W9-CTH-461

Quality and Performance Improvement in Healthcare

A Tool for Programmed Learning

Second Edition

Patricia Shaw, MEd, RHIA
Chris Elliott, MS, RHIA
Polly Isaacson, RHIA, CPHQ
and Elizabeth Murphy, MEd, RN

AMERICAN HEALTH INFORMATION
MANAGEMENT ASSOCIATION®

CHABOT COLLEGE LIBRARY

RA
399
.A1
Q27
2003

Copyright © 2003 by the American Health Information Management Association. All rights reserved. No part of this publication may be reproduced, stored in a retrieval system, or transmitted, in any form or by any means, electronic, photocopying, recording, or otherwise, without the prior written permission of the publisher.

ISBN 1-58426-116-1
AHIMA Product Number AB102703
Production Number IPC-2000-803

AHIMA Staff:
Marcia Bottoms, Acquisitions Editor and Director of Publications
Beth Hjort, RHIA, CHP, Content Reviewer
Katherine Kerpan, Project Editor
Jennifer Solheim, Assistant Editor
Julia Wixtrom, Production Manager

AHIMA strives to recognize the value of people from every racial and ethnic background as well as all genders, age groups, and sexual orientations by building its membership and leadership resources to reflect the rich diversity of the American population. AHIMA encourages the celebration and promotion of human diversity through education, mentoring, recognition, leadership, and other programs.

American Health Information Management Association
233 North Michigan Avenue, Suite 2150
Chicago, Illinois 60601-5800

http://www.ahima.org

CHABOT COLLEGE LIBRARY

Contents

About the Authors v
Acknowledgments for the First Edition vii
Acknowledgments for the Second Edition viii
Preface for Students ix
Preface for Educators and Practitioners x
Introduction xi

Part I   *A Performance Improvement Model*

Chapter 1   **Defining a Performance Improvement Model** 3
The model used for performance improvement activities is defined. How this model is
applied to organizationwide and team-based performance improvement is discussed.

Chapter 2   **Identifying Improvement Opportunities** 11
The four principal aspects of healthcare that are targeted for performance improvement
activities are defined. The QI toolbox techniques of brainstorming, nominal group
technique, and affinity diagramming are defined.

Chapter 3   **Applying Teamwork in Performance Improvement** 21
The effective use of teams in performance improvement activities is introduced. The
composition of the PI team and the use of mission, vision, and value statements are discussed.

Chapter 4   **Aggregating and Analyzing Performance Improvement Data** 31
Various data types, data display techniques, and aggregation and analysis methods to
support the performance improvement process are discussed.

Chapter 5   **Communicating Performance Improvement Activities and Recommendations** 53
Team communication tools such as minutes, quarterly reports, and storyboards are defined.
The importance of organizationwide communication of performance improvement activities
is emphasized.

Part II   *Continuous Monitoring and Improvement Functions*

Chapter 6   **Measuring Customer Satisfaction** 67
Identification of the customer is introduced. The difference between internal and external
customers is defined. The development and use of interviews and survey instruments are
discussed.

Chapter 7   **Optimizing the Continuum of Care** 87
Preadmission screening, utilization management, and discharge planning for the care
continuum are defined. The use of criteria sets, indicators, and Gantt charts is explored.

Chapter 8   **Preventing and Controlling Infectious Disease** 111
Management of the infection control function is discussed. The use of flowcharting
in the performance improvement process is introduced.

Chapter 9   **Decreasing Risk Exposure** 127
Risk management, incident reports, and sentinel event reporting are defined. The QI toolbox
techniques of cause-and-effect (fishbone) diagramming and root-cause analysis are discussed.

Chapter 10 Optimizing Patient Care **147**
The following are defined, discussed, and explored: the pharmacy and therapeutic
monitoring process; seclusion, restraint, and protective device use; the evaluation of
laboratory services and the use of blood products; patient care outcomes review; policy,
procedure, and documentation review; organizational standards of care.

Chapter 11 Improving the Care Environment and Life Safety **171**
The following steps to successful improvement activities for the care environment and
life safety are explored: security management programs, hazardous materials and waste
management programs, life safety management programs, medical equipment management
programs, building utility management and safety programs.

Chapter 12 Developing Staff and Human Resources **195**
Recruitment, retention, orientation, training, and performance appraisal are all critical
components of the staff performance improvement process in any healthcare organization.
All of these components, as well as the process for monitoring and managing physicians
and paraprofessional staff are discussed.

Part III   Management of Performance Improvement Programs

Chapter 13 Organizing for Performance Improvement **231**
The role of the healthcare organization's leaders in performance improvement activities is
outlined. An effective communication structure through the use of committees is discussed.

Chapter 14 Navigating the Accreditation, Certification, or Licensure Process **243**
Accreditation, certification, and licensure review processes are introduced. What to expect
during an on-site survey, as well as how to incorporate standards and regulations into
day-to-day operations for continuous improvement are explored.

**Chapter 15 Implementing Effective Information Management Tools
for Performance Improvement** **259**
The effective management of data and information collected for the performance
improvement function is described. The role of health information services managers,
information systems managers, and knowledge-based librarians in performance
improvement activities is introduced.

Chapter 16 Developing Effective Performance Improvement Teams **277**
The effective use of team charters, team roles, and ground rules is discussed. Effective
listening and questioning techniques within performance improvement team activities are
discussed.

Chapter 17 Managing Healthcare Performance Improvement Projects **289**
The application of project management principles in the performance improvement process
is discussed. The project life cycle and group dynamics of team life cycles are identified.

Chapter 18 Managing the Human Side of Change **303**
The three phases of change and the application of change management principles during
performance improvement projects are described.

Chapter 19 Developing the Performance Improvement Plan **313**
The components of a healthcare organization's performance improvement plan are
outlined and explored.

Chapter 20 Evaluating the Performance Improvement Program **329**
Appropriate review and evaluation of the performance improvement program
are discussed.

Chapter 21 Understanding the Legal Implications of Performance Improvement **337**
The legal aspects of performance improvement activities are explored.

Chapter 22 Predicting the Future of Performance Improvement in Healthcare **353**
Possible trends and evolving changes in theory and practice for the future of performance
improvement are offered.

Glossary **371**
Index **381**

About the Authors

Patricia Shaw, MEd, RHIA, earned her master's degree in education in 1997 and has been on the faculty of Weber State University for over twelve years, where she teaches in the Health Information Management and Health Services Administration programs. She has primary teaching responsibility for the quality and performance improvement curriculum in those programs. Pat maintains contact with practice settings as a consultant specializing in the areas of reimbursement and coding issues. Prior to accepting a position at Weber State University, Pat managed hospital health information services departments and was a nosologist for the 3M Corporation's Health Information Systems Division.

Chris Elliott, MS, RHIA, holds a master's degree in information systems and is pursuing a doctorate in medical informatics at the University of Utah. He is currently the director of Health Information Services at San Francisco General Hospital Medical Center. Prior to his current position, he had taught in health information and health administration programs at Weber State University in Ogden, Utah. Before becoming a full-time health professions educator, he worked for 20 years in various positions related to health information management and quality improvement in the San Francisco bay area.

Polly Isaacson, RHIA, CPHQ, is a founding partner with HealthCare Consulting Associates (since 1994) and HealthCare Pharmaceuticals (since 1997). She has twenty-four years of healthcare experience in health information management and performance improvement in a variety of healthcare settings. Polly has practiced as a full-time consultant throughout the Intermountain West since 1994, primarily in the capacity of facilitating healthcare organizations through the process of state licensure, Medicare certification, or JCAHO accreditation. Polly also works as an adjunct faculty member at Weber State University teaching the online performance improvement course and continues to consult in various positions related to health information management. Polly holds a registration in health information administration (RHIA), a certification as a professional in healthcare quality (CPHQ), and a license as a pharmacy technician.

Elizabeth Murphy, MEd, RN, holds a master's degree in education, with an emphasis in community counseling and addiction rehabilitation. She is a registered nurse and has worked in a variety of rehabilitation settings as a nurse administrator since 1987. Currently, she is employed by the University of Utah College of Nursing in Salt Lake City as a nurse supervisor for Youth Corrections. Prior to her current position, Elizabeth was the director of Clinical Services for Children's Comprehensive Services, Inc., which serves 4,500 children in psychiatric and rehabilitative settings across the nation. Her focus of practice in most of her positions has included accreditation and licensure, strategic planning, budgeting, performance improvement, and quality management.

Acknowledgments for the First Edition

The authors wish to thank the following individuals for their contributions to the first edition of this text:

Patrick Baggs for his real-life example regarding improving the environment of care.

Marcia Bottoms for her chapter on managing the human side of change and for editing the book so assiduously.

Melanie Brodnik, PhD, RHIA, for her review of the text and suggestions for improvement.

Shirley Eichenwald, MBA, RHIA, for her review of the text and excellent suggestions for improvement of it.

Danece Fickett, RN, for her real-life example regarding managing the continuum of care in chapter 7.

Beth Hjort, RHIA, for her review of the text and suggestions for improvements.

Merida Johns, PhD, RHIA, for reviewing the text and holding the authors to a higher standard.

Marie Kotter, PhD, for her case study on the organizational structure necessary for performance improvement programs.

Virginia Mullen, RHIA, for her real-life example regarding measuring customer satisfaction.

Leonard Gary Nielson, MS, MT (ASCP), for reviewing the infection control chapter and providing background materials on the Clinical Laboratory Improvement Amendments.

Karen Patena, MBA, RHIA, for her review of the text and excellent suggestions for its improvement.

Dorothy Grandolfi Wagg, JD, RHIA, (deceased) for her chapter on the legal aspects of performance improvement. The authors wish to remember Dorothy Wagg for the valuable contributions she made to this book and to the profession of health information management. She will be greatly missed.

Acknowledgments for the Second Edition

The authors wish to thank the following individuals for their contributions to this second edition of the text:

Melanie S. Brodnik, PhD, RHIA, for her review of the text and suggestions for improvement.

Lloyd R. Burton, DM, for his expert treatment of two key topics—management of performance improvement projects (chapter 17) and the future of performance improvement (chapter 22)—in the two new chapters that he wrote for this second edition of the book.

Jean S. Clark, RHIA, for her review of part II of the text and suggestions for improvement.

Beth Hjort, RHIA, CHP, for her very helpful review of the text and key suggestions for improvement.

Nadine Myers, RHIA, for her suggestions for improvement.

Leonard Gary Nielson, MS, MT (ASCP), for reviewing the infection control chapter and providing key suggestions for improvement of this chapter.

Margaret A. Skurka, MS, RHIA, CCS, for her review of the text and suggestions for improvement.

Tori Sullivan, RHIA, for the new text section on business associates in the legal issues chapter.

Cathy Turner, MBA, BSN, RN, BC, for her review of the chapter on change management and for her well-researched suggestions for improvement.

Vicki L. Zeman, MA, RHIA, for her review of the text and suggestions for improvement.

Preface for Students

You will soon be entering your chosen profession in the healthcare field. The issues involved in the management of quality in healthcare span the various clinical and administrative disciplines and must be approached from a variety of perspectives. Many improvements for healthcare services are developed through team-based activities. Employers will expect you to be able to apply performance improvement data analysis and presentation tools. You may also find some time in the future that you will be asked to facilitate a performance improvement team meeting.

The authors of this text hope that this tool for programmed, incremental learning of the PI process will prepare you well for the challenges you will face in your new career. If you use this text carefully, you will probably find yourself miles ahead of your fellow students in preparation for today's healthcare environment.

Preface for Educators and Practitioners

This textbook from AHIMA presents a comprehensive introduction to the theory, practice, and management of performance and quality improvement processes in healthcare organizations. Parts I and II are intended for use by students in technology-level programs of all kinds and provide a basic background in performance improvement philosophy and methodology for healthcare practice today. Each chapter has real-life examples and case studies from healthcare settings that bring home the importance of quality in healthcare services. QI toolbox techniques are presented both in theory and in practice so that your students can see how the techniques can actually be used in performance improvement activities. Healthcare information management students will find the textbook's unique step-by-step and case study–based approach to the subject easy to use and understand. Students will also gain hands-on practice applying the analytical and graphic tools used in performance and quality improvement. Student projects are integrated into the chapter discussions, which range from designing specific improvement projects, to ongoing quality monitoring, to managing quality improvement programs and staff.

Part III is intended for use by students in management-level programs in the health services. Its chapters focus on the issues inherent in the management of quality and performance improvement programs in healthcare. Each chapter presents the issues and their backgrounds and concludes with a case study to reinforce student learning and encourage critical thinking about the issues.

Instructor materials for this book include lesson plans (for students in both two-year and four-year programs), lesson slides, and other useful resources. The instructor materials are available in online format through the Assembly on Education (AOE) Community of Practice (CoP). Instructors who are AHIMA members can sign up for this private community by clicking on the help icon within the CoP home page and by requesting additional information on becoming an AOE CoP member. An instructor who is not an AHIMA member or a member who is not an instructor may contact the publisher at publications@ahima.org. The instructor materials are not available to students enrolled in college or university programs.

Introduction

People naturally expect their world to improve over time. This expectation affects everything people come into contact with: food, housing, cars, education, and healthcare. Such expectations stimulate general social progress. Progress may take considerable time to develop, and the desire for progress sometimes takes a counterproductive path, as during times of war and political upheaval. Still, the objective of making the human situation better is a constant in human endeavors.

Progress is commonly accomplished in one of two ways. First, progress can be achieved through an understanding of the scientific basis of the natural world and its constituent parts. Understanding the way the human body functions through biochemistry, for example, facilitated the development of the pharmaceuticals in use today. Second, progress can be achieved through improvements in the ways that people perform work. Understanding the procedures that healthcare professionals must perform to help people get well, for example, facilitated the development of one of the best healthcare delivery systems in the world. This textbook examines the second type of progress.

The focus of this textbook is healthcare quality and performance improvement in the United States and the means by which progress is accomplished in healthcare organizations. Every healthcare professional needs to understand the issues surrounding quality and performance improvement in healthcare because society expects that progress will result in better and better healthcare products and services. But are there other reasons why healthcare professionals should be concerned about quality improvement? For example, is there a tradition to be followed?

Early Quality Improvements in Healthcare

There is a long tradition of quality improvement in healthcare. (See figure I.1.) From colonial times until the present, healthcare in the United States has undergone a series of developments and reforms, from the creation of hospitals in the eighteenth century to the scientific discoveries of the nineteenth century, to the professionalization of medical and nursing practice in the early twentieth century, to the technological advances of the late twentieth and early twenty-first centuries. Healthcare institutions, professional associations, individual leaders, and political visionaries all laid the foundations of modern healthcare.

Figure I.1.	Historical Perspectives on Performance Improvement in Healthcare	
1700s	**1800s**	**1900s to Present**
Mid-1700s Pennsylvania Hospital becomes the model for the organization and development of hospitals. **1760** New York State begins the practice of medical licensure. **1771** New Jersey begins the practice of medical licensure.	**1837** Massachusetts General Hospital sets limitations on clinical practice in the first granting of clinical privileges. **1851** Massachusetts General Hospital establishes the first disease/procedure index by classifying patient disposition. **Mid-1800s** Medical licensure is deemed undemocratic and is stopped. **1872** New England Hospital for Women and Children organizes a general training school for nurses. **1874** American Medical Association encourages the creation of independent state licensing boards.	**1903** North Carolina passes the first nurse registration bill in the United States. **1910** Flexner Report indicates unacceptable variation in medical school curricula. **1917** American College of Surgeons (ACS) establishes the Hospital Standardization Program. **1920** Most medical colleges meet rigorous academic standards and are approved by the American Association of Medical Colleges. **1946** Hill-Burton Act establishes funding to build new hospitals. **1953** Joint Commission on Accreditation of Hospitals (JCAH) was formed by the AMA, the CMA, the AHA, and the ACP. **1965** Public Law 89-97 establishes Medicare and Medicaid. **1972** Local peer review organizations are formed. **1980s** Prospective payment system is established. State and regional peer review organizations contract with HCFA. **1990s** JCAH becomes the Joint Commission on Accreditation of Healthcare Organizations (JCAHO). Deming's total quality management philosophy begins to spread in U.S. healthcare. JCAHO integrates quality improvement into the accreditation process. **2001** Ambulatory payment classification system is initiated. **2002** HCFA becomes the Centers for Medicare and Medicaid Services (CMS).

Healthcare Institutions

During the mid-1700s, before the American colonies became a nation, the citizens of Philadelphia, Pennsylvania, recognized the need for a place to house the mentally ill and to provide relief to the sick and injured. They also recognized the need to sequester newly arrived immigrants, who often carried diseases they had contracted during their long voyages to America. Thousands of people immigrated to the Pennsylvania colony in an attempt to improve their lives. Although most healthcare was provided in people's homes at that time, established inhabitants, particularly the poor, sometimes required a place to rest and mend during times of illness and injury. Recognizing these needs, Dr. Thomas Bond, with the help of Benjamin Franklin, persuaded the Pennsylvania legislature to undertake the organization and development of a hospital for the community, the famous Pennsylvania Hospital, the first in the growing nation (Morton 1895).

Over the next 150 years, the Pennsylvania Hospital became a model for the development of hospitals in other communities. It even attempted to standardize its care processes by publishing rules and regulations for its physicians and staff (Morton 1895). These regulations represent early attempts at healthcare improvement.

The annals of Massachusetts General Hospital provide an early example of an action taken by a hospital board of trustees to ensure the quality of care provided in the institution. In 1837, the trustees became aware that the son of a resident surgeon (a surgeon who had not attained appointment to the hospital) had practiced in the hospital during his father's absence. The trustees reiterated to all of the medical staff the need for allowing only those accorded privileges at the institution to practice there (Bowditch 1872, p. 135):

> . . . the Trustees have recently seen with great pain, that a violation of the rules of the institution by one of its officers has become the subject of newspaper animadversion. In an institution like this, to which it is so difficult to attract, and in which it is so important to command, public confidence, the strictest and most scrupulous adherence to rules, of which the propriety is unquestioned, is required by a just regard as well to its usefulness to the public, as to the character of those who have any agency in its direction and control. Where many persons are connected in different departments, the reputation of all is more or less affected by the conduct of each; and all are therefore bound, by respect for others as well as themselves to conduct in such a manner as to give no reasonable ground of complaint.

It is also interesting to note that the trustees believed that the expectations of the members of their community—their customers—should be considered.

The annals of Massachusetts General Hospital include other examples of the hospital's concern about service quality. For example, in 1851, the hospital hired a watchman to guard against the danger of fire during the night (Bowditch 1872, p. 367). In 1853, the hospital commended one of its surgical staff for compiling an analytical index for the surgical records of the institution and reflected on the quality of the surgical services provided (Bowditch 1872, p. 483). In 1872, the trustees decided to regulate the use of restraints at the institution, and they identified each by type and set the conditions under which the restraint could be utilized (Bowditch, 1872, pp. 679–80). Throughout the history of the institution, the trustees received regular reports on the number of patients treated as well as the classification of each patient's outcome as discharged "well," "relieved," "not relieved," or "dead" (Bowditch 1872, p. 447).

Medical Practice

Human anatomy and physiology were not well understood before the twentieth century. At one time, it was believed that four basic fluids, called humors, determined a person's temperament and health and that imbalances in the proportion of humors in the body caused disease. The therapeutic bleeding of patients was practiced into the early twentieth century. Early physicians also treated patients by administering a variety of substances with no scientific basis for their effectiveness. The science of medicine began to evolve in the late nineteenth century but was not fully realized until the second and third decades of the twentieth century.

Early on, the medical profession recognized that some of its members achieved better results than others and undertook to regulate the practice of medicine. At first, the regulation took the form of licensure, beginning in New York in 1760 and New Jersey in 1771. The New Jersey law stated that ". . . no person whatsoever shall practice as a physician or surgeon, within this colony of New Jersey, before he shall have first been examined in physic and surgery, approved of, and admitted by any two of the judges of the supreme court." The examination was to be performed before a board of "medical men" appointed by the state medical society (Trent 1977, p. 91). Various states developed similar legislation over the following decades.

By the middle of the nineteenth century, however, medical licensure had been repudiated as undemocratic, and the penalties for practicing medicine without a license were removed in most of the states. Buyer beware was the rule of thumb because the title of doctor could be used by anyone who wanted to sell medical services (Haller 1981). During this period, medical education consisted primarily of an apprenticeship with an already-established practitioner of some kind. Following the apprenticeship, the new doctor could then hang out a shingle and begin to treat patients. Some trainees did attend schools that claimed to teach them how to become physicians, but there was no established medical curriculum. Many people received diplomas just by paying a fee. Many others with no education, apprenticeship, or license just hung out a shingle and began piling up the fees. Effectively, doctoring had become a commercial enterprise. Any man with sufficient entrepreneurial talents could enter the practice of medicine. The emphasis was on making a living rather than joining a true profession. The result was an overabundance of "medical men" who provided medical care based on all kinds of traditions, and at times, no tradition at all.

The American Medical Association (AMA) was established in 1840 to represent the interests of physicians across the United States. The organization was dominated by members who had strong ties to the medical schools and the status quo. The organization's ability to lead reform was limited until it broke its ties with the medical colleges in 1874. At that time, the association encouraged the creation of independent state licensing boards (Haller 1981).

In 1876, the Association of American Medical Colleges (AAMC) was established. The AAMC was dedicated to standardizing the curriculum of U.S. medical schools and to developing the public's appreciation of the need for medical licensure.

Together, the AMA and the AAMC pushed for medical licensing. By the 1890s, thirty-five states had established or reestablished a system of licensure for physicians. Fourteen states granted medical licenses only to graduates of reputable medical schools. The state licensing boards discouraged the worst medical schools, but the criteria for licensing continued to vary by state and were not fully enforced (Haller 1981).

By the early twentieth century, it had become apparent that promoting quality in medical practice required regulation through curriculum reform as well as licensure. The membership of the AMA, however, was divided on this issue. Conservative members continued to believe that the organization should stay out of the regulatory arena. Progressive members advocated the continuing development of state licensure systems and the development of a model medical curriculum.

The situation attracted the attention of the Carnegie Foundation for the Advancement of Teaching and its president, Henry S. Pritchett. Pritchett offered to sponsor and fund an independent review of the medical curriculum and the medical colleges of the United States. The review was undertaken in 1906 by Abraham Flexner, an educator from Louisville, Kentucky.

Over the following four years, Flexner visited every medical college in the country and carefully documented his findings. In his 1910 report to the Carnegie Foundation, the AMA, and the AAMC, he documented the unacceptable variation in curriculum that existed across the schools. He also noted that applicants to medical schools frequently lacked a knowledge of the basic sciences. Flexner also reported how the absence of appropriate hospital-based training limited the clinical skills of medical school graduates. Perhaps most important, he documented the huge number of graduates who were being produced by the colleges each year, most with unacceptable levels of medical expertise.

Several reform initiatives grew out of Flexner's report and recommendations made by the AMA's Committee on Medical Education. One of the reforms required medical college applicants to hold a baccalaureate degree. Another required that the medical curriculum be founded in the basic sciences. Reforms also required that medical students receive practical, hospital-based training. Most important, Flexner recommended the closing of most of the medical schools in the country. The former recommendations were instituted over the decade after the release of Flexner's report, but only about half of the medical colleges actually closed. By 1920, most of the colleges met rigorous academic standards and were approved by the AAMC.

Nursing Practice

During the nineteenth century and throughout the first part of the twentieth, over half of the hospitals in the United States were sponsored by religious organizations. Nursing care at that time was usually provided by members of religious orders. As the U.S. population grew and more towns and cities were established, hospitals were built to accommodate the healthcare needs of new communities. Older cities were also growing, and city hospitals became more and more crowded.

In the late nineteenth century, nurses received no formal education and training. Nursing staff for the hospitals was often recruited from the surrounding community, and many poor women who had no other skills became nurses. The nature of nursing care at that time was unsophisticated, and ignorance of basic hygiene often promoted disease rather than wellness. In 1871 at Bellevue Hospital in New York City, for example, 15 percent of patients died while hospitalized, and hospital-acquired infections were common (Kalisch and Kalisch 1995, p. 71). Even simple surgical procedures and maternity care often resulted in death due to infection.

In 1868, the president of the AMA, Dr. Samuel Gross, called the medical profession's attention to the need for trained nurses. During the years that followed, the public began to call for better nursing care in hospitals.

A small group of women physicians working in the northeast area of the country created the first formal program for training nurses. Dr. Susan Dimock, working with Dr. Marie Zakrzewska at the New England Hospital for Women and Children, organized a general training school for nurses in 1872 (Kalisch and Kalisch 1995). The school became a model for other institutions throughout the United States. As hospital after hospital struggled to find competent nursing staff, many institutions and their medical staffs developed their own nurse training programs to meet staffing needs.

The responsibilities of nurses in the late nineteenth and early twentieth centuries included housekeeping duties such as cleaning furniture and floors, making beds, changing linen, and controlling temperature, humidity, and ventilation. Nurses also cooked the meals for patients in kitchens attached to each ward. Direct patient care duties included giving baths, changing dressings, monitoring vital signs, administering medication, and assisting at surgical procedures (Kalisch and Kalisch 1995). Nurses generally worked twelve-hour shifts, seven days per week.

During this time, nurses were not required to hold a license to practice. Because licensure was not required and because it was difficult to attract women to nursing staff positions, many women who had no training at all continued to work in the nation's hospitals and as private-duty nurses.

In the years immediately following the turn of the twentieth century, nurses began to organize state nursing associations to advocate for the registration of nurses. Their goal was to increase the level of competence among nurses nationwide. Despite opposition from many physicians who believed that nurses did not need formal education or licensure, North Carolina passed the first nurse registration bill in the United States in 1903. Many other states initiated similar legislation in subsequent years.

Allied Health Professions

Other healthcare professions in the United States developed as specialized areas of practice over the course of the twentieth century. Each underwent periods of formalization in similar ways. Each became regulated either by the states or by national professional associations as membership and professional responsibilities grew and the public demanded that they document their professional competence. These developments made important contributions to the quality of healthcare delivered in this country. The allied health professions include radiologic technology, respiratory therapy, occupational therapy, and physical therapy, among others. For example, health information management professionals today are certified and registered by the American Health Information Management Association.

Contributions of Individuals

Many individual healthcare professionals have made significant contributions to the early improvement of healthcare delivery in the United States, including the development and implementation of a variety of improvement strategies. We have included a small sample of these individuals and their contributions here. It is important to recognize the progress that can be made when healthcare professionals care about the quality of their work.

Maude E. Callen, an African-American public health nurse/midwife, undertook the training of midwives in coastal South Carolina in 1926. A registered nurse, Callen recognized that the midwives' lack of training contributed to high infant and maternal mortality rates in the region, and she traveled extensively throughout the region to assist at deliveries and improve the expertise of midwives (Hill 1997).

Robert Latou Dickinson, an obstetrician and gynecologist practicing in New England around the turn of the twentieth century, developed a standardized patient questionnaire. He used the patients' answers on the questionnaire to structure his examinations. His questionnaire represents one of the first uses of a structured health assessment tool in the United States (Bullough 1997).

Lavinia Lloyd Dock, a nurse and early nurse educator, developed important approaches to disaster nursing at the end of the nineteenth century. After graduating from a nurse training program, Dock worked to institute appropriate nursing practices during the yellow fever epidemic in Jacksonville, Florida, in 1888, and during the aftermath of the Johnstown, Pennsylvania, flood in 1889 (Leighow 1997).

Roswell Park, a physician and surgeon during the late nineteenth century, helped to disseminate the principles of antisepsis during surgical procedures in the United States. On the basis of the findings of the English scientist Joseph Lister, Park advocated the use of antiseptic techniques and appropriate wound care in the treatment of surgical cases well before such approaches were common in the United States (Gage 1997).

Nicholas J. Pisacano, a physician who practiced in the middle and late twentieth century, recognized the need to upgrade the practice of the general practitioner as new technologies and treatments were developed. He worked tirelessly to develop and promote the specialty of family practice in the United States (Adams and Moore 1997).

Ernst P. Boas, a physician who practiced in New York City during the first half of the twentieth century, was among the first to call for the coordinated, interdisciplinary care of the chronically ill. Prior to his advocacy, the chronically ill were often considered incurable. He believed that the development of new therapeutics and restorative technologies could return people with chronic illnesses to better health and productivity. His work led to the establishment of the Goldwater Memorial Hospital for Chronic Diseases on Welfare Island in New York City (Brickman 1997, p. 21).

Mary Steichen Calderone, medical director of the Planned Parenthood Federation of America during the 1950s, launched a clinical investigation program to scientifically identify effective contraceptive methods. Hers was one of the first efforts to identify appropriate clinical practice through the use of scientific evidence in a controversial area (Meldrum 1997).

Hospital Standardization and Accreditation

In 1910, Dr. Edward Martin suggested that the surgical area of medical practice needed to become more concerned with patient outcomes. He had been introduced to this concept through discussions with Dr. Ernest Codman, a British physician who believed that hospital practitioners should track their patients for a significant time after treatment to determine whether the end result was positive or negative. Dr. Codman also advocated the use of outcome information to identify the practices that led to the best results.

Dr. Martin and others had been concerned about the conditions in U.S. hospitals for some time. Many observers felt that part of the problem was related to the absence of organized medical staffs in hospitals and to lax professional standards. In the early twentieth century, hospitals were used primarily by surgeons who required their facilities to treat patients with surgical modalities. Therapies based on medical regimens were not developed until later in the century. It was natural, therefore, for the impetus for improvement in hospital care to come from the surgical community.

In November 1912, the Third Clinical Congress of Surgeons of North America was held. At this meeting, Dr. Franklin Martin made proposals that eventually led to the formation of the American College of Surgeons. Dr. Edward Martin made the following resolution (Roberts, Coate, and Redman 1987, p. 936):

> Be it resolved by the Clinical Congress of Surgeons of North America here assembled, that some system of standardization of hospital equipment and hospital work should be developed to the end that those institutions having the highest ideals may have proper recognition before the profession, and that those of inferior equipment and standards should be stimulated to raise the quality of their work. In this way patients will receive the best type of treatment, and the public will have some means of recognizing those institutions devoted to the highest levels of medicine.

Through the proposal and the resolution, the American College of Surgeons and the hospital improvement movement became intimately tied. Immediately upon formation, however, officers of the college realized how important their work would be. They were forced to reject 60 percent of the fellowship applications during the college's first three years because applicants were unable to provide documentation in support of their clinical competence (Roberts, Coate, and Redman 1987, p. 937). Medical records from many hospitals were so inadequate that they could not supply information about the applicants' practice in the institutions. Because of this situation and many others of which they became aware, college officers petitioned the Carnegie Foundation in 1917 for funding to plan and develop a hospital standardization program.

In 1917, a committee on standards was formed by the college and met to consider the development of a minimum set of standards that U.S. hospitals would have to meet if they wanted approval from the American College of Surgeons. On December 20, 1917, the American College of Surgeons formally established the Hospital Standardization Program and published a formal set of hospital standards, which they called *The Minimum Standard.*

During 1918 and part of 1919, the college undertook the review of hospitals across the United States and Canada as a field trial to see whether *The Minimum Standard* would be effective as a measurement tool. In total, 692 hospitals were surveyed, of which only 89 met the standard entirely. Some of the most prestigious institutions in the United States failed to meet the standard. Brief and clear in its delineation of what was believed to promote good hospital-based patient care in 1918, *The Minimum Standard* (American College of Surgeons 1930, p. 3) stated:

1. That physicians and surgeons privileged to practice in the hospital be organized as a definite group or staff. Such organization has nothing to do with the question as to whether the hospital is "open" or "closed," nor need it affect the various existing types of staff organization. The word STAFF is here defined as the group of doctors who practice in the hospital inclusive of all groups such as the "regular staff," the "visiting staff," and the "associate staff."

2. That membership upon the staff be restricted to physicians and surgeons who are (a) full graduates of medicine in good standing and legally licensed to practice in their respective states or provinces; (b) competent in their respective fields; and (c) worthy in character and in matters of professional ethics; that in this latter connection the practice of the division of fees, under any guise whatever, be prohibited.

3. That the staff initiate and, with the approval of the governing board of the hospital, adopt rules, regulations, and policies governing the professional work of the hospital; that these rules, regulations, and policies specifically provide: (a) That staff meetings be held at least once each month. (In large hospitals the departments may choose to meet separately.) (b) That the staff

      review and analyze at regular intervals their clinical experience in the various departments of the hospital, such as medicine, surgery, obstetrics, and the other specialties; the clinical records of patients, free and pay, to be the basis of such review and analysis.

4. That accurate and complete records be written for all patients and filed in an accessible manner in the hospital—a complete case record being one which includes identification data; complaint; personal and family history; history of present illness; physical examination; special examinations, such as consultations, clinical laboratory, X-ray and other examinations; provisional or working diagnosis; medical or surgical treatment; gross and microscopical pathological findings; progress notes; final diagnosis; condition on discharge; followup and, in case of death, autopsy findings.

5. That diagnostic and therapeutic facilities under competent supervision be available for the study, diagnosis, and treatment of patients, these to include, at least (a) a clinical laboratory providing chemical, bacteriological, serological, and pathological services; (b) an X-ray department providing radiographic and fluoroscopic services.

      The adoption of *The Minimum Standard* marked the beginning of the **accreditation** process for healthcare organizations. A similar process is still followed today. (For more information, see chapter 14 of this textbook.) Basically, the process is based on the development of reasonable quality standards and a survey of the organization's performance on the standards. The accreditation program is voluntary, and healthcare organizations request participation in order to improve patient care (Roberts, Coate, and Redman 1987).

      The American College of Surgeons continued to examine and approve hospitals for three decades. By 1950, however, the number of hospitals being surveyed every year had grown unmanageable, and the college could no longer afford to administer the program alone. After considerable discussion and organizing activity, four professional associations from the United States and Canada decided to join the American College of Surgeons to develop the Joint Commission on Accreditation of Hospitals. These associations were the American College of Physicians, the American Medical Association, the American Hospital Association, and the Canadian Medical Association. The new accrediting agency was formally incorporated in 1952 and began accreditation activities in 1953. It continues its activities fifty years later as the Joint Commission on Accreditation of Healthcare Organizations.

Performance Improvement and Modern Healthcare

Until the Second World War, most healthcare was still provided in the home. Quality in healthcare services was considered a byproduct of appropriate medical practice and oversight by physicians. The positive and negative effects of other factors and the contributions of other healthcare workers were not given much consideration.

      In the 1950s, the number of hospitals grew to support developments in diagnostic, therapeutic, and surgical technology and pharmacology. Fueled by an expanding economy, the Hill–Burton Act of 1946 funded extensive hospital construction. A renewed insurance industry helped to pay for the new healthcare services provided to groups of individual beneficiaries.

      During this period, the Hospital Standardization Program was replaced by the Joint Commission on Accreditation of Hospitals. A whole new set of standards covered every aspect of hospital care. The intent was to ensure that the care provided to patients in accredited hospitals would be of the highest quality.

The construction of new facilities and the growth of the medical insurance industry did not guarantee access to services. As new treatments and "miracle" drugs such as antibiotics were developed, healthcare services became more and more costly. Many Americans, particularly the poor and the elderly, could not afford to buy healthcare insurance or to pay for the services themselves.

Medicare/Medicaid Programs

The idea of federal funding for healthcare services goes back to the 1930s, the Great Depression, and Franklin Roosevelt's New Deal. Harry Truman also supported a universal healthcare program in the late 1940s. But it was not until the 1960s and the presidency of Lyndon Johnson that the federal government developed a program to pay for the healthcare services provided to the poor and the elderly (American Hospital Association 1999).

In 1965, the United States Congress passed Public Law 89-97, an amendment to the Social Security Act of 1935. Title XVIII of Public Law 89-97 established health insurance for the aged and the disabled. This program soon became known as Medicare. Title XIX of Public Law 89-97 provided grants to states for establishing medical assistance programs for the poor. The Title XIX program became known as Medicaid. The objective of the programs was to ensure access to healthcare for citizens who could not afford to pay for it themselves. The Great Society, as the geopolitics of the United States was called in the 1960s, marshaled billions of federal tax dollars to fund care for millions of Americans.

During the 1970s, attempts were made to further standardize and improve the clinical services provided by physicians and hospitals. Under the authority of Medicare officials, hospital audits of physicians' medical records were mandated to identify physicians with substandard practice patterns or excessive patient care costs. Local peer review organizations (PROs), usually sponsored by local medical societies, then reviewed the findings at each local institution and developed recommendations for physician continuing education. Such retrospective **quality assurance** efforts were only partially successful and had little effect on the mounting cost to the government of the Medicare and Medicaid programs. As a result, utilization review (UR) programs were mandated to justify hospital admissions. The concept and practice of UR survives today (see chapter 7). Institutions must still provide payers a rationale for the level of services provided in order to be reimbursed.

The changes that most significantly improved patient outcomes in the 1970s involved the development and use of sophisticated medical technology and pharmaceuticals. The overall benefits of modern healthcare were evident in increased life spans and better medical outcomes. Americans had come to expect the best and the newest medical care available as a personal right not to be taken away.

By 1980, however, it was obvious that healthcare spending in the United States would consume more and more economic resources if left unchecked. The Medicare and Medicaid programs were on their way to becoming the most expensive government programs in U.S. history. At the same time, healthcare experts also began to understand that increased spending and technological advance did not automatically guarantee quality in healthcare.

In the early 1980s, a new nationwide system was developed to standardize reimbursement for hospital services provided to Medicare and Medicaid beneficiaries. Until 1983, Medicare/Medicaid reimbursement was based on a **retrospective payment system.** In a retrospective payment system, providers are paid for the services they provided to a patient in the past. Retrospective payment is also called fee-for-service payment. The patient goes to the doctor; the doctor cares for the patient; the doctor assigns charges and submits a bill to a payer; the doctor is reimbursed for his charges. The problem with this system is that there is no incentive for the doctor to hold down costs. If he provides more services, then he bills more, and he gets paid more. So, this type of arrangement did not help to rein in ever-increasing healthcare costs.

To slow the growth in cost of federal healthcare programs, a prospective payment system was developed. In a prospective payment system, providers (doctors and hospitals) receive a fixed, predetermined payment for the services they provide. The reimbursement amounts are determined annually by the Centers for Medicare and Medicaid Services (CMS), and billing of carriers and patients cannot exceed these assigned amounts. Because the amount of reimbursement is fixed and often lower than what the provider would otherwise charge, providers are motivated to use only those services absolutely necessary to the patient's care. In this way, costs are controlled and unnecessary services are avoided.

In the Medicare/Medicaid prospective payment system, reimbursement for hospital inpatient services is based on diagnosis-related groups (DRGs). The system assumes that similar diseases and treatments consume similar amounts of resources and therefore have similar total costs. Every hospital patient is assigned to an appropriate DRG on the basis of his or her diagnosis at discharge. Reimbursement levels for each DRG are updated annually and adjusted for the geographic location of the healthcare facility.

The Healthcare Common Procedure Coding System (HCPCS) was developed in 1983. HCPCS codes are used to report the healthcare services provided to Medicare and Medicaid beneficiaries treated in ambulatory settings. HCPCS includes three separate levels of codes: (level I) Current Procedural Terminology (CPT) codes, (level II) national codes, and (level III) local codes.

A prospective payment system for hospital outpatient and ambulatory surgery services provided to Medicare and Medicaid beneficiaries was implemented in 2001. This system is based on ambulatory payment classification groups (APCs). The APCs are generated on the basis of the CPT/HCPCS codes assigned for services such as outpatient diagnostic procedures and outpatient radiology procedures. A similar system has been implemented for the reimbursement of professional fees. This resource-based relative value scale (RBRVS) system takes into consideration the level of services provided by the physician in terms of time spent with the patient, complexity of physical exam and information gathering, as well as the diagnostic and procedural actions performed to arrive at the reimbursement amount.

U.S. hospitals and physicians provide billions of dollars worth of care to Medicare/Medicaid patients every year. The implementation of prospective payment systems has made it necessary for healthcare organizations to devise ways to control costs without endangering safe and effective patient care. It is necessary to recognize, however, that there are limits to the amount of dollars that can be extracted from the system by these methods. That will be the great debate and struggle engaging the nation in the first decade of the twenty-first century.

Managed Care Revolution

The growth of managed care in the United States has also had a tremendous impact on healthcare providers. *Managed care* is a broad term used to describe several types of managed healthcare plans. Health maintenance organizations (HMOs) are one of the most familiar types of managed care. Members of an HMO (or their employers) pay a set premium and are entitled to a specific range of healthcare services. HMOs control costs by requiring beneficiaries to seek services from a preapproved list of providers, by limiting access to specialists and expensive diagnostic and treatment procedures, and by requiring preauthorization for inpatient hospitalization and surgery.

Other types of managed care include preferred provider organizations (PPOs) and point-of-service (POS) plans. These types of managed care plans negotiate discounted rates with specific hospitals, physicians, and other healthcare providers. Many also restrict access to specialists and require preauthorization for surgery and other hospital services. In PPOs, enrollees are required to seek care from a limited list of providers who have agreed in advance to accept a discounted payment for their services. Enrollees in POS plans pay for a greater portion of their healthcare expenses when they choose to seek treatment from providers who do not participate in their plan.

Together, the Medicare/Medicaid programs and the managed care insurance industry have virtually eliminated fee-for-service reimbursement arrangements. At the same time, healthcare consumers are demanding more services and greater quality. Hospitals and physicians now find that they have no choice but to become more efficient and effective if they are to stay in business. Programs that promote **efficiency** and **effectiveness** have become the only way for providers to add value to the services they provide and the only way for them to ensure their financial viability.

Total Quality Management in Healthcare

In the 1980s, leaders in the healthcare industry began to take notice of a theory from general industry called **total quality management (TQM).** The concept of TQM was developed by W. Edwards Deming in the early 1950s as an alternative to authoritarian, top-down management philosophies. Philip Crosby and J. M. Juran adapted TQM and developed similar approaches. TQM mobilizes individuals directly involved in a work process to examine and improve the process with the goal of achieving a better product or outcome. It does not matter what the product or outcome might be. TQM is firmly based in the statistical analysis of objective data gathered from observation of the process being examined. The data are then carefully analyzed to identify the steps in the process that lead to a less-than-ideal product or outcome. Once the problematic steps in the process have been identified, individuals or teams can make recommendations for changing the process in order to get a better product or outcome. Key to Deming's philosophy was the concept that problematic processes, not people, cause inferior products and outcomes.

Total quality management revolutionized industrial production in Japan during the post–World War II period. When Japanese automobiles took over much of the U.S. car market in the 1960s and 1970s, American manufacturers began to take notice of TQM. They recognized that Deming's management philosophy might help them to create more efficient and effective manufacturing processes.

Avedis Donabedian was one of the first theorists to recognize that the TQM philosophy could be applied to healthcare services. Beginning in 1966, Donabedian advocated the assessment of healthcare from four perspectives: **structure, process, outcome,** and **cost.** Only in the 1990s, however, were his approaches becoming widely adopted. As the concept of TQM (or continuous quality improvement [CQI], as it became known in the U.S. healthcare system) was integrated into the healthcare industry's quest for improvement, the industry began using Donabedian's four perspectives to identify processes of providing care that could be improved. Using the team approach from Deming and his emphasis on objective data gathering to describe a process clearly, members of the industry began a self-examination that focused very specifically on the processes of care, rather than on the individuals who provided it. Many improvements were made for the nation's recipients of care in all types of healthcare organizations using this variant of Deming's TQM.

By the end of the 1990s, however, some individuals involved in the improvement of quality in healthcare had made a significant realization: quality in healthcare was tied very closely to the performance of individuals in the healthcare organization. Unlike manufacturing firms that utilize machinery to shape raw materials into physical products, the products of healthcare organizations were the services provided to patients by healthcare professionals who defined processes of practice. It was recognized that the performance of the professionals in care processes often determined the quality of the services. Quality improvement initiatives in healthcare organizations were renamed *performance improvement* initiatives, at least in those organizations affected by the JCAHO's performance improvement standards. Quality improvement was refocused to examine the performance of the people in the organization, rewarding those who obtained good outcomes or costs, and requiring those working in the healthcare industry to become more accountable for their patient or client outcomes.

Federal agencies began to emphasize quality improvement in the programs they sponsored during this time as well. The approach by federal Medicare and Medicaid programs retained the quality improvement terminology but focused largely on the same kinds of process issues. Today, contracted healthcare examiners—once called PROs, now called quality improvement organizations (QIOs)—retrospectively examine the care provided to beneficiaries and compare comparable providers' performance in different regions of the country. They are also now beginning to collect information on the most common diagnoses such as pneumonia or myocardial infarction in order to define the practices that achieve the best outcomes.

So, Why Care about Performance Improvement?

An individual working in the U.S. healthcare industry today will hear many terms reflecting the long development of quality improvement philosophy: *quality assurance, quality improvement, quality management,* and *performance improvement.* The differences in meaning are subtle, reflecting the time and place of their origins, as well as the individuals and philosophies that generated them. But they are all, in reality, focused on one thing: helping people with challenged health return to healthier, more productive lives and doing so by the most efficient and effective means possible. It is an evolving mission and one that is always seeking a better way.

Healthcare professionals must be concerned with performance improvement. As this introduction illustrates, there is a long tradition of seeking improvement in the healthcare industry. Today, performance improvement is the key to ensuring high-quality care and a performance improvement philosophy pervades leading healthcare organizations. To contribute to personal and organizational success, one must commit to participate in performance improvement. Today's patients increasingly are choosing their professional and institutional providers on the basis of quality. Furthermore, most contemporary payers prefer to negotiate with organizations that provide high-quality, yet cost-effective, services. Today's healthcare organizations must be able to back up their espousal of quality with reliable, objective data. Government-sponsored and commercial health plans, employers, and consumers are all now asking for more information on the quality of the healthcare services they receive and pay for. Additionally, a focus on quality is the key to meeting regulatory, licensure, and accreditation requirements. Demonstrating quality and improving performance are the definitive keys to success in the healthcare industry's mission to provide high-quality care.

Summary

Quality and performance improvement in healthcare have a long tradition in the United States. With the examination and licensure of physicians, the standardization of hospitals, and the adoption over time of a philosophy of continuous quality improvement, the healthcare industry has committed itself to improving the health of its customers in the most efficient and effective ways possible. Along the way, the government has stimulated the processes of improvement in the industry through innovative payment and review programs. Improvement initiatives at the end of the twentieth century concentrated on refining improvement methodologies and making improvement processes more scientific. Indeed, improvement initiatives have become one of the most important functions that healthcare organizations perform today to thrive in the healthcare marketplace.

References

Adams, D. P., and A. L. Moore. 1997. Nicholas J. Pisacano. In *Doctors, Nurses, and Medical Practitioners: A Bio-Biographical Sourcebook,* Lois N. Magner, editor. Westport, Conn.: Greenwood Press.

American College of Surgeons. 1930. *Manual of Hospital Standardization and Hospital Standardization Report.* Chicago: American College of Surgeons.

American Hospital Association. 1999. *100 Faces of Health Care.* Chicago: Health Forum.

Bowditch, N. I. 1872 (reprinted 1972). *History of the Massachusetts General Hospital.* Boston: Arno Press and New York Times.

Brickman, J. P. 1997. Ernst P. Boas. In *Doctors, Nurses, and Medical Practitioners: A Bio-Biographical Sourcebook,* Lois N. Magner, editor. Westport, Conn.: Greenwood Press.

Bullough, V. L. 1997. Robert Latou Dickinson. In *Doctors, Nurses, and Medical Practitioners: A Bio-Biographical Sourcebook,* Lois N. Magner, editor. Westport, Conn.: Greenwood Press.

Crosby, Philip B. 1980. *Quality Is Free.* New York City: Mentor Books.

Crosby, Philip B. 1984. *Quality without Tears.* New York City: Plume Books.

Donabedian, Avedis. 1988. The quality of care: how can it be assessed? *JAMA* 260(12):1743–48.

Donabedian, Avedis. 1980. *The Definition of Quality and Approaches to Its Management.* Volume 1: *Explorations in Quality Assessment and Monitoring.* Ann Arbor, Mich.: Health Administration Press.

Donabedian, Avedis. 1966. Evaluating the quality of medical care. *Milbank Quarterly* 44:166–203.

Gage, A. 1997. Roswell Park. In *Doctors, Nurses, and Medical Practitioners: A Bio-Biographical Sourcebook,* Lois N. Magner, editor. Westport, Conn.: Greenwood Press.

Haller, John S. 1981. *American Medicine in Transition 1840–1910.* Chicago: University of Illinois Press.

Hill, P. E. 1997. Maude E. Callen. In *Doctors, Nurses, and Medical Practitioners: A Bio-Biographical Sourcebook,* Lois N. Magner, editor. Westport, Conn.: Greenwood Press.

Kalisch, Philip A., and Beatrice J. Kalisch. 1995. *The Advance of American Nursing.* Philadelphia: J. B. Lippincott Company.

Leighow, S. R. 1997. Lavinia Lloyd Dock. In *Doctors, Nurses, and Medical Practitioners: A Bio-Biographical Sourcebook,* Lois N. Magner, editor. Westport, Conn.: Greenwood Press.

Meldrum, M. 1997. Mary Steichen Calderone. In *Doctors, Nurses, and Medical Practitioners: A Bio-Biographical Sourcebook,* Lois N. Magner, editor. Westport, Conn.: Greenwood Press.

Morton, Thomas G. 1895 (reprinted 1973). *The History of the Pennsylvania Hospital.* New York City: Arno Press.

Roberts, James S., Jack G. Coate, and Robert Redman. 1987. A history of the Joint Commission on Accreditation of Hospitals. *JAMA* 256(7):936–40.

Silin, Charles I. 1977. A state medical board examination in 1816. *Legacies in Law and Medicine,* C. R. Burns, editor. New York City: Science History Publications.

Trent, Josiah C. 1977. An early New Jersey medical license. *Legacies in Law and Medicine,* C. R. Burns, editor. New York City: Science History Publications.

Walton, Mary. 1986. *The Deming Management Method.* New York City: Perigee Books.

Walton, Mary. 1990. *Deming Management at Work.* New York City: G. P. Putnam's Sons.

Part I
A Performance Improvement Model

Chapter 1
Defining a Performance Improvement Model

Learning Objectives

- To explain the cyclical nature of performance improvement activities
- To introduce terminology and standards common to performance improvement activities
- To describe the distinction between organizationwide performance improvement activities and team-based performance improvement activities
- To outline the organizationwide performance improvement cycle
- To outline the team-based performance improvement cycle

Background and Significance

Efforts to ensure the quality of the healthcare services provided in the United States have gone by many names: QA (quality assurance), TQM (total quality management), QI (quality improvement), CQI (continuous quality improvement), QM (quality management), PI (performance improvement). The acronyms represent quality and performance improvement models and methodologies that have been used with varying degrees of success by healthcare organizations over the past thirty years. Many books and articles have been written on the subject, and new models and terminology will likely be developed in the future.

A professional just entering the healthcare field will probably work for many different organizations over his or her career and participate in many different quality and performance improvement projects. He or she will learn to use specific quality and performance improvement models and techniques as needed. With experience, he or she will develop the skill necessary to customize the models to specific organizations and healthcare services.

The goal of this chapter is to provide a general overview of quality and performance improvement as it is applied in healthcare organizations. The chapter describes a generic model of performance improvement. It also defines commonly used PI terms and explains the basic philosophy of continuous performance improvement.

Performance Improvement as a Cyclical Process

Accreditation organizations, groups of clinical professionals, quality management professionals, healthcare providers, and government regulatory and policy-making entities all have unique perspectives on quality in healthcare. Many have developed their own methodologies for quality and performance improvement. But most PI models being applied in healthcare today share one structural characteristic: they are cyclical in nature.

These cyclical models are based on the assumption that PI activities will take place continuously and that services, processes, and outcomes can always be improved. Quality is not to be treated as a goal to be accomplished and then forgotten. Rather, it is to be treated as an ongoing mission that guides everyday operations.

Accreditation and licensing agencies expect hospitals and other healthcare facilities to strive for the highest quality of care possible at all times. Healthcare leaders and their board of directors are responsible for the quality of the organization's services. Many large healthcare organizations employ experts in quality management who are responsible for organizing PI activities and reporting results to the leadership and the board of directors. At the same time, however, all employees are expected to have a basic understanding of PI principles and participate in PI activities.

The general PI model presented in this textbook includes two interrelated cycles. The cycle illustrated in figure 1.1 represents the organization's ongoing performance-monitoring

Figure 1.1.   Organizationwide Performance Improvement Processes

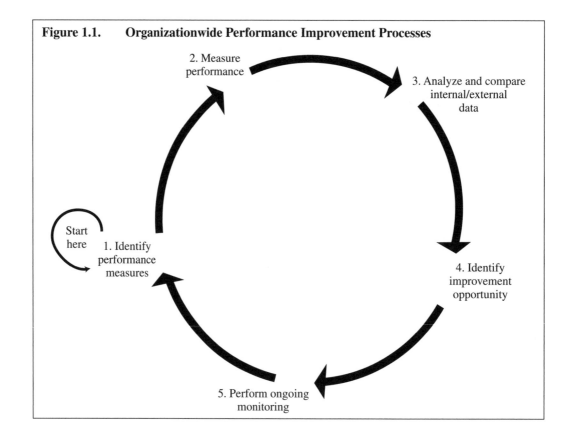

function. The cycle illustrated in figure 1.2 represents the activities of individual PI teams working on specific PI projects. Together, the two cycles make up the healthcare performance improvement model (figure 1.3).

Monitoring Performance through Data Collection

Performance monitoring is data driven. Monitoring performance through data collection is the foundation of all PI activities. Each healthcare organization must choose and prioritize which processes and outcomes (in other words, which types of data) are important to monitor based on its mission and the scope of care and services it provides. A logical starting point in identifying areas to monitor performance includes important organizational functions (addressed in part II of this text), particularly those that are high risk, high volume, or problem prone. Outcomes of care, customer feedback, and the requirements of regulatory agencies are additional areas organizations consider when prioritizing performance monitors. Once the scope and focus of performance monitoring are determined, the leaders define the data collection requirements for each performance measure.

As shown in figure 1.1, monitoring performance depends on the identification of **performance measures** for each process or outcome determined important to track. A performance measure is "a quantitative tool (for example, a rate, ratio, index, percentage) that provides an indication of an organization's performance in relation to a specified process or outcome" (JCAHO 2002). Monitoring selected performance measures can help an organization determine process stability, or can identify improvement opportunities. Specific criteria are used to define the organization's performance measures. Criteria for a performance measure include a documented numerator statement, a denominator statement, and a description of the population to which the measure is applicable.

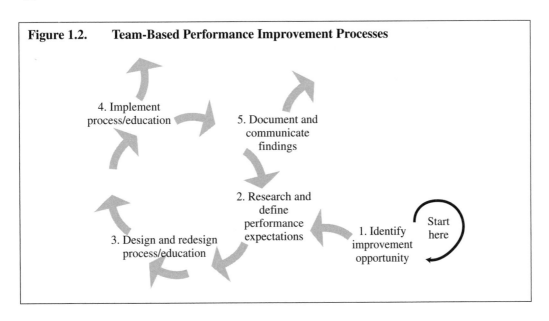

Figure 1.2.   Team-Based Performance Improvement Processes

4. Implement process/education

5. Document and communicate findings

2. Research and define performance expectations

3. Design and redesign process/education

1. Identify improvement opportunity

Start here

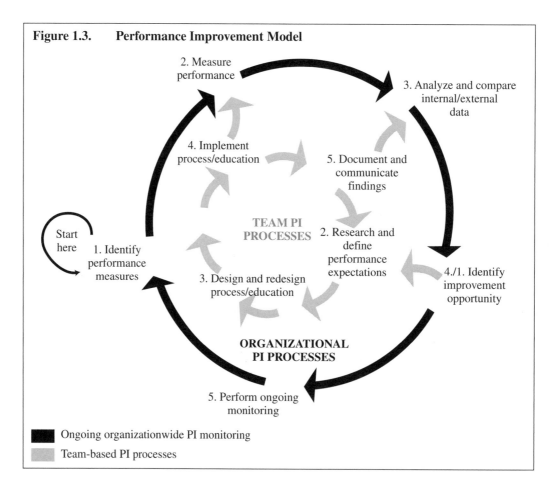

Figure 1.3.   Performance Improvement Model

One important outcome that hospitals are required to continuously monitor is the monthly delinquent medical record rate. The criteria used to establish this performance measure include:

$$\frac{\text{Number of incomplete medical records that exceed the medical}}{\text{staff established time frame for chart completion}}$$
$$\text{Average monthly discharges}$$

The populations included in this performance measure are the medical staff and inpatient medical records. Tracking this outcome allows the hospital to continuously monitor its rate or percentage of delinquent medical records. (See figure 1.1.) If the medical record delinquency rate exceeds the hospital's established performance standards (an internal comparison) or nationally established performance standards (external comparison), an opportunity for improvement has been identified. Following this, a team-based performance improvement process may be initiated. (See figure 1.2.)

When an organization compares its performance over time to its own internal historical data, or uses data from similar external organizations across the country, it helps establish a benchmark, also known as a standard of performance or best practice, for a particular process or outcome. The establishment of a benchmark for each monitored performance measure assists the healthcare organization in setting performance baselines, describing process performance or stability, or identifying areas for more focused data collection. The Joint Commission on Accreditation of Healthcare Organizations (JCAHO) is one available external resource hospitals use to establish their benchmark for medical record delinquency rates. The JCAHO will cite the healthcare organization with a Type 1 recommendation if the total average for medical record delinquency exceeds 50 percent of the average monthly discharges. It is common to see hospitals set the benchmark for their medical record delinquency rate at less than 50 percent.

Once a benchmark for each performance measure is determined, analyzing data collection results becomes more meaningful. Often further study or more focused data collection on a performance measure is triggered when data collection results fall outside the established benchmark. When variation is discovered through continuous monitoring, or when unexpected events suggest performance problems, members of the organization may decide that there is an opportunity for improvement. The opportunity may involve a process or an outcome that could be changed to better meet customer feedback, needs, or expectations.

An example of the PI model used as an improvement opportunity identified from ongoing data collection at Community Hospital of the West is shown in figure 1.4. The hospital had previously identified the employee turnover rate to be an important performance measure to monitor, and had a number of years of historical, internal data on this performance measure, as well as average external comparison data from other hospitals in the community and throughout the state. Using this comparison data, the hospital set its employee turnover rate benchmark at greater than five percent.

During the third year, employee turnover rates began to show a small, but steady, increase from three percent to six percent. After receiving third-quarter data that showed a continued increase in turnover, the Performance Improvement and Safety Council recommended further data analysis by job class. The findings from this analysis showed a pattern

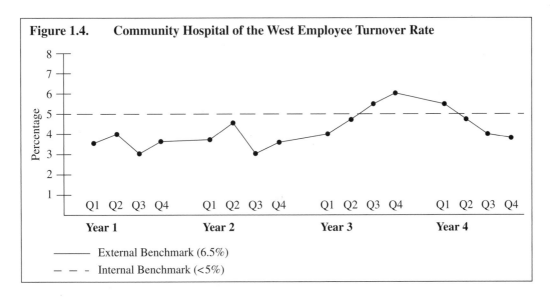

Figure 1.4. Community Hospital of the West Employee Turnover Rate

in employee turnover within nursing. Exit survey data received from nursing staff were also studied, with reasons for leaving linked to salaries and benefits. The council immediately recommended that the human resources department research community salary and benefit packages offered to nursing. The results of the research revealed that Community Hospital of the West's salary and benefit package had not remained competitive and that nursing personnel were being recruited away by hospitals with more attractive benefit packages. Once Community Hospital of the West redesigned, implemented, and advertised its benefit package for nurses, ongoing monitoring in year four showed that the turnover rate had decreased to below the established benchmark. Figure 1.5 shows how the PI model was applied in this situation.

It is a common practice in many healthcare organizations to appoint a leadership group to oversee organizationwide PI activities. This leadership group (sometimes named the PI/Safety Council or Executive Team) is responsible for defining the organization's PI program. Establishment of a PI program includes the following steps:

1. Define and implement the organizatiowide PI model.

2. Establish a staff education plan to train employees in performance improvement.

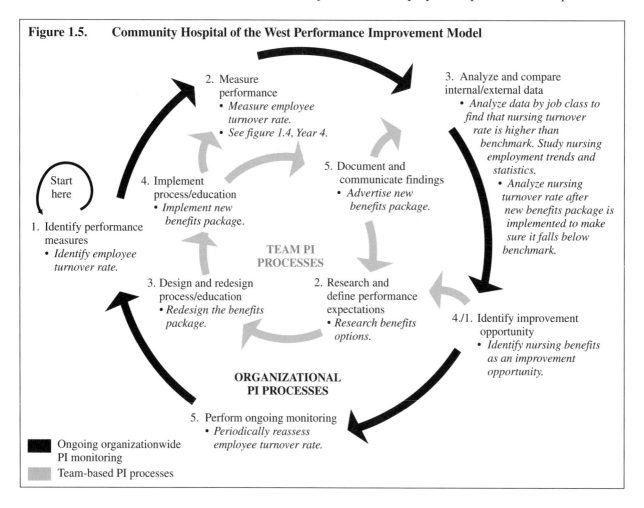

Figure 1.5.   Community Hospital of the West Performance Improvement Model

3. Prioritize PI monitors.

4. Define data collection and reporting responsibilities.

5. Appoint PI teams.

6. Maintain a process of reporting significant findings and corrective actions to the board of directors and other stakeholders.

Part III of this text covers PI leadership responsibilities in greater detail.

Who identifies opportunities for improvement? Participation depends primarily on leadership's commitment to establishing a culture of continuous improvement. Ideally, PI opportunities should be identified by those closest to care or service processes.

In some organizations, however, employees may not feel free to offer suggestions for improvement. Managers who have a traditional style of management may believe that only managers should direct change, and thus might feel threatened by PI processes. In such situations, managers assume responsibility for identifying opportunities for improvement. There is a wide spectrum of PI mind-sets in healthcare organizations.

Once an improvement opportunity is identified, the leadership group can respond in a variety of ways. When the improvement opportunity is believed to be the result of a lack of knowledge or experience, an educational program may be recommended. When the improvement opportunity is the result of inefficiency or ineffectiveness in a work process, the leadership group may convene a **performance improvement team** to examine the process. (The relationship between organizationwide performance monitoring and team-based PI processes is illustrated in figure 1.3.)

Team-Based Performance Improvement Processes

Once an improvement opportunity has been identified through performance monitoring and a team that consists of staff involved in the process under study has been assembled, the first task is to research and define performance expectations for the process targeted for improvement. The first steps may include the following:

1. Create a flowchart of the current process.

2. Brainstorm problem areas within the current process.

3. Research all regulatory requirements related to the current process.

4. Compare the current process to the organization's performance standards and/or nationally recognized standards.

5. Conduct a survey to gather customer input on their needs and expectations.

6. Prioritize problem areas for focused improvement.

Process redesign involves the following steps:

1. Incorporate findings or changes identified in the research phase of the improvement process.

2. Collect focused data from the prioritized problem areas to further clarify process failure/variation.

3. Create a flowchart of the redesigned process.

4. Develop policies and procedures that support the redesigned process.

5. Educate involved staff to the new process.

Performance improvement teams have a variety of tools that they can use to accomplish their goals. This textbook calls these quality improvement tools collected from traditional quality improvement practice and theory **QI toolbox techniques.** The tools make it easier to gather and analyze information, and they help team members stay focused on PI activities and move the process along efficiently. The subsequent chapters of part I of this textbook and all of the chapters in part II introduce at least one technique from the QI toolbox.

After implementation of a new process, continue to measure performance against customers' expectations and other performance standards. The team may need to redesign the process or product when measurements indicate that there is room for further improvement. When measurements indicate that the improvement is effective, ongoing monitoring of the process is resumed (as in figure 1.1). The team documents and communicates its findings to the leadership group and other interested parties in the organization. Results may also be communicated to interested groups in the community.

The team is usually disbanded at this point in the cycle, and routine organizational monitoring of the performance measures is resumed. When another opportunity for improvement arises, the team-based improvement process may be reinstituted.

Summary

Performance improvement in healthcare is a cyclical process. Healthcare professionals are expected to continuously look for opportunities to improve the quality of processes, services, and/or outcomes. Many performance improvement methodologies can be applied in healthcare organizations. Most methodologies, however, follow a model of continuous performance monitoring, ongoing identification of improvement opportunities, and team-based improvement processes.

References

Gaucher, Ellen J., and Richard J. Coffey. 2000. *Breakthrough Performance: Accelerating the Transformation of Health Care Organizations.* San Francisco: Jossey-Bass Publishers.

Joint Commission on Accreditation of Healthcare Organizations. 2002. *2003 Hospital Accreditation Standards.* Oakbrook Terrace, Ill: JCAHO.

Kazandjian, Vahe A., and Terry R. Lied. 1999. *Healthcare Performance Measurement: Systems Design and Evaluation.* Milwaukee, Wis.: ASQ Quality Press.

McLaughlin, Curtis P., and Arnold D. Kaluzny. 1994. *Continuous Quality Improvement in Health Care: Theory, Implementation, and Applications.* Gaithersburg, Md.: Aspen Publishers.

Meisenheimer, Claire G., editor. 1997. *Improving Quality: A Guide to Effective Programs.* Gaithersburg, Md.: Aspen Publishers.

Tacket, S. A. 1991. The quality council: a catalyst for improvement. *Journal of Quality Assurance* 13(5):30–36.

Van Belle, Gerald. 2002. *Statistical Rules of Thumb.* San Francisco: Jossey-Bass Publishers.

Chapter 2
Identifying Improvement Opportunities

Learning Objectives

- To explain the four principal aspects of healthcare that are targeted for performance improvement
- To describe the significance of outcomes and risk reduction in performance improvement methodology
- To explain how brainstorming and the nominal group technique can be used in performance improvement activities

Background and Significance

The American healthcare system is extremely complex. The idea of improving even a tiny element of the system may seem daunting to students new to the concept of performance improvement (PI). Where does the process begin? How are potential areas for improvement identified? To answer these questions, it is important, first, to develop a general understanding of the areas of healthcare services that are the focus of quality improvement efforts.

Most healthcare quality improvement philosophies focus on four areas of healthcare:

- **Structures:** The foundations of care giving, which include buildings (environmental services), equipment (technical services), professional staff (human services), and appropriate policies (administrative systems); also known as service systems
- **Processes:** The interrelated activities in healthcare organizations, which promote effective and safe patient outcomes across services and disciplines within an integrated environment
- **End products:** The final results of healthcare services in terms of the patient's expectations, needs, and quality of life, which may be positive and appropriate or negative and diminishing
- **Cost:** The financial investment of individuals, healthcare organizations, managed care entities, and society in providing healthcare services and its relationship to the quality of care provided

All PI programs in healthcare organizations should address these four aspects of patient care (Donabedian 1988). Of course, not every improvement project can address all four areas, but every improvement project should address quality in at least one of these areas.

Continuous Improvement Builds on Continuous Monitoring: Steps to Success

In order to discover what structures (or systems), processes, end products, or costs need to be improved, a healthcare organization must first find out what is and what is not working with respect to the needs and expectations of its customers. (See the discussion of performance measurement and monitoring in chapter 1.) Most improvement methodologies recognize that the organization must continuously analyze the care environment to identify aspects of the organization that can be improved. (See figure 2.1 for an illustration of the process.)

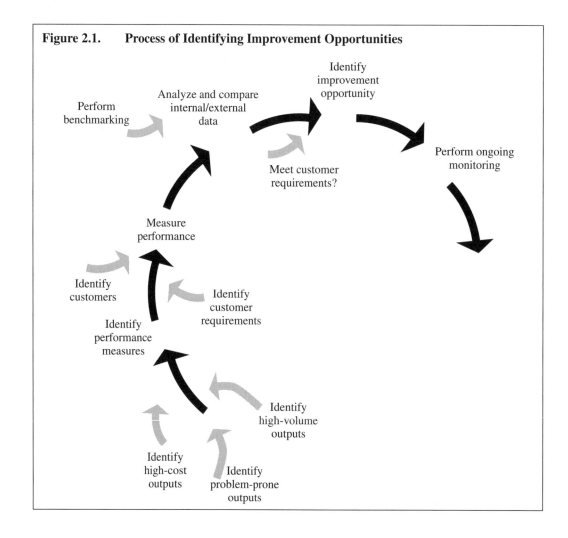

Figure 2.1.   Process of Identifying Improvement Opportunities

Step 1: The organization identifies its end products

End products are the measurable products of the organization's work. In healthcare, end products include:

- The products and services provided to customers, patients, and clients

- The effects services have on customers, patients, and clients (end products such as timeliness, cost, efficiency, safety, and effectiveness)

Many facilities develop measurable criteria to determine if the effects of services on customers are desirable or undesirable related to the end products. The organization must then develop a data collection process or tool within the facility to track information related to these services.

In addition, the processes of every organizational unit have end products that affect other organizational units. This means that the members of the organization are each other's customers. For each process, the end products must first be identified. End products received by customers, patients, and clients, as well as end products received by other members of the organization, must be considered. (See the discussion of internal and external customers in chapter 6.)

Benchmarking is another means through which systems, processes, end products, and cost can be identified for improvement. **Benchmarking** is the systematic comparison of the products, services, and end products of one organization with those of a similar organization. Benchmarking comparisons can also be made using regional and national standards if the data collection processes are similar. This process of benchmarking against an established norm helps the agency to determine whether its processes fall within the acceptable standard deviations of the norm or whether they are outside the norm of these standards. Items that fall outside the norm may be appropriate for PI projects.

Finally, organizations sometimes receive dramatic information about the ineffectiveness of a care process through a sentinel event. **Sentinel events** usually involve the significant

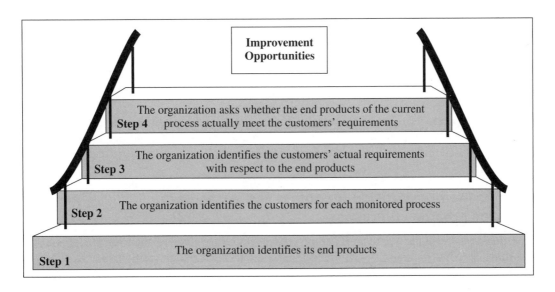

injury or death of a patient or an employee through avoidable causes. The Joint Commission on Accreditation of Healthcare Organizations (JCAHO) defines a sentinel event as an unexpected occurrence involving death or serious physical or psychological injury, or the risk thereof. Serious injury specifically includes loss of limb or function. The phrase "or risk thereof" describes any process variation for which a recurrence would carry significant chance of an adverse reaction. Reviewable sentinel events would include suicide of a patient in a setting where the patient receives 24-hour care, infant abduction or discharge to the wrong family, rape, hemolytic transfusion reaction involving administration of blood or blood products having major group incompatibilities, or surgery on the wrong patient or wrong body part. Such problem-prone end products get the organization's attention very quickly. An analysis of the causes of a sentinel event usually allows the organization to make significant improvements in a process. JCAHO maintains a publication, "Sentinel Event Alert," for information related to sentinel events and outcomes. This information outlines risk reduction strategies for many common sentinel events based on root-cause analysis from different organizations. The review of this information in a credible root-cause analysis allows the organization to measure its own performance against others. (See chapter 9.)

The most important end products are those identified as strategically important services to the organization's overall mission. Some organizations use the criteria of high/low volume, high cost, and problem prone (high risk) to identify the end products that should receive the most scrutiny:

- Processes related to high-/low-volume end products affect numerous customers. High-volume processes increase the risk of incidents through sheer numbers of processes, while low-volume processes increase the risk because the process may be unfamiliar or utilized very infrequently, creating the opportunity for increased errors.

- Processes related to high-cost end products affect the financial health of the organization, especially if the client is underinsured or not insured or if the process requires expensive employee training or certification.

- Processes related to problem-prone (high-risk) end products can result in patient injury or negative outcomes that might open the organization to malpractice suits or other legal actions. Either of these items increases the long-term cost to the patient via increased insurance fees or costly malpractice fees for physicians performing processes.

The end products related to strategically important services and to the organization's overall mission become the organization's performance measures or indicators—those end products by which the quality of the organization and its work units will be measured by internal and external customers.

Step 2: The organization identifies the customers for each monitored process

The question to be answered in this step is who is the receiver of the end products? The list must be exhaustive and should include internal and external customers (as described in chapter 6).

Step 3: The organization identifies the customers' actual requirements with respect to the end products

Identifying actual requirements must be done from the customers' perspective. Identifying the factors most valued by customers is the objective, whether those customers are internal or external to the organization (as described in chapter 6).

Step 4: The organization asks whether the end products of the current process actually meet the customers' requirements

When the answer is yes, another process within the organization is selected and examined. When the answer is no, a PI team may be formed to examine the process in greater detail. (See chapter 3.) Alternatively, an educational program may be developed to fine-tune the organization members' ability to execute the process effectively.

Real-Life Examples

Some ways that healthcare work units have identified improvement opportunities using the techniques introduced in this chapter are discussed in the following paragraphs.

Registration for Day Surgery

At one hospital, many patients had complained about having to come to the facility two and three times for preoperative testing. The PI team assigned to explore the problem identified the following end products, customers, and customers' needs:

Step 1 **End products?**
Registration information provided
Preoperative workup completed
Surgery scheduled

Step 2 **Customers for monitored process?**
Patient and patient's family
Physician and office staff
Surgery staff
Registration staff

Step 3 **Customers' actual requirements?**
One-time communication of registration information
Preoperative workup completed before surgery and coordinated in one visit
Surgery date and time confirmed during the same visit

Step 4 **Customers' requirements met?**
Patient satisfaction surveys showed only 76 percent satisfaction with same-day surgery registration and preoperative workup processes.
Registration staff and physician office staff were hearing complaints from patients that registration process was cumbersome and not user-friendly.
Duplicate data collection occurred between physician office registration and same-day surgery registration.

The data collection process indicates dissatisfaction with services and clearly identifies several problems that could lead to an improved outcome for patients through a performance improvement process with a combined team approach.

Business Office and Health Information Management Department

In another hospital, the number of accounts waiting to be billed had increased over the past six months. The business office and health information management (HIM) department decided to look at the timeliness, appropriateness, and effectiveness of their information-processing procedures. A PI team assigned to examine the process developed the following information:

Step 1   **End products?**
Patient insurance and benefits information
Complete health record documentation
Clinical codes
Billed accounts, revenue increase

Step 2   **Customers?**
Patients
Physicians and other clinical staff
Business office staff
HIM staff
Administration

Step 3   **Customers' requirements?**
Unbilled accounts continuously below $1 million
Health record delinquency rate less than 50 percent
HIM health record completion standards met
Business office benefits verification standards met

Step 4   **Customers' requirements met?**
Unbilled accounts greater than $3 million
Health record delinquency rates greater than 50 percent
HIM backlogs in all chart completion areas
Business office benefits verification at admission occurring only 48 percent
of the time

QI Toolbox Techniques

The most common toolbox techniques used to identify performance improvement opportunities include brainstorming, the nominal group technique, and affinity diagrams.

Brainstorming

Brainstorming can be conducted in a structured or an unstructured way. In *structured* brainstorming, the leader solicits input from team members by going from one to the next around the table or room. Each team member comments on the issue in turn or passes until the next round. This process continues until participants have no new ideas to suggest or until the time period set in the meeting's agenda has elapsed. In *unstructured* brainstorming, members of the team offer ideas as they come to mind. Some members may have no ideas to offer and others may contribute a number of ideas.

In either method of brainstorming, several general rules are followed:

- Everyone agrees on the issue to be brainstormed.

- All ideas are written down on a white board or flip chart in the team members' own words.

- Ideas are never criticized or discussed during the brainstorming period.

- The process is limited in the time allotted, five to fifteen minutes at most.

Nominal Group Technique

The **nominal group technique** gives each member of the team an opportunity to select which ideas under discussion are the most important. This technique allows groups to narrow the focus of discussion or to make decisions without getting involved in extended, circular discussions during which more vocal members dominate. All of the ideas obtained during an earlier brainstorming session are written in a place where everyone can see them. Teams usually use white boards or flip charts for this purpose.

Next, team members vote on the various issues or ideas to determine which should be considered first. The facilitator then writes a numeral 1 by the idea chosen by each team member in turn. This process continues until all of the team members have ranked all of the issues or ideas on a numerical scale. For example, if there were five issues listed, each team member would rank the five issues from 1 to 5, with 1 being the most important and 5 being the least important. Then the facilitator adds up the rankings for each issue. The issue with the lowest sum is selected as the team's choice for most important. The team then works on this issue first, followed by the other four in ranked order.

Another way to vote is to use adhesive dots in various colors. Each team member chooses a color, and the members affix their dots to the five issues they feel are most important. The issue that has the most dots is the most important. This method works well because members can be influenced by other members' placement of votes, thus allowing consensus to begin to develop during the voting process.

Affinity Diagrams

Affinity diagrams are used to organize and prioritize ideas after the initial brainstorming session. This type of diagram is useful when the team generates a large amount of information. The team members agree on the primary categories or groupings from the brainstorming session and then secondary ideas are listed under each primary category. This process allows the team to tackle a large problem in a more manageable way (ASQ 1992). (See figure 2.2 for an example.)

Adhesive notes are an easy organizational tool for this process. They are easily transferred from one category or grouping to another as team members work to group or rate ideas.

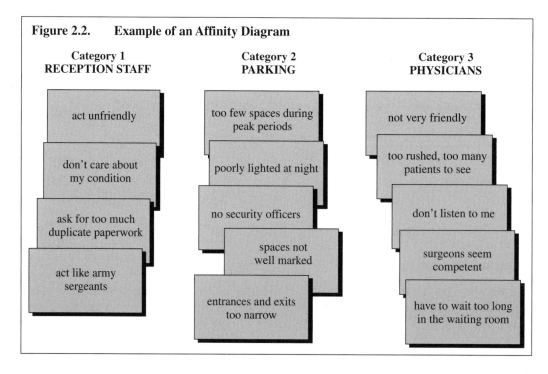

Figure 2.2.   Example of an Affinity Diagram

**Category 1
RECEPTION STAFF**

- act unfriendly
- don't care about my condition
- ask for too much duplicate paperwork
- act like army sergeants

**Category 2
PARKING**

- too few spaces during peak periods
- poorly lighted at night
- no security officers
- spaces not well marked
- entrances and exits too narrow

**Category 3
PHYSICIANS**

- not very friendly
- too rushed, too many patients to see
- don't listen to me
- surgeons seem competent
- have to wait too long in the waiting room

Case Study

Students should watch the movie *The Doctor,* with William Hurt, or *Patch Adams,* with Robin Williams. In class, students should brainstorm positive and negative healthcare quality issues as portrayed in the film they watched. Then they should create an affinity diagram by writing the brainstormed issues on adhesive notes and group similar quality issues together on a white board or flip chart.

Project Application

Students should brainstorm issues or opportunities for improvement that they have observed in their academic environment, general community, work settings, or personal activities. The issues can be educational processes such as course sequencing, professional practice issues, or customer service issues in relation to the bookstore or registration area. Students who are working in a healthcare setting may have issues that student teams could work on and then offer recommendations to healthcare administrators. Governmental processes that do not seem to be working might be good opportunities for improvement. Processes needing improvement at the grocery store or at church might also be evident.

Summary

Four aspects of healthcare are considered the most important areas for performance improvement: structure (systems), process, end products, and cost. Healthcare organizations usually focus on these areas. Many organizations emphasize the need to monitor and evaluate services that are high/low volume, high cost, or problem prone (high risk). Identifying end products for each organizational work unit helps define processes and products produced by the unit and valued by customers.

References

Abdelhak, Mervat, et al. 2001. *Health Information: Management of a Strategic Resource.* Philadelphia: W. B. Saunders.

Albright, J. M., et al. 1993. Reporting tools for clinical quality improvement. *Clinical Performance Quality in Health Care* 1(4):227–32.

American Society for Quality. 1996. *Total Quality Tools: PQ Systems.* Milwaukee, Wis.: ASQ Quality Press.

Donabedian, Avedis. 1988. The quality of care: how can it be assessed? *JAMA* 260(12):1743–48.

Joint Commission on Accreditation of Healthcare Organizations. 2003. *The Measurement Mandate.* Oakbrook Terrace, Ill.: JCAHO.

Chapter 3
Applying Teamwork in Performance Improvement

Learning Objectives

- To identify the effective use of teams in performance improvement activities

- To describe the composition of performance improvement teams

- To outline the advantages that agendas lend to performance improvement team processes

- To enumerate the differences between the roles of the leader and the members in performance improvement teams

Background and Significance

Performance improvement (PI) teams involve a number of people working over long periods of time. They are expensive in terms of both time and money. Therefore, PI teams should be considered an organizational resource to be used appropriately and prudently.

The early quality improvement philosophies of Deming, Juran, and Crosby were based on team processes. Therefore, when quality improvement methodologies were first applied to healthcare organizations, it was assumed that PI activities were best accomplished by PI teams. Experience with PI processes in healthcare settings, however, has shown that PI teams are not always required.

Today, when an improvement opportunity is identified, the organization's leadership or performance improvement council can initiate one of three approaches:

- Establishing a **blitz team**

- Disseminating information or developing an educational training program

- Developing a cross-functional PI team

Sometimes the leaders of an organization may decide that they already have all of the facts they need about an improvement opportunity. In such cases, they may decide to establish a blitz team. Blitz teams do not spend a lot of time gathering data and reengineering processes. These teams are usually composed of individuals in a similar group who are very familiar with the processes or end products that need improvement (Caldwell 1998). For example, the members of the health information services (HIS) department may want to increase response time to chart sign-off. The HIS department may plan several ways to make charts more available in their department for all other staff or they may create a more conducive environment for privacy during documentation. The blitz team constructs relatively simple fixes that improve work processes without going through the whole PI process and without the need to involve other departments. Thus, an improvement can be implemented without a major investment of time, personnel, and resources.

Some improvement opportunities involve only the dissemination of information or better individual training. Again, this approach to implementing an improvement can be accomplished without expending large amounts of resources.

When, however, the improvement opportunity is complex and involves multiple departments or multiple work units within a department, the team-based approach is required. A **PI team** is instituted to research, plan, and implement the improvement. For example, a multidisciplinary team may be needed to address a shortage of blood or blood products in a hospital that cares for emergency, surgical and obstetrics patients, any of who may need blood at differing times during their hospital stay. Individuals from all areas including the laboratory, administration, surgery, obstetrics, and emergency room may need to help set priorities and explain the particular needs of their own areas.

The goal of this chapter is to discuss the composition of PI teams and the roles of individual members. The purpose of team mission and vision statements is also explained.

Team Composition

When convening a cross-functional PI team is appropriate, the leadership of the organization must determine the composition of the team. The PI team should be made up of individuals close to the process to be improved, because they are best qualified to accomplish the process review. This may mean that some teams will include more staff than managers. Some questions that should be asked to identify individual team members include the following:

- Which departments or disciplines are involved in the process?

- Who are the customers of the process? In other words, who will receive the product or service that the process produces?

- Who supplies the process? In other words, who provides materials or services for use in the process under investigation?

Limitations on the number of people who can participate on a team and other factors sometimes mean that all of the people involved in a process to be improved cannot participate directly on the PI team. When this happens, the team must make provisions to contact the other individuals affected by the improvement initiative. Because their perspectives and information are of critical importance, some means for obtaining their input must be developed.

It is important to keep teams small and manageable. The general rule of thumb is that teams should include no more than ten members. Once the team has been formed, the team members should determine which individuals, departments, and disciplines are key players in the process. Then they can bring other individuals, departments, and disciplines into the process on an ad hoc basis when necessary.

Team Roles

After the team has been selected, a **team leader** should be chosen. Having a leader is necessary to get the team organized in the most basic ways. The organization's leaders may select the team leader, or the team itself may select its own leader.

The team leader should be someone whom the team respects as well as someone who is organized and will take the initiative to see the team through the process. The team leader is primarily responsible for championing the effectiveness of the process in meeting customers' needs. He or she is also responsible for the *content* of the team's work. The team leader is responsible for the following specific activities:

- Preparing for and scheduling meetings (standard meeting day, time, and location)

- Sending out announcements of meetings and any other necessary materials

- Conducting meetings (the importance of following an agenda is discussed later in the QI toolbox techniques section of this chapter)

- Focusing the group's attention on the task at hand

- Ensuring group participation and asking for facts, opinions, and suggestions

- Providing expertise in the organization's performance improvement methodology and PI tools and techniques

- Overseeing data collection

- Making task assignments

- Facilitating implementation of action plan items

- Conducting critiques of the meetings

- Serving as the primary spokesperson and presenter for the team

- Keeping attendance records

- Contacting absent members personally to review the results of the meeting and provide any materials that were distributed during the meeting

The role of the team member includes the following functions:

- Participating in decision making and plan development for the team

- Identifying opportunities for improvement

- Gathering, prioritizing, and analyzing data

- Sharing knowledge, information, and data that pertain to the process being investigated

The role of the secretary is primarily that of record keeper.

- Keeping accurate minutes from each meeting, including attendance
- Ensuring participating team members receive minutes in a timely manner along with scheduled meeting times
- Producing an agenda for new meetings with assignments for team members from the previous meeting

A variety of other roles may be implemented in performance management teams. (See chapter 16.)

Mission and Vision Statements

Healthcare organizations use mission statements, vision statements, and values statements at many different levels. The corporation as a whole may have a mission and vision statement, as may the separate divisions of the corporation. Departments within facilities may also have mission and vision statements.

To be effective, mission, vision, and values statements should be developed in concert with the organization's overall goals as developed in the organization's strategic plan. The statements should reflect the mission and vision of the overall organization as well as the goals of the individual PI team. The mission statement for a PI team identifies its name, what it does, and whom it serves. For instance, the mission of the health information PI team is to provide quick accurate billing to all clients and third-party payers in an honest, efficient, user-friendly manner. The vision statement for the team may be "better service to all" and its values will be honesty, efficiency, and user-friendliness.

After the team has been assembled and the leader has been chosen, the team should establish its mission. Developing a mission statement can help both the team and the larger organization identify the goals and purpose of the performance improvement initiative. The team's mission statement should answer the following questions:

- What process is to be improved?
- For whom is the process performed?
- What products does the process produce?
- What is not working with the current process?
- How well must the process function?

(See the example in figure 3.1.)

Figure 3.1.   Sample Mission Statement

Evaluate the HIM lab in regard to accessibility, resources, library access, Internet access, quality of equipment, and adequacy of equipment while maintaining 95 percent HIM student satisfaction with these services.

The team should also articulate its vision for the process. A vision statement is a description of the ideal end state or a description of the way the process should function. (See the example in figure 3.2.)

The team's vision of the way a process should function may not be validated by existing data and observations of the process. In such cases, a disharmony between the vision and reality becomes apparent. This disharmony represents the team's opportunity for improving the process. Humans have a natural tendency to resolve disharmony. The more clearly the group maintains its focus on "what should be" and acknowledges "what is," the more powerfully the team will be drawn toward implementing its vision of the desired outcome.

Quite often, organizations give up their vision because of the discomfort connected with the schism between their vision and the current reality. Some organizations focus on their vision and ignore the way things are or believe the current situation is better than it really is. In either case, the natural tendency for resolution or change is dissipated.

The team that focuses on "what should be" and at the same time maintains an accurate description of the current state of the process takes a powerful step toward creating the results it envisions. When the PI team focuses clearly and consistently on its mission and vision, it naturally and almost effortlessly gets a sense of what still needs to be done. New processes seem to announce themselves, and the group becomes increasingly aware of additional opportunities for continuing improvement.

Real-Life Examples

Three examples of the effectiveness of PI teams follow. The first example focuses on the triage process in a small metropolitan hospital. In the second example, health information services and business office department processes are explored. Finally, in the third example, safety issues in the laboratory of a hospital are considered.

Continuum of Care Team

In a small metropolitan hospital, the triage process in the emergency department was inadequate. Communication was fragmented. Precertification was not taking place in a timely manner. Intake processing time had increased. The main patient waiting area was not private, and referral volume was increasing.

To address these problems, a PI team was instituted. The team included the following people:

- Team leader: Emergency department intake coordinator

- Team members: Representatives from the business office, health information services, administration, utilization review, and finance

- Ad hoc members: Representatives from regulatory affairs, reception, nursing, case management, and an emergency department physician.

Figure 3.2.   Sample Vision Statement

The HIM lab provides access to a variety of application software resources, library knowledge bases, and the Internet.
A convenient, comfortable work environment exists.

After analyzing the situation, the team's vision was to "Design a centralized clinical assessment center to facilitate patient privacy, data collection, and staff communication."

Health Information Services and Business Office Services

An organization's health information services department was experiencing an increase in physician chart delinquencies, delays in diagnosis and procedure coding, and delays in processing requests for patient information. In addition, the business office was experiencing delays in the billing process due to incomplete insurance information and an increase in accounts payable due to the chart completion problems occurring in the HIS department.

To address these issues, a PI team was formed. The team was composed of the following individuals:

- Team leader: Health information services (HIS) consultant

- Team members: Representatives from both the health information services department and the business office, registration, administration, and the physician chairman of the HIS committee

The HIS department's mission was to "contribute and provide support to the effective organizationwide management of information." The business office's mission statement was to "give support to organizationwide financial viability and provide accurate and timely exchange of financial information that allows for an efficient and effective billing and collection process."

This team made a recommendation to the HIS committee to change the physician chart completion policy from thirty days following discharge to seven days following discharge. This improved the coding turnaround time and requests for patient information, which in turn improved the billing cycle. Additional process change was to include a business office representative interviewing the patient upon admission. This eventually improved the accuracy of financial information needed for billing and collection.

Clinical Laboratory Services

Safety issues and other problems concerning the laboratory department at one hospital increased 207 percent over a period of one year. The objectives of the PI team were to identify specific problem areas within laboratory services, conduct a baseline survey to assess each area, analyze the survey results, develop an action plan, implement improvements, and evaluate the results of the changes. The team was made up of the following people:

- Team leader: Chief of pathology

- Team members: The laboratory manager, an emergency department physician, the director of nursing, a laboratory technician

The mission of this team was "to provide reliable, timely diagnostic services for the clinical staff." The survey indicated that inappropriate techniques for specimen collection were occurring and that the reference laboratory was slow to return results to the organization.

Team recommendations included education and training on specimen collection and a change in the contract reference laboratory used by the organization.

QI Toolbox Technique

For a meeting to be effective, the team must operate with a common purpose and specific goal. Communication of the meeting's common purpose and specific goal is usually accomplished by establishing an **agenda.** An agenda is a list of the tasks to be accomplished during a meeting. Using an agenda ensures that every team member knows what items will be discussed or worked on. The agenda should be sent to all team members before the meeting. This allows them to prepare ahead of time to discuss specific agenda items. The agenda should also include an indication of how long the team will spend on each item. (See the sample agenda in figure 3.3.) Setting time frames for agenda items helps the team leader to keep the group focused on the process and moving forward.

Standard agendas begin with a review and approval of the last meeting's minutes. Once this has been accomplished, the PI team should review the agenda for the current meeting and approve the time frames that have been set. This allows the individual team members to have input into how long a certain agenda item should be discussed. Next, the specific PI steps for current discussion should be listed and discussed.

As a closing item of business, many teams find it helpful to evaluate the meeting itself in terms of its effectiveness. Asking the following questions may be helpful:

- Did the team accomplish what it set out to accomplish during the meeting?

- Is the PI process moving forward?

- Does the team need to ask additional people to sit in on the process meetings?

- Did members participate appropriately, listen effectively to other members' suggestions, and stay focused on the agenda?

Figure 3.3.   Sample Agenda

AGENDA

Date: January 15   **Team:** Registration Process
Time: 10:00 a.m.   **Place:** Conference Room B

Time Allotted: **Agenda Item:**

Time Allotted	Agenda Item
5 minutes	1. Review and approve minutes from last meeting
5 minutes	2. Review agenda and time frames
15 minutes	3. PI step: Registration process discussion on how the computer system affects the registration process
15 minutes	4. PI step: Registration process discussion on what happens now when the computer system is "down"
15 minutes	5. Brainstorm possible ideas to improve the computer system
10 minutes	6. Process (evaluate) meeting
10 minutes	7. Plan next steps and agenda for next meeting

Finally, the next meeting's agenda should be agreed upon and tied to the current meeting's evaluation process and minutes. In other words, the next meeting should be designed in light of the accomplishments of the current meeting.

Case Study

"Well, why does it have to take so darn long?" the nurse shouted into the telephone receiver. "We've got to be able to order tests for our patients! They can't wait until next Christmas for their meds!" She punched the receiver onto its cradle and turned to the rest of the staff collected at the nursing station. "436 was transferred here from ICU two hours ago and I still can't enter any orders. How do those idiots in Admitting think we're supposed to get our work done? I will never understand why it has to take so long to get a patient transferred in that darn system."

"That witch!" exclaimed the admissions clerk to her supervisor as she hung up the phone. "If someone would let us know once in a while what they're doing with patients in this place, maybe we could do our jobs! Evidently, they transferred Mr. Campbell to 436 from ICU hours ago!"

The clerk walked over to the report printer. A long, wide ribbon of paper hung from it onto the floor where it curled in a short pile. She yanked at the hanging pages and ripped them along a perforation. Then she sat down at a desk, where she began to sort the half-sheet messages from the hospital network communication system into different piles. When she was finished sorting, she began entering the changes in status and location for each of the patients on each message into the patient accounting and order-entry system.

"But NO, instead of giving us a call and telling us that the patient's been transferred, they'd rather wait 'til a couple of hours later and the patient doesn't have his meds and then call up and rag on us to death like it was OUR fault!" She entered "Campbell, Roy" from one of the half-sheets, and the patient's location data came up on screen. She keyed 4-3-6 in the location field and pressed Enter.

University Hospital employees had been confronted by this same kind of conflict for two years, ever since management purchased a new patient accounting and order-entry system from ABS Company. XYZ Software was used by clinicians to look up outpatient histories, laboratory reports, and other diagnostic data for both inpatients and outpatients. It also provided departments with information about the current location of the patient so that they would know where to send the final paper-based reports of blood tests, X rays, and other diagnostic reports. But the nursing staff had never been very careful about transferring or discharging patients in the system. It was one of those tasks that took secondary importance to the other patient-oriented duties that they had to perform in their hectic schedules.

University Hospital was the principal tertiary care hospital in a community of 200,000, and patients were referred to it from all over the state. The census never dipped below 70 percent. As a result, the nursing units were almost always full, and the patients had high acuity levels.

The admitting and patient accounting departments had been requesting a new information system for many years when administration finally agreed to make the investment in the ABC system. The new system allowed them to tie order-entry functions to the geographic location of the patient in the hospital. Thus, the system provided more accurate information to lab and other diagnostic testing personnel such as ECG and X ray to inform them of the patient's whereabouts when a test was ordered. This feature would also ensure that the patient's account was charged correctly for room rate, which had never been the case when they were relying on the XYZ system. Nursing staff had never seemed to get changes in room location entered into the system in a timely fashion. And the census from the XYZ system was never accurate—everyone knew that.

The managers of the admitting and patient accounting departments had stood their ground when the new work procedures around the ABC system were developed. They were adamant that nursing was not going to be in charge of patient census management. It had been obvious for years that nursing could not take it seriously. The admitting department staff would be making all changes in patient location in the system. Finally, their system would generate an accurate daily inpatient census. They had IS programmers add a short utility in XYZ so that whenever a patient was discharged or transferred in the XYZ census management application, a notification message of same would print out on the admitting department printer. Periodically, during the day, the admitting clerks would take the messages and enter them into the ABC system. If for some reason they were not notified, they had the authority to go into the system and change the dates and times of the discharge or transfer so that the accounting files would be correct and the patients' bills would reflect correct charges. One additional feature of the ABC system, however, was that orders on a patient could only be entered from the nursing unit on which the system currently had the patient located.

Case Study Questions

1. In your opinion, is there an opportunity for improvement in this system? Why or why not?

2. If there is, is a PI team appropriate in this context?

3. From your knowledge of hospital organizational structure, who should be on the team? What is your rationale for including each individual?

Student Project Application

Students should form teams of two to four members. Each team should select a process or product to evaluate for improvement opportunities. After each team has identified a process or service to evaluate, each should develop a mission statement for the team and a vision statement for its process. This will help each team to identify its goals and purposes for the PI process as well as to clarify what it is trying to accomplish. Mission and vision statements can be written on newsprint flipchart paper similar to that used in healthcare organizations. The statements should then be displayed on the wall so that the students can share their initial mission and vision statements and introduce their proposed projects.

Summary

Performance improvement teams often undertake PI activities in healthcare organizations. The team process allows the members to represent the varied perspectives of departments and other entities in the organization. At a minimum, a team should include a team leader and several team members. Mission and vision statements and team meeting agendas help keep the team focused on team activities.

References

Caldwell, Chip, editor. 1998. *Handbook for Managing Change in Health Care.* Milwaukee, Wis.: ASQ Quality Press.

Carboneau, C. E. 1999. Achieving faster quality improvement through the 24-hour team. *Journal of Healthcare Quality* 21(4):4–10.

DeMarco, Tom, and Timothy R. Lister. 2000. *Peopleware: Productive Projects and Teams,* 2nd edition. New York City: Dorset House Publishing.

Fast Company Magazine. 2001. *What Makes Teams Work: The Teamwork Syllabus (eBook).* New York City: Gruner + Jahr USA Publishing.

LaVallee, Rebecca, and Curtis P. McLaughlin. 1999. Teams at the core. In *Continuous Quality Improvement in Health Care, 2nd edition,* Curtis P. McLaughlin and Arnold D. Kaluzny, editors. Gaithersburg, Md.: Aspen Publishers.

Lencioni, Patrick. 2002. *The Five Dysfunctions of a Team: A Leadership Fable.* San Francisco: Jossey-Bass.

Lynch, Robert F., and Thomas J. Werner. 1992. *Continuous Improvement: Teams and Tools.* Atlanta: QualTeam.

MacMillan, Pat. 2001. *The Performance Factor: Unlocking the Secrets of Teamwork.* Nashville: Broadman & Holman.

O'Malley, S. 1997. Total quality now! Putting QI on the fast track. *Quality Letter for Healthcare Leaders* 9(11):2–10, December.

Parisi, Lenard L. 1994. Quality improvement teams and teamwork. In *Improving Quality: A Guide to Effective Programs,* Claire G. Meisenheimer, editor. Gaithersburg, Md.: Aspen Publishers.

Chapter 4
Aggregating and Analyzing Performance Improvement Data

Learning Objectives

- To differentiate between internal and external benchmark comparisons
- To identify common healthcare data collection tools
- To introduce the concept of data aggregation in support of data analysis
- To describe the various data types
- To recognize the correct graphic presentation for a specific data type
- To design graphic displays for a given set of data
- To analyze the data for changes in performance displayed in graphic form

Background and Significance

After the performance improvement team has administered its survey or collected data by abstracting it from other sources, the team is ready to aggregate and analyze the data. Other sources of data may include medical records, organizationwide incident reports, annual employee performance evaluations and staff competency results, and so forth. Abstracted data and survey results can provide invaluable information about a process and thereby point the team in a specific direction for improvement. Internal and external data comparisons, also known as **benchmarking,** can provide additional information about why and how well the process works—or does not work—in meeting the customers' expectations.

It is best to conduct an internal data comparison with data collected over a period of time. Collecting data over a three- to six-month time frame, for example, establishes an internal baseline for benchmark purposes. To establish the baseline, the organization averages all collected data. This internal baseline becomes the organization's benchmark to maintain or improve upon when external benchmark comparisons are not available.

Comparing an organization's performance to the performance of other organizations that provide the same types of services is known as external benchmarking. The other organizations need not be in the same region of the country, but they should be comparable in terms of patient mix and size. The use of external benchmarks can be instructive when comparisons are made to an organization that is doing an outstanding job with a process similar to the process on which the PI team is focusing. For example, if the national standard for the average number of adverse drug reactions were *X*, then comparing an organization's number to the national average would give the team information about the effectiveness of the organization's medication program. The JCAHO's ORYX data is another external benchmark that permits rigorous comparison of the actual results of care across hospitals.

In figure 4.1, note the comparison of Western States University Hospital's community-acquired pneumonia data to those of similar hospitals using the same ORYX data vendor. This report allows the organization to do both internal and external comparisons. Internal comparisons can be performed on the mean length of stay observed between specialties of practice (Mean LOS Obs). For instance, note the difference in mean length of stay between the Family Practice Service and the General Internal Medicine Service. External benchmarking can also be performed comparing the mean length of stay observed and mean length of stay expected (Mean LOS Exp). Benchmarking on mortality can be performed using the columns % Deaths Obs and % Deaths Exp. Where might Western States University Hospital want to focus attention on its pneumonia mortality rate?

Several tools can be used by PI teams for data aggregation, analysis, and presentation. The tools are discussed later in this chapter.

Data Collection Tools

Certain types of data and information need to be accumulated over time to support clinical and management functions. The organization must assess its need for aggregate data and information, and define the types of required data and information to support individual care and care delivery, decision making, management and operations, analysis of trends over time, performance comparisons over time within and outside the organization, and performance improvement. The common types of data collection tools used in healthcare organizations include: incident reports, safety and infection surveillance reports, employee performance appraisals, staff competency examinations, restraint use logs, adverse drug reaction reports, surveys, diagnosis and procedure indices, and peer review reports. Individually, these tools do not describe the quality of care being provided at the healthcare organization. The data from these reports must then be aggregated to provide useful information on the organization's performance in key areas.

To aggregate data, the values of one data element over a set period of time (such as a month or quarter) are added together. These aggregated data are then compared to previous months or quarters to determine if there is a variance from the established benchmark. Chapter 15 discusses the transformation of data into knowledge in more detail and figure 15.1 shows an example of aggregate data over time between quarters.

When an organization needs to collect a new data element that is not already a part of the existing data collection tools, a check sheet may be used.

Check Sheets

A **check sheet** is used when one needs to gather data based on sample observations in order to detect patterns. When preparing to collect data, a team should consider the four *W* questions:

- *Who* will collect the data? For example, using the JCAHO's Core Measure for heart failure, determining who will collect the data is vital to data accuracy.

Figure 4.1.   LOS Summary by Attesting Physician Specialty

Physician Specialty	Cases	Mean LOS (Obs)	StDev LOS (Obs)	Mean LOS (Exp)	LOS Index	Savings Opp (Days)	% 30 Day Readmit	% With Comp's	% Deaths (Obs)	% Deaths (Exp)	% Early Deaths
Cardiology	1	6.00		6.36	0.94	0	0.00	0.00	0.00	0.75	0.00
Endocrinology	1	3.00		3.93	0.76	−1	0.00	0.00	0.00	0.83	0.00
Family Practice	17	8.47	11.44	6.26	1.35	38	11.76	0.00	0.00	4.27	0.00
Infectious Disease	4	4.00	2.94	4.35	0.92	−1	0.00	0.00	0.00	1.13	0.00
General Internal Medicine	34	3.82	3.00	4.89	0.78	−36	0.00	0.00	5.88	4.39	2.94
Nephrology	3	3.67	0.58	4.99	0.73	−4	0.00	0.00	0.00	4.03	0.00
Neo/Perinatal Med	1	4.00		3.69	1.08	0	100.00	0.00	0.00	0.92	0.00
Prev/Occ Med	6	4.83	4.83	6.37	0.76	−9	16.67	0.00	33.33	16.57	16.67
General Pediatrics	7	3.43	2.15	3.55	0.97	−1	0.00	0.00	0.00	0.18	0.00
Pulmonary/Crit Care	2	17.50	21.92	14.65	1.19	6	0.00	0.00	50.00	61.46	50.00
Gen Diag/Interv Radiology	1	1.00		12.02	0.08	−11	0.00	0.00	100.00	80.25	100.00
Rheumatology	1	2.00		3.40	0.59	−1	0.00	0.00	0.00	0.39	0.00
Internal Med Heme/Onc	1	3.00		3.71	0.81	−1	0.00	0.00	0.00	0.90	0.00

Legend:

Comp's = Complications
Crit = Critical
Diag = Diagnostic
Exp = Expected
Gen = General
Heme = Hematology

Interv = Interventional radiology
LOS = Length of Stay
Med = Medicine
Neo = Neonatal
Obs = Observed
Occ = Occupational

Onc = Oncology
Opp = Opportunity
Prev = Preventive
Readmit = Readmission
StDev = Standard deviation

More often than not, the *who* collecting the heart failure measures will be a clinician with a background in cardiology. A nonclinical person can be trained to look for specific documentation of the measure in the health record, but the complexity of the heart failure data makes a cardiology clinician the best candidate for this type of data collection.

- *What* data will be collected? Here the JCAHO Core Measure is defined for the organization. The data elements for this core measure include discharge instructions, left ventricular failure, treatment with ACE inhibitors for left ventricular systolic dysfunction, and adult smoking cessation counseling.

- *Where* will the data be collected? Data for the heart failure core measure will be abstracted or collected most often from the individual patient health record. However, some measures may have to be collected from other data sources. There could conceivably be core measures that required the abstractor to go to other electronic systems like those used in the ICU that might not be printed for the paper-based record.

- *When* will the data be collected? This question is defined by time parameters. The data for the JCAHO Core Measures are collected on patients for a three-month period (such as July, August, and September) and must be reported to JCAHO four months from the end of the last month of the reporting quarter (for reporting on July, August, and September, this is the end of January) (JCAHO 2003b).

Once the team has the answers to the four *W* questions, it can develop a check sheet to collect the data. (See figure 4.2.) Check sheets make it possible to collect a large volume of data systematically. It is important to make sure that the data are unbiased, accurate, properly recorded, and representative of typical conditions for the process.

A check sheet is a simple, easy-to-understand form used to answer the question, "How often are certain events happening?" It starts the process of translating opinions into facts. Constructing a check sheet involves the following steps:

1. The PI team determines who is the responsible person to collect the data: clinician, technician, or other person.

2. The PI team comes to an agreement on which event it wants to observe.

Figure 4.2.   Example of a Check Sheet

Problem	Day			
	1	**2**	**3**	**TOTAL**
A	II	III	II	7
B	I	I	I	3
C	IIII	II	IIII	10
TOTAL	7	6	7	20

3. The team decides on the time period during which the data will be collected. The time period can range from hours to weeks.

4. The team determines the appropriate source from which the data will be collected. This may be health records, charge slips, and so forth.

5. The team designs a form that is clear and easy to use. The team should make sure that every column is clearly labeled and that there is enough space on the form to enter the data.

6. The team collects the data consistently and honestly. Enough time should be allowed for this data-gathering task.

Check sheets can also be used to tally responses on surveys. (See survey design and administration in chapter 6.) For example, if a survey included a question that asked for the days of the week on which patients had surgery, the results could be tabulated by using a check sheet that included each day of the week.

Data collection in healthcare organizations does not have to develop new methods for every PI project. Organizations must determine what they are already collecting and how those data can be folded into the PI measurement process for use in the future.

Types of Data

Before decisions can be made on how to display performance data, a determination of what type of data has been collected must be made. The four data categories are nominal, ordinal, discrete, and continuous.

Nominal data, also called categorical data, include values assigned to name-specific categories. For example, gender can be subdivided into two groups, "male" and "female," or two categories, "1" and "2." Nominal data are usually displayed in bar graphs (figure 4.8) and pie charts (figure 4.11).

Ordinal data, also called ranked data, express the comparative evaluation of various characteristics or entities, and relative assignment of each, to a class according to a set of criteria. Many surveys use a Likert scale to quantify or rank statements. A Likert scale, in which the respondent may state the degree to which he disagrees or agrees with the statement, typically ranges from 1 to 5. This type of scale allows the PI team to determine how respondents feel about issues. Ordinal or ranked data, like nominal data, are also best displayed in bar graphs and pie charts.

Discrete or count data are numerical values that represent whole numbers, for example, the number of children in a family or the number of unbillable patient accounts. Discrete data can be displayed in bar graphs.

Continuous data assume an infinite number of possible values in measurements that have decimal values as possibilities. Examples of continuous data include weight, blood pressure, temperature, and so on. Continuous data are displayed in histograms (figure 4.9) or line charts (figure 4.12).

Two terms often used in data analysis are **absolute frequency** and **relative frequency.** Absolute frequency refers to the number of times that a score or value occurs in the data set. For example, in the data set shown in table 4.1, the frequency of the score or value 25 is 3. Relative frequency is the percentage of the time that the characteristic appears. In the example, the percentage of observations on which the respiration rate was 25 was 3:7, or 42.9 percent, its relative frequency.

Statistical Analysis

Methods of statistical analysis that are necessary to the graphic display of data include the mean and standard deviation. Although an in-depth discussion of these techniques can be found in any elementary statistics text, a brief review is provided here before discussion of graphic display.

The **mean** is also known as the arithmetic average of a distribution of numerical values. The values may be discrete, count, or continuous in nature. If the data are discrete or count, the mean should be rounded to the nearest whole value. If the data are continuous, whole numbers or numbers with decimal fractions can be reported.

To calculate the mean, the various observed values are first summed, or added together. There may be repetitions of specific values, all of which are included. Then, the sum is divided by the number of observations made. For example, fifteen observations of a person's systolic blood pressure revealed the following data set:

| 122 | 124 | 116 | 115 | 120 | 128 | 126 | 122 | 121 | 121 | 124 | 120 | 117 | 116 | 121 |

Note there were three 121s and two 122s in the set. All are summed. The sum of those observations is 1,813, which is then divided by the number of observations made (15). This equals 120.866666666666 . . . . Next, round to a whole number, because systolic blood pressure is commonly expressed as whole numbers. The reported mean, or average systolic measurement, is 121.

The **median** is derived usually without calculation. Instead, the observed values are placed in ascending or descending order. Then, the value that is in the very middle of the set is taken as the median. There has to be an odd number of observations in order for there

Table 4.1. Sample Data Set	
Date and Time	**Respiration Rate Recorded**
Jan. 5, 8:00 a.m.	22/min
Jan. 5, 12:00 noon	25/min
Jan. 5, 4:00 p.m.	25/min
Jan. 5, 8:00 p.m.	22/min
Jan. 5, 12:00 midnight	21/min
Jan. 6, 4:00 a.m.	23/min
Jan. 6, 8:00 a.m.	25/min

to be a value in the middle. In the data set above, the values are rearranged in ascending order as follows:

115	116	116	117	120	120	121	121	121	122	122	124	124	126	128

Note again that the repeated values are retained. The middle value is 121, as indicated by the arrow above. Thus, 121 is the median. If, however, there were an even number of values, the middle falls between the two values at place 7 and 8 in the row.

115	116	116	117	120	120	121	121	121	122	122	124	124	126	

The values in places 7 and 8 are added together and divided by 2. Here that would be 121 + 121 = 242, which divided by 2 = 121. If the values in place 7 and 8 had been 122 and 125, adding them together = 247, which divided by 2 = 123.5.

It is sometimes better to use a median value in displaying some graphic representations of data, particularly if the observed values have a lot of variation in them or are skewed to one side or the other. Skewing means there are a lot of very high or very low values in the observations that distort the calculated mean. Because the median is not calculated, if the data set is greatly distorted by the extreme values, it can help in defining a more true picture of the middle of the set.

The **standard deviation** (SD) is a more complex analysis technique used in developing control charts for the display of some PI data. (There is a discussion of control charts later in this chapter.) The standard deviation is most easily calculated using the statistical analysis feature of a spreadsheet application. To do so, enter and highlight the column of data and select the standard deviation function from the menu bar button "fx." Alternately, one can choose a cell to contain the SD and enter =STDEV( ) and record the range inside the parentheses.

Although the SD is easily calculated using a spreadsheet application, what this reveals about a data set is more difficult to understand. When a PI team begins to make observations of a continuous measure, the observations are plotted along the x- and y-axis with very little clustering, or with no discernable pattern or trend. (See figure 4.3.) This can be

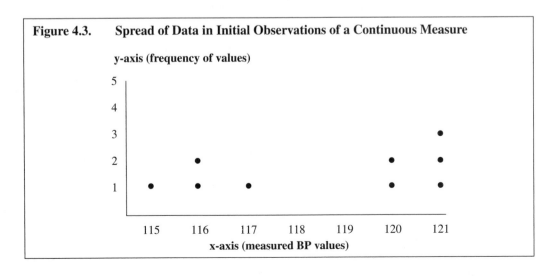

Figure 4.3.   Spread of Data in Initial Observations of a Continuous Measure

graphed as in figure 4.2 on an x-y axis graph where the x-axis is the values of the measure and the y-axis is the absolute frequency of that value in the set. Taking the systolic blood pressure data set from above, it could be graphed as depicted in figure 4.3.

Taking the first nine values, we see 115 with an absolute frequency of 1, 116 an absolute frequency of 2, 117 an absolute frequency of 1, and so on. As the measure is observed more and more times, however, the observed values begin to congregate more often around the mean as in figure 4.4.

As the number of observations increases into the hundreds or thousands, they begin to form a graph that looks like figure 4.5, the typical bell-shaped curve of what is called a *normal distribution*. Almost all measures, when graphed, will take on this bell-shaped or "normal" appearance as the number of observations increases and increases.

At the center or vertex of the normal distribution is the calculated mean. For the small data set above, 121 was calculated as the mean of the observed blood pressures. Many people know that the commonly accepted "normal" or mean value for systolic

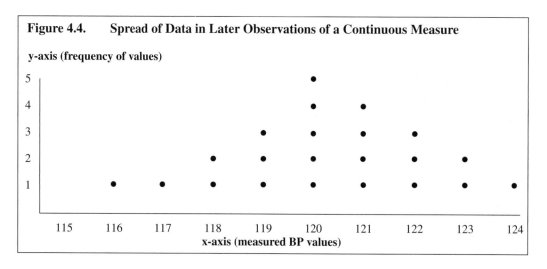

Figure 4.4.   Spread of Data in Later Observations of a Continuous Measure

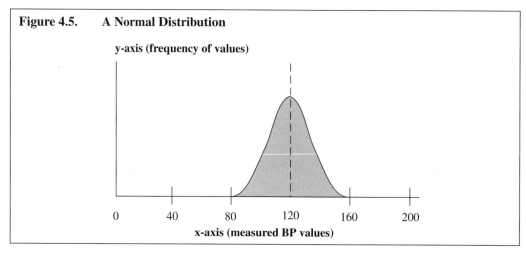

Figure 4.5.   A Normal Distribution

blood pressure is 120, so that calculation is relatively close. It has the greatest absolute frequency in this data set as denoted by the apex of the curve residing at that 121 value. Then, as the absolute frequencies of the measured values of the data set are examined to the left and right sides of the mean, they are seen to decrease and decrease until there are no observations below 80 or above 160 in relatively "normally" functioning human beings.

For the purposes of control chart construction, the standard deviation can be defined by the percentage of the frequencies contained beneath various portions of the normal distribution. In the interval under the curve from −1SD from the mean TO +1SD from the mean, there are approximately 68 percent of the observations. (See figure 4.6.)

In the interval under the curve from −2SD from the mean TO +2SD from the mean, there are approximately 95 percent of the observations. (See figure 4.7.)

Finally, in the interval under the curve from −3SD from the mean TO +3SD from the mean, there are approximately 99 percent of the observations.

Figure 4.6.   One Standard Deviation from Mean

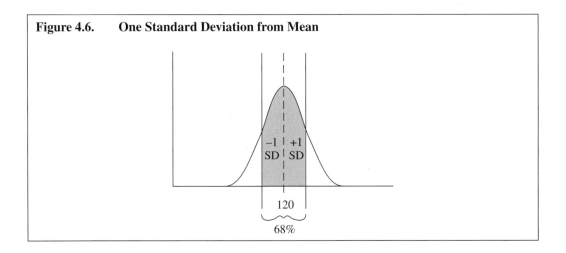

Figure 4.7.   Two Standard Deviations from Mean

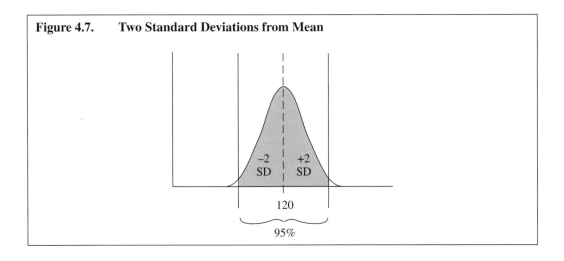

This is useful for performance improvement in healthcare. It helps define the interval in which 95 percent of the observations should be observed. Any observation that occurs outside the interval defined by ± 2SD from the mean can then be identified as a variant. If many observations fall outside the ± 2SD interval, it could be that the process under examination is "out of control" and contributing negatively to the provision of healthcare services.

Data Display Tools

The display tools most commonly used in PI activities include bar graphs, histograms, Pareto charts, pie charts, line charts, and control charts.

Once the data have been collected, the PI team should sort the data and identify any significant findings. Charts and graphs of the data make it easier to identify trends and significant relationships. Graphs can be used to compare data sets from different years or over time to visually illustrate a trend in the data or a change in performance. The performance improvement measures reported to the board of directors often highlight variances in data using graphs. (See figure 15.1.) For example, in a bar graph, a change in the height of a bar would indicate an increase or decrease in the data represented by the bar. The PI team would then have to determine whether this increase or decrease was a significant change in the performance of the related process.

When constructing charts, graphs, and tables, the team must provide explanatory labels and titles. Data display should be simple and accurate. Team members must report all of the data, even when the data appear to have positive or negative implications for the organization. Sometimes what appears to be a negative trend may actually turn out to be a positive trend after the team fully analyzes the data.

Bar Graphs

Bar graphs are used to display discrete categories, such as the gender of respondents, or the type of health insurance respondents have. Such categories are shown on the horizontal, or x, axis of the graph. The vertical, or y, axis shows the number or frequency of responses. The vertical scale always begins with zero (0). Most spreadsheet software programs can be used to "draw" a bar graph from a given data set. An example of a bar graph is in figure 4.8.

Histogram

A **histogram** is a bar graph that displays data proportionally. Histograms are used to identify problems or changes in a system or process. They are based on raw data and absolute frequencies, which determine how the graphs will be structured. Unlike a **Pareto chart** (described in the following section), the data remain in the order of the scale against which they were obtained. The horizontal axis measures intervals of continuous data such as time or money. The scale is grouped into intervals appropriate to the nature of the data. For example, time might be grouped into intervals of six hours and money into groups of $5,000. The vertical axis shows the absolute frequency of occurrence in each of the interval categories. Because of its visual impact, a histogram is more effective for displaying data than a check sheet of raw data, particularly when the frequencies are large. A histogram can be used in place of a pie chart for continuous data, which should not be displayed in pie charts.

Figure 4.8.   Example of a Bar Graph

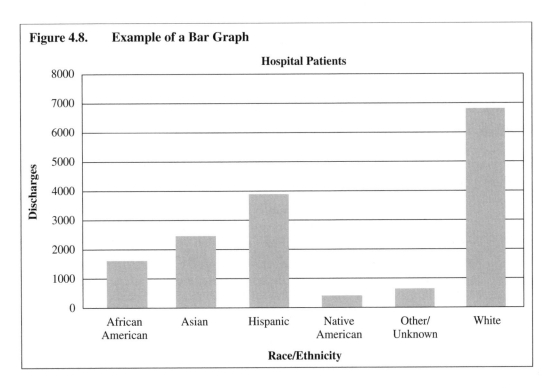

Histograms have the following characteristics:

- They display large amounts of continuous data that are difficult to interpret in lists or other nongraphic forms.

- They show the relative frequency of occurrence of the various data categories indirectly using the height of the bars.

- They show the distribution of the absolute frequencies of the data in the grouped intervals.

The first step in creating a histogram is to gather the data. The data can be collected on check sheets or gathered from department logs or other resources. A histogram should be used in situations in which numerical observations can be collected. Cost in dollars for a surgical procedure is one example. Once the data have been gathered, the team can begin to group data into a series of intervals or categories. A check sheet can be used to count how many times a data point appears in each interval grouping. For the cost of a surgical procedure, for example, to each patient might be grouped into the following intervals: $0 to $24,999, $25,000 to $49,999, $50,000 to $74,999, and $75,000 to $99,999.

To create the actual histogram, one should set up the horizontal, or x, axis with the interval groupings and the vertical, or y, axis with the absolute frequencies, always beginning at 0 for both. Each bar should be drawn upward as it relates to the tabulated frequencies from the check sheet.

To analyze a histogram, one should look for things that seem suspicious or strange. The team should review the various interpretations and write down their observations.

An example of a histogram is provided in figure 4.9. The example shows data related to the number of patients and their total charges. The horizontal, or x, axis shows the interval groupings indicating the "number of patients" and the vertical, or y, axis lists the "revenue/stay" by dollar amounts.

Pareto Charts

A **Pareto chart** is a kind of bar graph that uses data to determine priorities in problem solving. Using a Pareto chart can help the team to focus on problems and their causes and to demonstrate which are most responsible for the problem. Following these steps will result in a Pareto chart:

1. Use a check sheet to collect the required data. Figure 4.10 shows the top ten major diagnostic categories (MDC) by total charges. A check sheet is used to collect cases by MDC and for totaling the hospital charges for that MDC.

2. Arrange the data in order, from the category with the greatest frequency to the category with the least frequency. In the example, the data are arranged from the MDC with the highest charges to the MDC with the lowest charges.

3. Calculate the totals for each category. Figure 4.10 lists MDC 10 first with charges totaling approximately $2,500,500, then MDC 6 with charges totaling $2,500,000, MDC 4 $2,300,000 and so forth to MDC 18 with charges totaling approximately $300,000.

4. Compute the cumulative percentage. This is accomplished by calculating the percentage of the total for each category. Then you add the percentage for the greatest frequency to the percentage for the next greatest frequency and so on. Using the example in figure 4.9, the charges for MDC 10 ($2,600,000) represents 19.3 percent of all charges listed, while the charges for MDC 6 ($2,500,000) represent 18.6 percent of all charges listed. The cumulative percentage then is obtained by adding the percentages together, MDC 10's 19.3 percent to MDC 06's 18.6 percent for a total of 37.9 percent. This cumulative percentage is then calculated for all categories until 100 percent is obtained.

5. Draw horizontal and vertical axes on graph paper. The horizontal, or x, axis is the MDC and the vertical, or y, axis is the total charges.

6. Scale the vertical axis for absolute frequency (0 to the total calculated above).

7. Working from left to right, construct a bar for each category, with height indicating the frequency. Start with the category with the largest value and add categories in descending order.

8. Draw a vertical scale on the right side of the graph, and add a percentage scale (0 to 100 percent).

9. Plot the cumulative percentage line, as shown in figure 4.10.

Pie Charts

Pie charts are used to show the relationship of each part to the whole, in other words, how each part contributes to the total product or process. The 360 degrees of the circle, or pie,

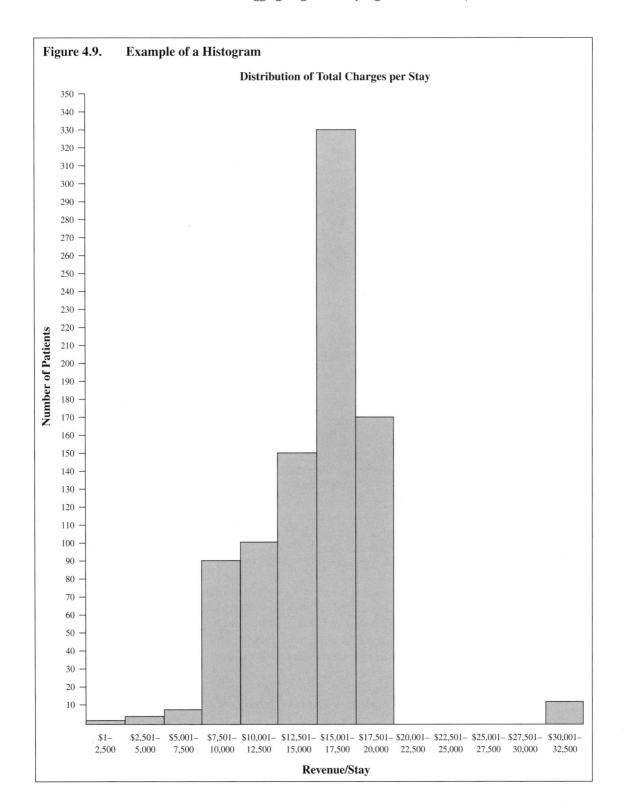

Figure 4.9.   Example of a Histogram

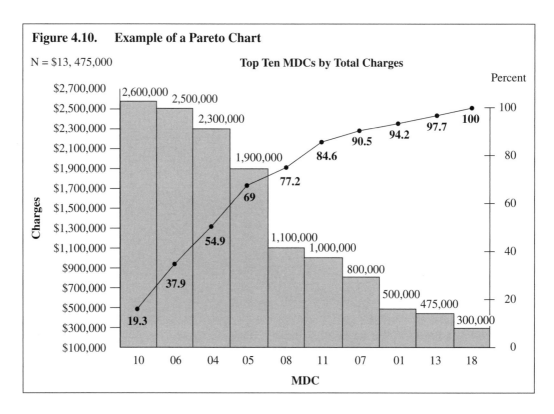

Figure 4.10.   Example of a Pareto Chart

represent the total, or 100 percent. The pie is divided into "slices" proportionate to each component's percentage of the whole. To create a pie chart, one should first determine the percentages for each data element of the total population and then draw the slice accordingly. Creating pie charts by hand requires the use of a protractor. For example, if a slice were to represent 45 percent of the whole, one would multiply the 360 degrees of the circle by .45 to find that 45 percent of the pie equals 162 degrees. Then, using the protractor, one could mark off 162 degrees on the pie and draw lines to the center to configure the slice. Spreadsheet programs can automatically create pie charts from a given data set. (See figure 4.11 for an example of a pie chart.)

Line Charts

A **line chart** is a simple, plotted chart of data that shows the progress of a process over time. By analyzing the chart, the PI team can identify trends, shifts, or changes in a process over time. The chart tracks the time frame on the horizontal axis and the measurement (the number of occurrences or the actual measure of a parameter) on the vertical axis. The data are gathered from sources specific to the process that has been evaluated. Each set of data (measurement/number of occurrences and time frame) must be related.

A line chart can be created by going through the following steps:

1. Select a time frame.

2. Identify the data that are to be tracked.

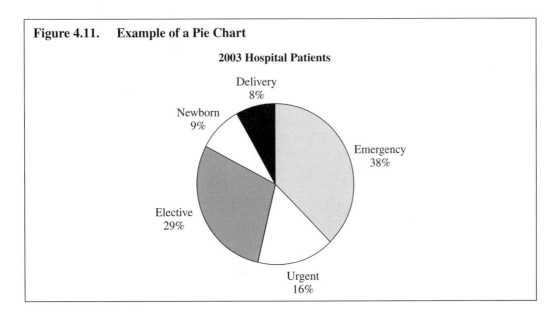

Figure 4.11.   Example of a Pie Chart

2003 Hospital Patients

- Delivery 8%
- Newborn 9%
- Emergency 38%
- Elective 29%
- Urgent 16%

3. Use a check sheet to collect frequency data or a tally sheet to collect measurement data.

4. Draw the chart and place the measurement or frequency on the vertical axis and the time frame on the horizontal axis.

5. Label the chart with specific details.

6. Plot the data in the sequence they appear.

7. Connect the points to form a relationship line.

8. Take the average of the data points collected.

9. Finally, draw a line parallel with the horizontal axis to represent the average.

To analyze the chart, the team should look for peaks and valleys that indicate that there may be a problem with the process.

Periodically redoing the line charts for a process helps the team to monitor changes over time. A line chart is a good way to display trends in the data. For example, a line chart could be displayed on a large plastic graph that could be updated month by month. When the team evaluates the results, they should look for seasonal peaks and valleys. For example, summer vacation times may show a change in a chart plotting staff productivity. Figure 4.12 provides an example of a line chart.

Control Charts

Control charts can be used to measure key processes over time. Using a control chart focuses attention on any variation in the process and helps the team to determine whether that variation is normal or a result of special circumstances. Normal variation may also be called *common cause variation,* or the expected variance in a process due to the fact that the process will not or cannot be performed in exactly the same manner each and every time. When a special circumstance or unexpected event occurs in the process, this will result in what is called *special cause variation*. It is this special cause variation that the performance improvement process needs to investigate.

The specific statistical calculations used to determine the upper and lower limits of a control chart are dependent on the type of data collected. The calculations used in calculating statistical process control are the data mean, the median of the range between data points, and the standard deviation. (See discussion of the mean and standard deviation above and Meisenheimer 1997.) The appearance of the control chart is like turning the classic bell curve on its left tail and running it along horizontally left to right as the time on the x-axis goes by. (See figure 4.13.)

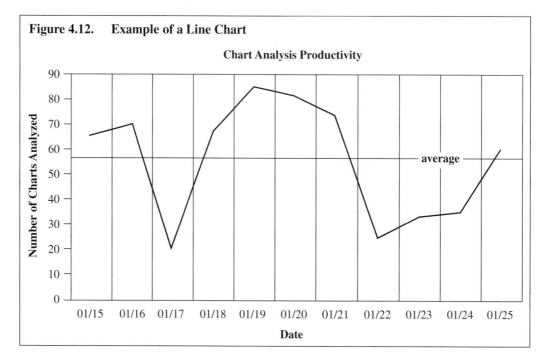

Figure 4.12.   Example of a Line Chart

Chart Analysis Productivity

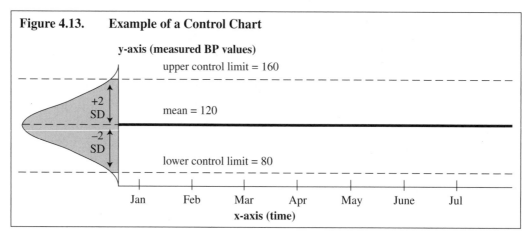

Figure 4.13.   Example of a Control Chart

The upper and lower control limits are always ±2 standard deviations from the mean. As each successive month of data is added to the chart, the standard deviations are recalculated and may fluctuate in value. As the process is tightened up and improved over time, the standard deviation should become smaller and smaller as variation is driven out of the process. If the standard deviation expands, the process becomes less controlled, taking on variation and most likely lesser quality. The latest calculated standard deviation is always used in the current display of the chart. Data points that lie outside upper or lower control limits may signal special cause variation that should be examined.

The following list shows the steps in developing a control chart:

1. Determine which process and what data are to be measured.

2. Collect about twenty observations of the measure.

3. Calculate the mean and standard deviation for the data set. Various spreadsheet software programs can be used to make this calculation. The mean, or average, becomes the center line for the control chart.

4. Calculate an upper control limit and a lower control limit. The upper control limit is usually represented by a dashed line two standard deviations above the mean, and the lower control limit is usually two standard deviations below the mean.

The resulting control chart becomes the standard against which the team can compare all future data for the process. For example, figure 4.13 displays the incidence of nosocomial infections for a facility ICU. Commonly, the rate is computed against 1,000 ICU inpatient census days, because there is no way to derive a denominator otherwise. The mean is calculated at 0.01, and the standard deviation is calculated at 0.012. The upper control limit (UCL) is two standard deviations above the mean, or 0.034. Some organizations, however, use a standard for nosocomial infection rate of 0.03, and so the UCL for these data is 0.03. The lower control limit (LCL) is two standard deviations below the mean, or −0.014. Because the calculated LCL is a negative number, 0 becomes the lower control limit.

Advanced Statistical Analysis

More advanced forms of statistical analysis are sometimes used in performance improvement activities. Examples include ANOVA, linear regression, and correlation. Information on these methods is available in the suggested readings for this chapter.

Real-Life Example

Table 4.2 shows a portion of a patient profile for one hospital in the years 1998 and 2003. When the 1998 data are compared to the 2003 data in a bar graph (figure 4.14), it can be concluded that the hospital has experienced an increase in the number of Asian patients in its customer base. The hospital must look at how this increase affects its processes. For example, what changes might need to be made in the dietary area? What staffing changes might need to be made to accommodate patients who have religious and cultural differences?

Table 4.2.	Data Set for Bar Graph		
Profile of Hospital Patients			
1998		**2003**	
Race/Ethnicity	**Discharges**	**Race/Ethnicity**	**Discharges**
White	6,254	White	6,874
Black	1,859	Black	1,763
Hispanic	4,251	Hispanic	3,954
Native American	254	Native American	301
Asian	1,352	Asian	2,514
Other/Unknown	750	Other/Unknown	594

Table 4.3 shows another set of data from the hospital's patient profile. Comparison of the two pie charts created from the data (figure 4.15) shows that the number of emergency admissions has increased 14 percent over five years and that the number of urgent admissions has decreased by about 12 percent over five years. This information indicates that either the admissions are being categorized incorrectly or the incidence of trauma is increasing. The facility may need to look at its emergency department capacity and procedures, or consider changes to its admission criteria.

Case Study

From the data provided in table 4.4, students should select three sets of data and create an appropriate graph to represent the data. They should keep in mind that the type of data in the table and choose the best graphic display tool for those data.

Project Application

Using the data collected from their QI projects, students should select data to be displayed and then determine the best graphic presentation. They should use a spreadsheet software program to design graphs for their student projects.

Summary

Using data analysis tools and data aggregation are important skills for PI teams to learn. Making improvement decisions on the basis of actual experience is much better than making decisions on the basis of intuition, or gut feelings. Graphic depictions of process outcome are also easier to track over time and translate data into meaningful information when the team can clearly see the magnitude of changes. The most commonly used graphic tools include bar graphs, pie charts, Pareto charts, histograms, line charts, and control charts.

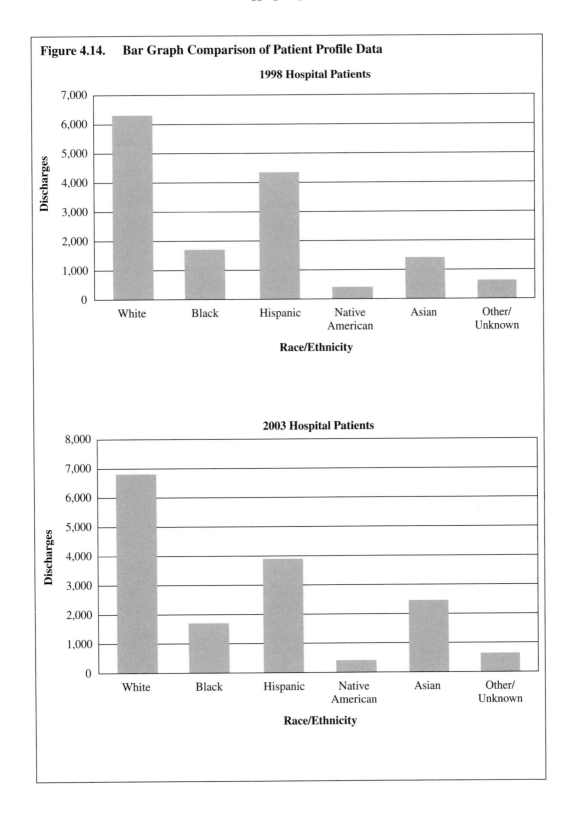

Figure 4.14.   Bar Graph Comparison of Patient Profile Data

Table 4.3.   Data Set for Pie Chart

Profile of Hospital Patients			
1998		**2003**	
Admission Type	**Discharges**	**Admission Type**	**Discharges**
Emergency	2,163	Emergency	5,987
Urgent	4,325	Urgent	2,478
Elective	5,784	Elective	4,458
Newborn	1,659	Newborn	1,342
Delivery	1,478	Delivery	1,270

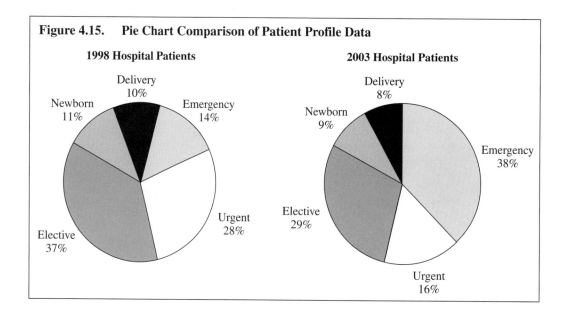

Figure 4.15.   Pie Chart Comparison of Patient Profile Data

Table 4.4. Data Set for Case Study

HSA 11 – HFPA 933 – LAKEWOOD REGIONAL MED CTR – SOUTH ST

PROFILE OF HOSPITAL PATIENTS, CALIFORNIA   ID# 106190240   SEE VOLUME III FOR LTC REPORTS

AGE	DISCHARGES MALE	FEMALE	TOTAL	AVER STAY	ADJUSTED TOTAL CHARGES TOTAL $	$/DAY	$/STAY
<29 DAYS	546	502	1,048	1.6	781,716	476	746
29-364 DY	6	5	11	1.9	29,432	1,402	2,676
1-4 YRS	14	10	24	1.9	74,596	1,622	3,108
5-14 YRS	28	33	61	2.3	280,570	1,990	4,600
15-18 YRS	24	121	145	2.1	641,857	2,077	4,427
19-44 YRS	616	1,672	2,288	3.0	17,251,826	2,528	7,540
45-64 YRS	835	798	1,633	4.9	24,390,763	3,064	14,936
65-69 YRS	291	397	688	5.8	12,545,685	3,159	18,235
70-74 YRS	381	437	818	6.2	15,700,837	3,078	19,194
75-84 YRS	417	700	1,117	6.6	21,994,942	2,998	19,691
85+ YRS	128	278	406	6.4	6,490,931	2,499	15,988
TOTAL	3,286	4,953	8,239	4.4	100,183,155	2,787	12,160

UTILIZATION & CHARGES FOR MOTHERS & BABIES

	DISCHRGS	DAYS	ALOS	ADJUSTED TOTAL CHARGES # TOTAL	$/DAY	$/STAY
NORMAL NEWBRN (1)	910	1,379	1.5	623,095	452	685
DELVRY-VAGINL (2)	853	1,275	1.5	2,594,805	2,035	3,042
DELVRY-C.SECT. (3)	168	490	2.9	1,073,141	2,190	6,388

DEFINITIONS: (1) DRG 391; (2) DRGS 372 THRU 375; (3) DRGS 370 & 371.

MAJOR DIAGNOSTIC CATEGORY	DISCHARGES	%	AVER STAY	ADJUSTED TOTAL CHARGES # TOTAL $	$/DAY	$/STAY	PERCENTILES OF $/STAY # 10TH	50TH	90TH
01 NERVOUS SYSTEM	433	5.3	5.4	6,037,289	2,587	13,943	3,462	8,682	27,240
02 EYE	11	0.1	3.1	91,369	2,687	8,306			
03 EAR, NOSE, ETC	63	0.8	3.5	494,246	2,247	7,845	1,413	5,497	13,293
04 RESPIR SYSTEM	783	9.5	6.7	14,830,414	2,807	18,941	4,905	13,088	40,052
05 CIRCU SYSTEM	1,728	21.0	5.1	33,469,870	3,788	19,369	4,218	10,712	48,831
06 DIGESTIV SYSTM	598	7.3	5.0	8,158,578	2,723	13,643	3,160	7,335	22,872
07 HEPATO SYSTEM ETC	209	2.5	4.7	3,208,921	3,268	15,354	3,976	8,740	27,930
08 MUSCULOSKEL SYSTEM	656	8.0	5.5	9,884,375	2,760	15,068	4,153	11,050	26,487
09 SKIN, BREAST, ETC	174	2.1	5.3	2,008,206	2,195	11,541	3,096	7,001	19,284
10 ENDOCRIN, NUTRI, ETC	269	3.3	4.8	4,547,948	2,017	9,697	2,533	5,973	17,435
11 KIDNEY/URINARY	257	3.1	4.8	2,713,015	2,222	10,556	2,999	8,177	18,577
12 MALE REPRODUCTV	148	1.8	4.1	1,664,124	2,742	11,244	5,793	9,014	19,010
13 FEMALE REPRODUCTV	132	1.6	3.6	1,311,264	2,738	9,934	5,587	8,815	14,482
14 PREGNCY/CHILDBIRTH	1,104	13.4	1.8	4,058,449	2,092	3,676	2,214	3,101	8,245
15 NEWBORN/NEONATE	1,047	12.7	1.6	781,465	476	746	438	567	1,177
16 BLOOD, ETC.	35	0.4	4.6	362,063	2,249	10,345	3,632	6,887	26,568
17 MYELOPROLIF, ETC	54	0.7	5.6	731,165	2,413	13,540	2,517	8,806	29,828
18 INFEC & PARASIT	142	1.7	8.5	3,225,324	2,668	22,714	4,442	13,583	52,218
19 MENTAL DISORDERS	23	0.3	3.3	162,376	2,165	7,060	2,886	5,201	15,823
20 ALC/DRUG USE ETC	43	0.5	4.0	398,022	2,341	9,256	1,990	3,381	22,500
21 INJURY/POISON/DRUG	104	1.3	4.1	1,389,320	3,223	13,359	2,733	5,074	17,604
22 BURNS	3	0.0	13.0	68,008	1,744	22,669			
23 OTHER FACTORS	14	0.2	4.7	127,105	1,926	9,079			
24 MULTI SIG TRAUMA	1	0.0	5.0	14,540	2,908	14,540			
25 HIV INFECTIONS	8	0.1	20.3	445,699	2,751	55,712			
UNGROUPABLE	0	0.0	.0	0	0	0			
TOTAL	8,239	100.0	4.4	100,183,155	2,787	12,160	886	6,684	25,358

EXPECTED SOURCE OF PAYMENT	DISCHARGES	%	AVER STAY	ADJUSTED TOTAL CHARGES TOTAL $	$/DAY	$/STAY
MEDICARE	2,862	34.7	6.5	52,618,588	2,836	18,385
MEDI-CAL	1,836	22.3	3.2	12,966,166	2,194	7,062
INSURANCE CO	812	9.9	3.0	6,189,201	2,512	7,622
HMO/PHP	2,020	24.5	3.3	21,706,087	3,269	10,746
BLUE X/SHIELD	31	0.4	3.2	291,884	2,919	9,416
SELF PAY	423	5.1	2.9	3,139,734	2,551	7,423
MED INDIGENT	0	0.0	0.0	0	0	0
OTHER GOVMT	0	0.0	0.0	0	0	0
WORKERS COMP	3	0.0	3.3	32,663	3,266	10,888
OTHER	251	3.0	4.1	3,236,576	3,124	12,895
NONGOVMT	0	0.0	0.0	0	0	0
NO CHARGE	0	0.0	0.0	0	0	0
TITLE V	1	0.0	1.0	2,256	2,256	2,256
UNKNOWN	0	0.0	0.0	0	0	0
TOTAL	8,239	100.0	4.4	100,183,155	2,787	12,160

RACE/ETHNICITY	DISCHARGES	%	STAY
WHITE	5,235	63.5	4.8
BLACK	772	9.4	4.3
HISPANIC	1,597	19.4	3.1
NATIVE AMER	98	1.2	3.8
ASIAN	482	5.9	3.8
OTHER & UNKNOWN	55	0.7	3.3

ADMISSION SOURCE	DISCHARGES	%	STAY
ROUTINE	3,585	43.5	4.2
EMERGENCY ROOM	3,299	40.0	5.2
HOME HEALTH SVC	1	0.0	3.0
SHRT TERM ACUTE HOS	117	1.4	4.5
SNF/ICF	167	2.0	8.4
OTHER FACILITY	38	0.5	4.7
NEWBORN	1,032	12.5	1.6
OTHER & UNKNOWN	0	0.0	0

DISPOSITION	DISCHARGES	%	STAY
ROUTINE	6,710	81.4	3.6
SHRT TERM ACUTE HOS	318	3.9	3.6
OTHER FACILITY	187	2.3	7.8
HOME HLTH SRVC	277	3.4	9.1
SNF/ICF	417	5.1	10.2
DIED	245	3.0	8.2
LEFT AGNST MED ADV	85	1.0	2.9
OTHER & UNKNOWN	0	0.0	.0

ADMISSION TYPE	DISCHARGES	%	STAY
EMERGENCY	606	7.4	7.2
URGENT	4,704	57.1	5.2
ELECTIVE	868	10.5	4.3
NEWBORN	1,032	12.5	1.6
DELIVERY	1,022	12.4	1.7
OTHER & UNKNOWN	7	0.1	1.7

EXCLUDES DISCHARGES WHERE TOTAL CHARGES WERE REPORTED AS UNKNOWN

References and Suggested Readings

Drennan, David, and Steuart Pennington. 1999. *12 Ladders to World Class Performance.* London: Kogan Page Ltd.

Fields, Willa L., and Dale Glaser. 1997. Using statistical process control tools in the quality process. In *Improving Quality: A Guide to Effective Programs,* Claire Meisenheimer, editor. Gaithersburg, Md.: Aspen Publishers.

Joint Commission on Accreditation of Healthcare Organizations. 2003a. *Comprehensive Accreditation Manual for Hospitals.* Oakbrook, Ill.: JCAHO.

Joint Commission on Accreditation of Healthcare Organizations. 2003b. Accessed online (5/22/03) http://www.jcaho.org/pms.index.htm.

Lighter, Donald, and Douglas C. Fair. 2000. *Principles and Methods of Quality Management in Health Care.* Gaithersburg, Md.: Aspen Publishers.

McLaughlin, Curtis, and Arnold D. Kaluzny. 1999. *Continuous Quality Improvement in Health Care: Theory, Implementation, and Applications, 2nd edition.* Gaithersburg, Md.: Aspen Publishers.

Meisenheimer, Claire. 1997. *Improving Quality: A Guide to Effective Programs, 2nd edition.* Gaithersburg, Md.: Aspen Publishers.

Osborn, Carol. 2000. *Statistical Applications for Health Information Management.* Gaithersburg, Md.: Aspen Publishers.

Rudman, William. 1997. *Performance Improvement in Health Information Services.* Philadelphia: W. B. Saunders.

Chapter 5
Communicating Performance Improvement Activities and Recommendations

Learning Objectives

- To apply communication tools such as minutes, quarterly reports, and storyboards in performance improvement processes
- To recognize the key elements in a storyboard and critique a storyboard layout

Background and Significance

The effective communication of information about the activities of performance improvement (PI) teams is vital to the PI process in healthcare organizations. All PI activities should be reported using the committee or meeting structure defined by the healthcare organization. This structure may include medical staff standing committees, PI team workgroups, or department meetings. This communication process is further defined in chapter 13.

The PI and Patient Safety Council then receives committee reports of PI activities throughout the organization. In turn, this council reports significant findings to the leaders of the organization—typically, the executive committee and board of directors.

Evidence of PI activity is also required by various regulatory and accreditation agencies. Such organizations require that process improvement activities take place within every healthcare organization, and they look for PI compliance during the survey process.

Common methods of communicating PI activities include minutes, quarterly reports, and storytelling. This chapter focuses on these three basic communication tools. Figure 5.1 shows the flow of information from the Chartered PI Teams to the Performance Improvement and Safety Council regardless of the reporting method.

Minutes

It is important that the committee or team keep track of its progress and activities. Documentation of these activities is often recorded in the form of minutes.

Organizations should select the method they will use to set agendas for their meetings and allocate meeting time for action items and discussions. (Information on setting agendas

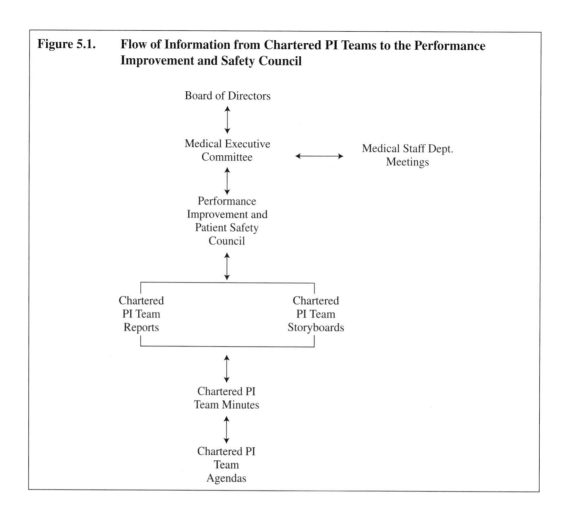

Figure 5.1.   Flow of Information from Chartered PI Teams to the Performance Improvement and Safety Council

can be found in chapter 3.) The responsibility for distributing the agendas for upcoming meetings and the minutes from past meetings should be discussed and assigned. The role of recorder, or secretary, should be assigned to an individual trained in the process of recording minutes.

Many different formats can be used to record the minutes of meetings. One method that is particularly helpful for PI documentation is the CRAF method. By using the CRAF format, the recorder can avoid getting distracted from the discussion. The format allows the recorder to focus only on necessary documentation. CRAF stands for the following categories of recordable information:

- *Conclusions* of group discussion

- *Recommendations* made by the committee or team

- *Actions* that the committee, team, or individual members decide to take

- *Follow-up* activity (See figure 5.2 for an example.)

Figure 5.2.   Sample Minutes from a PI Meeting

Committee Name: HIM Laboratory   **Approval:**

Attendance/Name and Title: HIM 3320 Class

Date: January 15   **Beginning/Ending Time:** 10:05 a.m.–11:20 a.m.

Recorder: John Smith

Leader: Sue Jones

Facilitator: Kathy Anderson

Conclusions	Recommendations	Actions	Follow-up
Reviewed and finalized customer survey tool.	Additions/changes to customer survey: • Add "WSU" and class name to survey title. • Add A, B, C, . . . to question responses on survey (all questions as applicable). • Add "on average" to question #2 on survey x2. • Change question #5 wording on survey from "do you have knowledge" to "are you aware." • Add to question #6, "If not applicable to you, circle NA." • Change under question #6 "CPU" to "computer." • Add "NA" column for each response in #6. • Add month and year (1/00) to last page of survey—bottom left corner to indicate design date.	Terry to update survey per team recommendations by 1/20. Request HIM department secretary to copy survey once changes have been made.	Consensus reached on survey tool questions and administration of same.

(Continued on next page)

Figure 5.2. *(Continued)*

Conclusions	Recommendations	Actions	Follow-up
Administer survey next week during Stats class and HIM 3010 classes.		Lori and Terry to administer survey to Stats class next week. Michelle to administer survey to HIM 3010 class next week.	
Tabulate results of survey during meeting next week.	Break out in teams, compile data, decide on best QI tool(s) to use in presenting data and which software applications to use.		
Individual teamwork completed on outputs, i.e., identifying customer requirements, possible measure(s) for each requirement, and flowcharting the process.	Break out in teams, assign leader, facilitator, and recorder per each team. Document and report all meeting decisions/actions.	Teamwork completed on following outputs: • Transcription • Chart analysis • Lab/resource accessibility	Assignment complete. Team roles practiced.
Additional assignments given in preparation for next week's meeting: • Read handout "Developing an Information Management Plan." • Be prepared to discuss JCAHO IM standards, which one related to HIM Laboratory.			
Evaluate meeting. Trouble with amount of time allotted per each agenda item. Adjourn.			

The conclusion section of the minutes should document the end result of the discussion and any decisions made by the group about future actions. The recorder should be sure to clarify the final conclusions at the end of a discussion if there is any ambiguity.

The recommendations section of the minutes should capture the plan for putting its decision into effect. The approach to solving the problem at hand should be listed as a recommendation.

The recorder then documents the actions planned for the different process steps. In figure 5.2, the actions section documents who was assigned to accomplish which activities during the next period of work. Activities between meetings might include gathering data, talking with persons collateral to the process being examined, or doing a literature search.

The follow-up section documents the assessment of the actions: if they were accomplished, for example, and whether the group is ready to make decisions and recommendations for future activities. Documenting the assessment of progress ensures that the process plan is followed appropriately and that sufficient analysis of previous actions has been conducted.

Reports

In addition to documenting meeting activities, the committee or team must provide regular reports to the organization's PI and Patient Safety Council. The frequency of reporting is usually based on the committee or team meeting schedule. At a minimum, quarterly reports should be submitted to the organization's PI and Patient Safety Council. The quarterly report is based on the documented meeting minutes. The quarterly report should include information about PI activities, which would include summaries of data collection, conclusions, and recommendations. (See sample reporting form in figure 5.3.)

Storyboards

A **storyboard** is another effective method of reporting PI activities by chartered teams. Storytelling has always been a powerful method of teaching and learning, and storyboards (or electronic presentations) help teams explain their work to those who may not be familiar with the PI process. The purpose of storyboarding is to summarize an entire PI project in a single graphic presentation. The team uses words, pictures, and graphs to tell the story of the project in a fashion that permits listeners to grasp the thought process of the team and to understand its specific applications of PI tools. Storyboards also demonstrate a growing knowledge of customer needs and the understanding of gathered statistical data. Electronic presentation software has streamlined storytelling significantly because such software provides standard formatting and key symbols to display the PI tools. However, healthcare organizations often use the storyboard as well because it provides the organization with a valuable teaching tool for its customers and staff. In addition, prominent display of the storyboard recognizes the accomplishments and participation of the organization's staff and employees.

Who started PI storytelling? In the early 1990s, Kaoru Ishikawa called attention to the importance of relaying the PI process in a structured manner to support learning and organizationwide performance improvement (Ishikawa 1986). Since the technique was introduced, elaborate rules that outline this storytelling process have been developed. Leaders in continuous quality improvement and total quality management continue to emphasize the importance of these rules.

Figure 5.3.   Example of a Quarterly Report

Committee Name: HIM Laboratory
Leader: Sue Jones
Facilitator: Kathy Anderson
Date: January 15

Opportunity Statement	Student complaints have been received about the quality of resources and technology available in the HIM Laboratory.
Mission	Evaluate the HIM lab in regard to accessibility, resources, library access, Internet access, quality of equipment, and adequacy of equipment for HIM students.
Vision	The HIM lab provides access to a variety of application software resources, library knowledge bases, and the Internet. A convenient comfortable work environment exists.
Performance Measure	Survey student satisfaction with HIM Laboratory.
Sample	All HIM and HIT students
Summary of Progress to Date	Survey will be administered to the students during the next week. Data will be tabulated and recommendations will be made. PI team should wrap up its activities by the end of February.
Conclusions	Not applicable at this time
Recommendations	In progress

Reported To: _____   Date: _____

Signature: _____   Date: _____

Who benefits from PI storytelling? Actually, very few people do not benefit from learning in a clear, concise manner how a performance improvement effort proceeded. Examples of how individuals can benefit from the storytelling process include the following:

- Team accomplishments are documented over an extended period of time in an organized and succinct way.

- Presentations are focused, and the presenter gains practice in sharing the pride that comes from working on PI projects.

- Team members crystallize their thinking about the process of improvement.

- Teamwork is tracked in a succinct and focused manner, thus facilitating communication while reducing the accumulation of paper.

- Team members receive the public recognition that they are due and learn how they might contribute more in the future.

- Other groups or departments learn how they can think in new ways about their work and improvement of the systems that they manage. Questioning each other helps create clearer awareness of how much people can learn from their associates.

- The professional staff can see clearly the impact of their role on overall care of their patients. They also learn how they can actively participate in PI work themselves.

- The board of directors can learn a great deal about the organization by studying the application and effect of PI processes on outcomes. It can also better meet its responsibility to ensure that the organization provides high-quality patient care.

- The administrative team can prepare itself to lead and teach the process of management and improvement throughout the organization. In addition, it can recognize the contributions of the staff and encourage everyone to do just a little bit more. It can also provide regular opportunities to celebrate gains made in the continuing journey of performance improvement.

- Storytelling provides an outstanding forum for new employees to be introduced to what PI is all about.

- Guests, suppliers, and others can learn about PI without imposing a significant additional burden on the presenters.

- The whole organization learns the habit of making improvements in everything it does. The dominant culture of the organization becomes one of continuous improvement in every facet of the facility's life.

There are several keys to successful storytelling, including:

- Organization
- Structure
- Timeliness
- Frequency
- Connection
- Celebration
- Feedback

To create effective storyboards, keep the following rules in mind:

- Map the board in advance with labels for each section.
- Prepare clean boards for group presentation and display.
- Keep detailed information in a team record binder for reference.
- Plan the presentation to fit the size of the storyboard (36 x 48 inches is standard) and the general size of the panels.
- Use large fonts (24 point or greater) so that people can read the boards from a distance.

The basic storyboard format is illustrated in figure 5.4.

Figure 5.4.	Sample Storyboard Layout	
Storyboard Title		
• Opportunity statement (mission/vision) • Team members • Customers • Relevant dimension of performance	• Key team activities in process steps • Flowcharts • Cause-and-effect (fishbone) diagram • Benchmarks	• Data gathered and analyzed (baseline and during PI activities) • Gantt chart • Future plans and goals

Once a PI project is complete, the organization may decide to communicate the outcome to its communities of interest, such as patients, medical staff, or employers in the region. Organizations that have improved their services often want to market this.

Several approaches can be used to communicate information on performance improvements. For example, many healthcare organizations have Web sites. Some large healthcare corporations routinely present performance data on their Web sites to show customers and other stakeholders how they are progressing with important performance measures.

Many healthcare organizations also publish information about care quality initiatives in their annual reports. Such reports are an excellent vehicle for communicating performance information, emphasizing an organization's mission within its community, and communicating ongoing efforts to provide the community with the best healthcare possible.

In some segments of the healthcare industry, such as long-term care, report cards provide consumers and other stakeholders with information on the performance of individual facilities. Report cards usually present data on an organization's performance with respect to a preestablished set of criteria relevant to the organization's service segment. The Department of Health and Human Services publishes report cards for every licensed long-term care facility in the country (http://www.medicare.gov/nhcompare/home.asp). The report cards show how each facility performed with respect to meeting the state's established licensing criteria.

Real-Life Example

The storyboard in figure 5.5 represents a student project. The PI project examined services at the campus student health center. Students were concerned about customer satisfaction in terms of hours of operation, quality of care, and confidentiality. So the student PI team developed mission and vision statements, and identified both the center's customers and those customers' requirements. Then the group conducted a survey of students, faculty, and staff to assess customer satisfaction with the center.

Case Study

Using the criteria listed below, students should critique the storyboard shown in figure 5.5.

Figure 5.5.   Example of a Student Team's Storyboard Presentation

WSU STUDENT HEALTH CENTER
AND PHARMACY:
GOOD OR BAD?

MISSION STATEMENT:
Weber State University's
student health center
and pharmacy are
dedicated to providing
high-quality healthcare
to students, faculty, and staff.

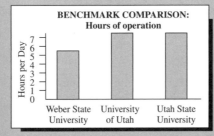

BENCHMARK COMPARISON:
Hours of operation

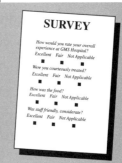

SURVEY

VISION STATEMENT:
To provide high-quality
care in a patient-
focused environment

BENCHMARK COMPARISON:
Drop-in versus scheduled appointments
Weber has drop-in appointments.
The University of Utah and
Utah State University schedule
appointments.

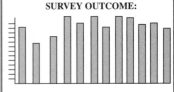

SURVEY OUTCOME:

All areas rated 3.5 or better on a 5-point
scale except for "Hours of operation
adequate" and "Length of time
waiting to be seen"

ACTION PLAN:
Determine whether there are
any problems with the services
provided by Weber State University's
student health center and pharmacy
by conducting a survey.

CUSTOMERS:
Students, faculty, and staff

TEAM MEMBERS:
Davis
James
Mary
Anne

RECOMMENDATIONS:
Student health center hours
should be extended.
The drop-in system should be
replaced with scheduling
by appointment.

CUSTOMERS' REQUIREMENTS:
Patient focused
Respect and compassion for individuals
Responsible use of resources
Responsive and user-friendly

Case Study Questions

1. Is the storyboard pleasing to look at, colorful, and easy to read?

2. Is the storyboard set up logically?

3. Is each of the process improvement steps taken into account?

4. Are the steps in the process easy to read and understand?

5. Are all of the elements of a good mission statement present?

6. Are all of the elements of a good vision statement present?

7. Are the external and internal customers of the process identified?

8. Does the storyboard identify the requirements of the customers? If so, how well? If not, why?

9. Does the storyboard display the team's findings in the improvement process?

10. Are the team's recommendations based on the data they collected? Are the recommendations sound? Are there other recommendations that the team did not identify?

Project Application

Students should develop a storyboard or electronic presentation for their projects. The presentation or storyboard should include each of the steps in the PI process, as well as the students' findings and recommendations. Electronic presentations should include no more information than a storyboard would include.

Summary

Communication among the various constituencies involved in healthcare PI activities is of paramount importance. Team members must keep the organization's leadership and/or PI and Patient Safety Council informed of their progress. The team must also track its activities carefully so that it stays focused on the issues for which it was constituted. The use of meeting minutes facilitates this communication and tracking. When the team has completed its work, communication of its activities to the organization as a whole informs everyone of changes in work processes and allows the people not involved in the project to see how the team arrived at its conclusions. Storyboards or electronic presentations are effective vehicles for this internal communication. Sometimes the organization may want to communicate performance improvement information to external stakeholders. Annual reports, information on Web sites, and report cards are tools that facilitate external communication.

References

Bemowski, Karen, and Brad Stratton, eds. 1999. *101 Good Ideas: How to Improve Just About Any Process.* Milwaukee, Wis.: American Society for Quality.

Denning, Stephen. 2000. *The Springboard: How Storytelling Ignites Action in Knowledge-Era Organizations.* Woburn, Mass.: Butterworth-Heinemann.

Drennan, David, and Steuart Pennington. 1999. *12 Ladders to World Class Performance.* London: Kogan Page Ltd.

Ishikawa, Kaoru. 1986. *Guide to Quality Control.* Milwaukee, Wis.: ASQ Quality Press.

McLaughlin, Curtis, and Arnold D. Kaluzny. 1999. *Continuous Quality Improvement in Health Care: Theory, Implementation, and Applications,* 2nd ed. Gaithersburg, Md.: Aspen Publishers.

Meisenheimer, Claire. 1997. *Improving Quality: A Guide to Effective Programs, 2nd edition.* Gaithersburg, Md.: Aspen Publishers.

Part II
Continuous Monitoring and Improvement Functions

Chapter 6
Measuring Customer Satisfaction

Learning Objectives

- To identify the differences between internal and external customers

- To outline the reasons why customers' perspectives are important to the performance improvement process

- To describe the difference between surveys and interviews

- To outline the characteristics that make surveys and interviews effective

- To critique a survey or interview format

Background and Significance

As discussed in chapter 1, researching and defining performance expectations includes an investigation of what the customers of an organizational process expect from that process. Because there are many types of organizational processes, there are also many types of customers. Their **expectations** must be identified and incorporated into the design or redesign of an effective process.

Customers receive a product or service as a result of an organizational process. Just as one can identify the customers of a dress shop or an auto dealership, one can also identify the customers of a healthcare process. For example, when a nurse inserts a catheter into an artery to administer medication, the patient is receiving a service from the nurse. Similarly, when a pharmacist dispenses a medication to a patient, the patient is receiving a product from the pharmacist.

Identifying the patient, client, or long-term care resident as a customer should seem fairly straightforward. But customers can be identified for all kinds of healthcare processes. The families and friends of patients are the customers of volunteer services when they ask for a patient's room number. Emergency care physicians are the customers of central supply services when they request sterile suturing trays to close a patient's laceration. Surgeons are the customers of the pathology laboratory when they request frozen-section examination of tissue in the operating room during resection of a breast lesion.

Types of Customers

Customers can be placed in one of two categories: **internal customers** and **external customers.** Within the healthcare organization setting, internal customers are individuals within the organization who receive products or services from an organizational unit or department. In the preceding examples, surgeons are the internal customers of the pathology laboratory, and emergency care physicians are the internal customers of central supply services.

External customers are individuals from outside the organization who receive products or services from within the organization. In the preceding examples, patients, family members, and friends of patients are external customers.

In determining the customers of a process, however, the organizational frame of reference must also be taken into consideration. Sometimes the frame of reference modifies the customer type, as graphically displayed in figure 6.1. Figure 6.2 graphically displays an overhead view of figure 6.1. The large oval is Western States University Hospital as a whole organization. Patients are external customers because their frame of reference comes totally

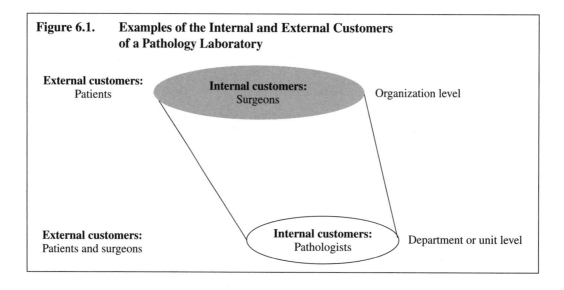

Figure 6.1.   Examples of the Internal and External Customers of a Pathology Laboratory

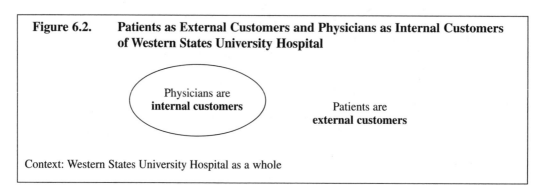

Figure 6.2.   Patients as External Customers and Physicians as Internal Customers of Western States University Hospital

from outside the organization. When one thinks of the organization as a whole, surgeons, on the other hand, are identified as internal customers because they are members of an organizational unit, the medical staff.

Now, if the frame of reference is focused on the departmental level, as depicted in figure 6.3, with the pathology laboratory taken as the frame of reference, then the surgeons might be identified as the external customers of pathology lab processes and the pathologists might be identified as the internal customers of the laboratory technicians who process the specimens.

It is important to recognize that internal and external customers must be identified in relation to the organizational process under consideration. Each process has a unique set of customers whose needs and expectations must be recognized.

The opinions of internal and external customers regarding the effectiveness of a healthcare process should be of primary importance to healthcare organizations. No one is a better judge of products and services than the customer. Additionally, a dissatisfied customer is said to tell ten times as many people about a negative experience than the satisfied customer is to relate a positive experience.

There are several ways for an organization to obtain information about its customers' perceptions of its products and services. With internal customers, simply ask them. Many internal customers never get the opportunity to express their expectations in a positive context. Often the only time a department representative hears about the expectations of internal customers is when a process has been mismanaged or has resulted in a negative outcome. Giving internal customers the opportunity to vocalize their expectations increases their overall satisfaction.

With external customers, particularly patients, identifying expectations about service quality is more complicated. Patients' expectations are multifaceted and often based upon the condition for which the patient is being treated. Assessments of patient expectations must be undertaken judiciously.

Administrative subjects such as parking, hours of operation, room decor, and so forth can be assessed using an anonymous patient satisfaction survey. Clinical subject matter pertaining to the patient's condition and medical and nursing treatment, however, may need to be assessed from the viewpoint of the clinicians involved in the patient's care and the outcomes achieved through that care. (See chapter 10.)

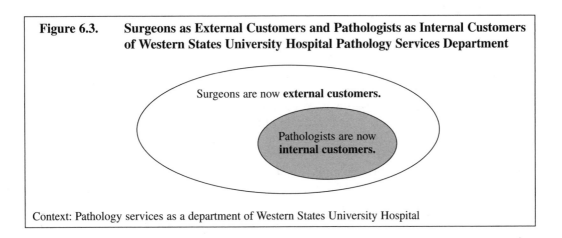

Figure 6.3.   Surgeons as External Customers and Pathologists as Internal Customers of Western States University Hospital Pathology Services Department

Surgeons are now **external customers.**

Pathologists are now **internal customers.**

Context: Pathology services as a department of Western States University Hospital

Subjective patient expectations, such as pain management, are completely different from one patient to the next. Another highly subjective area that influences a patient's expectations is his or her return to a satisfactory quality of life after treatment. One of the most important factors that influences how patients assess their care is the care and respect they are given by caregivers (Atlantic Information Services 1995, pp. 5–6).

Because of this inherent complexity, assessment of patient satisfaction might best be performed by multidisciplinary teams using carefully tailored assessment tools. Nationally recognized vendors offer patient satisfaction surveys. (See table 6.1.) Most of these surveys also provide healthcare organizations with important benchmarking feedback, which measures their customers' satisfaction against that of other, similar organizations.

Monitoring and Improving Customer Satisfaction: Steps to Success

In order to monitor and improve customer satisfaction, an organization must know:

- Exactly who its customers are

- What its customers want and value

- What improvements could be made to better meet its customers' needs

Step 1: Identify internal and external customers

Assessing whether a process has met the expectations of its customers is difficult when some customers have not yet been identified. To identify customers, the PI team should list everyone who comes in contact with the process and takes away a product or a service. (See figure 6.4.)

Step 2: Identify products and services important to customers

The PI team should develop a list of the products and services used by internal and external customers. Not every product is tangible; it is not always seen and evaluated as an object in the environment. Commonly, the tangible aspects of products include such things as facilities, equipment, and appearance of personnel. For example, information could be collected in a database and used by many customers, but only rarely might the database be printed out on paper in its tangible form.

Outcomes of care are not necessarily tangible, either. In order to report outcomes of care, the care service recipient needs some established means of describing the outcome, or clinical staff may have to use healthcare monitoring instruments to determine improvements in the patient's condition. An example of this would be the use of heart-monitoring equipment to determine if a drug has had the appropriate effect on a patient's arrhythmia.

Another aspect of quality that customers may emphasize in their evaluations of healthcare performance is reliability. *Reliability* is the level at which an organization can provide an offered product or service when requested and as advertised. For example, one hospital in the western U.S. advertises emergency department visits to occur within fifteen minutes. To be judged reliable by customers, that fifteen-minute goal must be met when they show up for emergency services.

Table 6.1.   Vendors of Patient Satisfaction Surveys

Vendor Name/ Contact Information	Product Description
National Research Corporation Gold's Galleria 1033 O Street Lincoln, NE 69508 www.nationalresearch.com	• Provides customizable patient satisfaction surveys • Provides SF-36-based health assessment surveys • Conducts surveys through both telephone and direct mail • Maintains national databank for benchmarking • Provides extensive analysis of results
Dey Systems, Inc. 230 Executive Park Louisville, KY 40207 www.dey-systems.com	• Developed Patient Satisfaction Information System (PSIS) for data analysis • Uses time-dependent Functional Status Outcomes Measurement System (FSMS) developed from SF-36 questionnaire for health assessment • Offers consulting services for aid in survey design • Maintains processing center for mailing, collection, scanning, comment typing, coding, and reporting • Offers on-line access to benchmarking information
Press, Ganey Associates 1657 Commerce Drive South Bend, IN 46628 www.pressganey.com	• Specializes in patient satisfaction measurement via direct-mail surveys • Produces quarterly reports detailing hospitalwide performance comparing individual units, noting satisfaction trends and national benchmarking statistics • Offers on-line access to completed surveys • Produces special annual reports containing detailed demographic analyses of individual hospital data and national comparative data
Parkside Associates, Inc. 205 West Touhy Avenue Suite 204 Park Ridge, IL 60068 www.femf.org	• Designs and distributes customizable surveys for patient satisfaction • Performs direct-mail surveys • Performs data entry, analysis, and reporting • Offers individual departmental analysis • Provides national benchmarking data • Offers consultation on result application
The Gallup Organization The Gallup Building 47 Hulfish Street Princeton, NJ 08542 609-924-9600	• Conducts telephone surveys from lists of discharged patients • Uses patient satisfaction survey that is customized by specialty • Uses customizable, time-dependent telephone surveys developed from SF-36 questionnaire to assess functional outcomes • Analyzes collected data and creates quarterly reports • Generates recommendations on successful application of results • Provides national benchmarking results provided within quarterly reports
The Picker Institute 1295 Boylston Street Suite 100 Boston, MA 02215 www.pickereurope.org	• Designs surveys that target dimensions of care that patients are most concerned about • Offers SF-36 surveying and analysis services only in conjunction with patient satisfaction surveying programs • Conducts continuous survey and analysis • Provides a priority matrix that helps clients to focus on areas needing improvement • Provides national benchmarks and comparisons with peer institutions
Professional Research Consultants 11326 P Street Omaha, NE 68137 www.prconline.com	• Uses proven methodology for patient satisfaction telephone interviewing • Incorporates patient satisfaction and expectation assessments with outcomes research (not health assessment outcomes) • Provides statistically valid measurement of perceptions of "quality" with open-ended response capabilities
California Institute for Health Systems Performance 1215 K Street Sacramento, CA 95814 www.cihsp.org	• Conducts PEP-C patient satisfaction survey for selected hospitals in California • Allows hospitals to compare their performance on key dimensions with other participating institutions • Web-based analysis and reporting tools, allowing hospitals immediate access to results and flexibility in conducting data analyses

Source: Adapted from Atlantic Information Services. 1995. *A Guide to Patient Satisfaction Survey Instruments*, 2nd ed. Washington, D.C.: Atlantic Information Services, pp. 5–6.

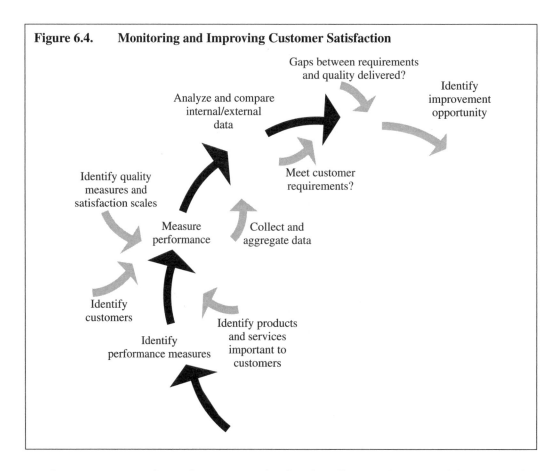

Figure 6.4.   Monitoring and Improving Customer Satisfaction

Responsiveness refers to how an organizational staff responds to unanticipated service needs. This includes staff willingness to continuously monitor both the customer's condition and satisfaction with services. Currently, pain management is an important function as defined by the Joint Commission on Accreditation of Healthcare Organizations (JCAHO). Pain management is a continuous process; patients' pain must be monitored, and medication dosage must be adjusted, when necessary, to keep pain within their individual bounds. Customers judge responsiveness to pain management based on the staff's willingness to go the extra distance to attend to this often continually changing need.

The healthcare PI term *assurance* describes the knowledge and courtesy of the staff who provide the goods and services. This aspect of care quality generates customer trust and confidence in the individuals providing the products and services as well as the products and services themselves. For example, in a patient's selection of a physician, assurance is key. If patients do not gain trust in the physician's judgement as new patients, they will probably question the outcome of all visits or attempt to find another provider.

Empathy can be defined here as the willingness of the staff to relate to customers as fellow human beings who have feelings and emotions about the provider–customer relationship. Particularly in healthcare, this can be an important aspect because when people are ill, most have an emotional reaction to the illness. When major illness strikes, or when fragile individuals such as children or the elderly are involved, the emotional response can be intense. Healthcare customers expect that staff and providers will understand these kinds of feelings and help them cope with these trying situations.

Features are the aspects of healthcare services that distinguish one organization from another or that add particular value in the customer's evaluation of an organization. Examples of this include an organization's renown in its community for providing exceptionally successful heart disease treatment with few negative outcomes, while employing the latest in diagnostic imaging and therapeutic modalities. Another example is the recent influx of birthing centers as a feature of organizations' service provision.

Finally, *perceived quality* is also an important aspect of healthcare services. This encompasses the organization's reputation for high-quality service and consumer reaction to services and products based on their experiences with the organization.

It may or may not be important to look at the quality of a healthcare product or service from all of these aspects. The PI team must identify which aspects have relevance to the process or service under examination, then complete the cycle of measurement and improvement, as necessary, regarding each aspect.

Step 3: Identify quality measures and satisfaction scales for each product and service

For each quality aspect of a product or service, **performance measures** must be identified. For example, if one were assessing the patient's experience at a physician's office for a medical appointment, waiting time would be a relevant performance measure. After waiting time has been identified as a performance measure, a satisfaction scale for waiting time must be developed. The team decides whether it would be best to measure waiting time in seconds, minutes, or hours.

Step 4: Collect and aggregate data on each performance measure

Next, the best methods for collecting data on each performance measure currently under assessment should be determined. In assessments of customer satisfaction, the principal methods of data collection are **survey tools, interviews,** and **direct observation.**

The construction of an effective survey tool requires a significant investment of time. (See the discussion of survey tools later in this chapter.) Interviews are often easier to use because they consist of a series of open-ended questions. Data aggregation from a survey is easier than aggregation of information from an interview because surveys are often composed of structured responses. The responses to interview questions, on the other hand, must be analyzed to identify common themes and perceptions.

The healthcare organization's institutional review board (IRB) should preapprove the use of any data-gathering tool. Institutional review board approval is mandated by federal regulations on the use of human subjects in biomedical and health services research. Although every institution's policies and procedures will vary, PI teams using such data-gathering methodologies should recognize their responsibility to obtain IRB approval in order to maintain the highest standards respecting their human subjects. If the organization has no IRB, then review of all tools used should be undertaken by either the quality council or the committee involved in the approval of PI projects. A statement such as "Your completion and return of this survey implies your consent to the use of your feedback for quality improvement purposes" should be prominent in the survey introduction so that respondents are aware that their responses will be used individually and collectively.

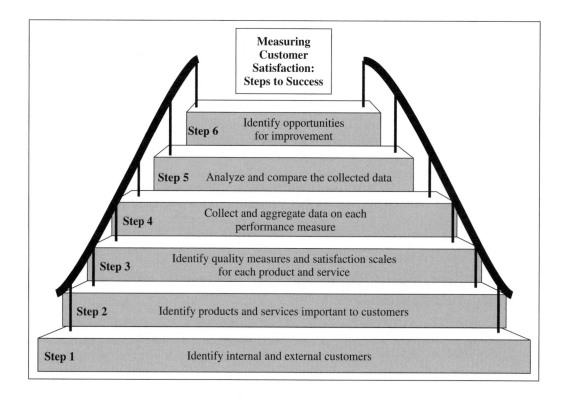

Step 5: Analyze and compare the collected data

The PI committee will next compare aggregate satisfaction ratings on performance measures with previous trends within the organization or with satisfaction levels achieved by other organizations. Such comparisons may be conducted using patient satisfaction data collected within the enterprise or satisfaction ratings published by state departments of health, national accreditation agencies, or other national organizations specializing in healthcare quality assessment. (The vendors of patient satisfaction surveys noted in table 6.1 can provide national data.)

Step 6: Identify opportunities for improvement

Finally, the committee develops improvement areas based on the comparisons with aggregate satisfaction ratings on quality measures. Performance improvement teams should be implemented to define appropriate performance expectations and design or redesign processes.

Real-Life Examples

The following two examples will help illustrate the process of measuring customer satisfaction. The first example looks at a PI measure involving external customers. The second example addresses both internal and external customers.

External Customers of a Food Services Department

Virginia Mullen, RHIA, is the director of quality and service excellence for a large U.S. healthcare corporation. She was working with one of her corporation's smaller facilities, a small acute care hospital with eighty beds. In evaluating the service excellence of the facility, Mullen and the administration had decided to use a satisfaction survey. The survey was to be mailed to patients after they had been discharged from the hospital.

One part of the survey asked for patient feedback on inpatient food services. The quality of food services was rated on a five-point scale, with 5 indicating very satisfied and 1 indicating very dissatisfied. The survey contained the following items:

- Taste of the food

- Temperature of the food

- Appearance of the food

- Variety of menu items

- Overall satisfaction with the food

After administering the survey for several months, Mullen and the administration noted that only 10 to 30 percent of patient ratings of food services fell within the excellent category.

Mullen considers the percentage of responses in the excellent category to reflect the loyalty of a facility's customers. Customers who ranked food service as less than excellent may or may not return to the facility for care in the future. In a highly competitive environment, developing market loyalty is extremely important to the survival of the facility.

Historically, institutional food is not an area that gets rave reviews. Upon review of the literature for institutional nutrition services, the investigators found that the research noted that assessment of food quality is highly subjective. Everyone has his or her own idea about what is high quality. The fact that the respondents are hospitalized further complicates the assessment because illnesses and medications often significantly affect taste sensation. Food services have traditionally received the lowest ratings across the healthcare industry. Yet despite this historic precedent, administration and management of the food services decided to see if they could improve ratings in this area.

Using PI methodology, the PI team developed an entirely new approach to menu selection for inpatients. They focused on lunch and dinner, leaving breakfast as it was, with a standard provision of the usual breakfast items. The directors had read about a few hospitals using a method similar to hotel room service. They contacted a hospital in Washington State that had implemented a similar program and invited the staff to give a presentation about the pros and cons of the program. Following the presentation, the team did additional research to locate other facilities that used the system and identified the systems' successes and challenges. It became apparent that every facility needed its own process for designing this service. The service features needed to be based on the specific organization's patient population, staffing, equipment, and financial resources. The pilot hospital assembled a multidisciplinary team to identify possible issues that would arise with such a change. The food services team needed the support of the administration, staff, ancillary staff, and especially nursing.

The program was implemented as planned. Patients on regular diets had a three-and-a-half-hour window during which they could order from a menu similar to a hotel room service menu. Patients on therapeutic diets were visited by the dietician, who helped them develop appropriate diets for their medical conditions.

Over the following months, Mullen and the administration of this hospital saw the ratio of very satisfied ratings increase from the 10 to 30 percent range to over 50 percent. The ratio of ratings in the satisfied and very satisfied responses rose to over 80 percent. Annual projected costs in food services decreased $20,000 thanks to a decrease in the amount of wasted food. Patients could now eat what they wanted, when they wanted it, and they were more likely to eat all of it.

Long term, the system has worked extremely well. Food services staff members who were initially skeptical would not go back to the old way of doing things. Patient satisfaction scores have remained high. There is less food waste, and the nursing staff keeps all liquid diet products on the units so that patients have access to them whenever they want them. Nursing staff also knows when patients are moving from a liquid to solid diet, and they can help their patients order appropriate food.

Food services at the hospital has become a revenue-generating department because the room service menu has been expanded to serve visitors. Families are thrilled to be able to have meals with the patients who are in the hospital and are willing to pay for the meals. Recently, this service has been expanded to the hospital staff as well. The hospital cafeteria is closed on weekends, and the staff can call for room service at any time. They love it, and so the program has been of benefit with respect to employee satisfaction as well.

A copy of a patient satisfaction survey is provided in figure 6.5.

Internal and External Customers of a Pathology Laboratory

Dr. Lagios had been chief pathologist at Western States University Medical Center for two months. He moved to the university setting from a large, tertiary care, private hospital in a metropolitan community because he wanted to expand the scope of his responsibilities. As do many managers when they come into new positions, he began an inventory of the pathology laboratory's functioning and examined each of its processes.

In general, the lab appeared to be functioning well, particularly in the area of processing surgical specimens. But in the area of autopsy report dictation, he found a huge backlog of reports that had not been dictated and finalized. The backlog of forty reports extended back more than two years. When he asked his secretary for an explanation of the backlog, she reacted with amazement and said she was totally unaware of the situation. No one had been monitoring the status of the autopsy reports on a continuing basis.

Dr. Lagios began his investigation by **interviewing** his fellow pathologists. Each of them told a similar story. Autopsies were usually performed the day following the death of a patient, and the remains were then picked up by a mortician. Organ specimens were harvested during the procedure, and microscopy slides were processed during the immediately following days. During the autopsy procedure, the pathologist would dictate notes about the findings to the diener (the nonphysician assistant who managed the remains and the specimens). The diener would write the notes on a photocopied form that outlined the order in which the autopsy was performed. But then the reports never got dictated because the pathologists had so many surgical specimens to process that they never made time to finalize the autopsy reports. When a physician needed the autopsy findings to fill out a death certificate, he had to call the assigned pathologist, who would verbally report the results.

Figure 6.5.   Example of a Patient Satisfaction Survey

Patient Satisfaction Survey

Instructions:
- Use a pencil or black pen to fill in your answers to the questions on the survey.
- Mark your answers in the circles provided.
- Answer only the questions that apply to your stay in the hospital.

Please choose one of the responses provided for the following questions:

1. Was this the first time you came to this hospital for inpatient care?
   ○ Yes   ○ No

2. Would you recommend this hospital to a friend or family member who needed inpatient care?
   ○ Yes   ○ No

3. Where were you admitted into the hospital?
   ○ Registration Area   ○ Emergency Department   ○ Other

4. How long did you wait before you were taken to your room?
   ○ Less than 20 minutes   ○ 21 to 30 minutes   ○ 31 to 60 minutes   ○ More than 1 hour

5. Did you have surgery while you were hospitalized?
   ○ Yes   ○ No

6. Were you in an intensive care unit at any time during your stay?
   ○ Yes   ○ No

Fill in the circle to the right of each statement that best describes how satisfied you were with the care and services you received while you were in the hospital.

Registration Process	**Very Dissatisfied**	**Somewhat Dissatisfied**	**Neutral**	**Somewhat Satisfied**	**Very Satisfied**
7. Courtesy and friendliness of the registration staff	○	○	○	○	○
8. How well the registration staff answered your questions	○	○	○	○	○
9. Amount of time needed to complete the registration process	○	○	○	○	○
10. Overall satisfaction with registration procedures	○	○	○	○	○
Nursing Staff					
11. Caring and concern of the nurses who cared for you	○	○	○	○	○
12. Skill of the nurses who cared for you	○	○	○	○	○
13. Time it took for nurses to respond to your calls	○	○	○	○	○
14. Willingness of your nurses to listen to your concerns	○	○	○	○	○
15. Amount of time your nurses spent with you	○	○	○	○	○
16. Overall satisfaction with nursing staff	○	○	○	○	○

PLEASE COMPLETE THE SURVEY ON THE REVERSE SIDE OF THIS PAGE.

(Continued on next page)

Figure 6.5.   *(Continued)*

Medical Staff	Very Dissatisfied	Somewhat Dissatisfied	Neutral	Somewhat Satisfied	Very Satisfied
17. Caring and concern of the doctors who cared for you	○	○	○	○	○
18. Availability of your doctors	○	○	○	○	○
19. Ways your doctors worked together and with your nurses	○	○	○	○	○
20. Information your doctors provided about your condition	○	○	○	○	○
21. Amount of time your doctors spent with you	○	○	○	○	○
22. Overall satisfaction with medical staff	○	○	○	○	○
Housekeeping Services					
23. Cleanliness of your room	○	○	○	○	○
24. Overall cleanliness of the hospital	○	○	○	○	○
25. Overall satisfaction with housekeeping services	○	○	○	○	○
Food Services					
26. Taste of the food	○	○	○	○	○
27. Temperature of the food	○	○	○	○	○
28. Appearance of the food	○	○	○	○	○
29. Variety of menu items	○	○	○	○	○
30. Overall satisfaction with food services	○	○	○	○	○
Other Hospital Services					
31. Overall satisfaction with X-ray services	○	○	○	○	○
32. Overall satisfaction with respiratory therapy services	○	○	○	○	○
33. Overall satisfaction with rehabilitation services	○	○	○	○	○
34. Overall satisfaction with emergency department services	○	○	○	○	○
35. **Overall satisfaction with the care and services you received**	○	○	○	○	○

36. What did we do really well? (Please be specific.) _____
_____
_____

37. What do we need to improve? (Please be specific.) _____
_____
_____

REMEMBER: ALL OF YOUR RESPONSES ARE CONFIDENTIAL.
THANK YOU FOR PARTICIPATING.

There was definite room for improvement in this process. Dr. Lagios decided to convene a PI team to examine the situation and come up with a better process for the process for dictating and finalizing autopsies. He asked two of his fellow pathologists who had been on staff longer than he, as well as the diener, to serve on the team. In addition, he requested the participation of the director of health information services, who managed pathology and autopsy report transcription processing, and two internal medicine physicians, whose deceased patients had been autopsied most frequently in the past two years.

Using the interview technique of information gathering (see discussion of interview design below) at the first meeting, the two internists talked the most. There was clearly a fair amount of suppressed frustration about the current situation. If the two internists on the team were unhappy, then other physicians were probably unhappy, too. The internists had morticians after them for cause of death to put on the death certificates, and tracking down the pathologists to get the findings from an autopsy considerably delayed the death certificate process.

Mrs. Castle, director of health information services, reported that it was becoming difficult to tell which autopsies had been done and which had not because there were so many of them. She was very willing to try and find a solution if the pathologists were willing to accept some change in procedures. She felt that as this was an information management problem, there might be an information management solution available.

The group decided to reconvene to determine whether new procedures could be developed. Mrs. Castle volunteered to contact the local dictation equipment vendor representative to see whether he had any suggestions. The representative introduced her to the concept of simultaneous dictation—the dictation of the details of a procedure during the procedure itself. The representative proposed installing a planetary microphone, which is designed to pick up recording circumferentially around the entirety of a room, above the autopsy table where the gross dissection is being performed. With a planetary microphone installed above the dissection table, the pathologist could move around the table as necessary during the dissection, unhampered by cords or handheld recording devices. Foot pedals under the table would allow him or her to turn the recording device on and off as he or she proceeded through the autopsy. The dictation equipment vendor representative proposed a similar setup for the pathologists' offices. Microphones were to be mounted to the desktops in such a way that they would capture the dictation while the pathologist was looking at the prepared slides on the microscope. Foot pedals under the desk would allow the pathologist to turn the recording device on and off during the review. This would keep the pathologist from having to take notes regarding findings on the slides and dictate using the traditional handheld microphone attached to the dictation machine.

At the next meeting of the PI team, Mrs. Castle reviewed her findings with the team members. All of the pathologists were delighted with the proposal from the dictation equipment vendor representative. They immediately saw that simultaneous dictation would streamline their report production operations. They also recognized that the benefits would extend to their dictation of the surgical specimen reports as well, which was accomplished completely at the desk microscope. The cost of the installation of the equipment was not exorbitant, so the team decided to move forward with the proposed solution.

Using the time freed with the use of simultaneous dictation systems, and with the commitment of the pathologists, the backlog of autopsy dictations was cleared in a few weeks. The pathologists were all happy with the new processes, and the attending physicians now routinely get their autopsy preliminary diagnoses in time to prepare death certificates.

QI Toolbox Techniques

Surveys and interviews are two commonly used data collection techniques to measure customer satisfaction. As mentioned earlier in this chapter, direct observation of behavior can be used as well, but behavior is difficult to analyze because it often changes when people realize that they are being observed. However, all three methods can be used to measure outcomes and processes. A discussion of some design considerations for surveys and interviews follows.

Survey Design

When designing a survey, the PI team must define the goal or goals of the survey in clear and precise terms. The purpose and audience of the survey must be kept in mind during the design phase as well. Careful consideration of the questions asked on the survey is imperative. The team must have a reason to include every item on the survey. It should avoid asking for information that is interesting, but not necessary for measuring process capabilities.

Survey items should be arranged from the general to the specific. For example, demographic data should be followed by process-specific questions. It is helpful to identify the broad categories of necessary information and then determine the order for these categories. The first question should not attempt to elicit information that is emotionally charged or sensitive. For example, if physicians are surveyed about the quality of the transcription system, the first question should not ask whether they are happy with the transcription system. The initial questions should ask how much they use the system, what types of reports they generate, and so on. The next set of questions can be used to determine their level of satisfaction with the system.

After the team has determined the broad categories of information it needs, it should think about individual questions or items. The single most important factor in item construction is clarity (Jagger 1982). Item format and content should be consistent. Similar questions should be formatted in the same way using the simplest sentence structure possible. (See figure 6.6.)

Figure 6.6.   Examples of Consistent and Inconsistent Format

Inconsistent Format	Consistent Format
What is your ZIP code? _____	Check which ZIP code you live in:
	___ 84065
Sex (circle one): Male Female	___ 84070
	___ 84092
	___ 84094
	___ Other (specify): _____
	What is your sex?
	___ Male
	___ Female

The survey should be written at the reading level of the respondents. The average reader in the United States reads at the sixth-grade level, so vocabulary should be simple, rather than using sophisticated medical or technical terminology that the average person would not understand. Furthermore, the items should be written in an objective manner so that they do not imply that any particular response is either desired or correct. (See figure 6.7.)

Surveys may incorporate a variety of question types. Open-ended questions allow respondents to construct a free-text answer in their own words. Responses to open-ended questions, however, are difficult to score, and response data are difficult to aggregate, because there is no defined scale of responses. Responses may show no clear connection or pattern.

Open-ended questions should only be used at the end of the survey, and they should not be used to elicit information that could be more easily collected in a structured format. (See figure 6.8.) When the researcher wants specific information about a particular area of investigation, the items must be worded precisely so that comparable data can be collected.

The use of structured questions on a survey limits the number of possible responses and thus standardizes the data collected. Care should be taken to include all possible responses to each question. Respondents must be able to select their answers from the choices provided. One method of ensuring this is to include a choice of "Other (specify)" so that respondents can write in an answer when the desired answer is not among the choices offered. Even then, it is important to provide as many of the common answers as possible to minimize the number of write-in responses. When items include categories of response, the categories must be mutually exclusive; that is, categories should not overlap. (See figure 6.9.)

Figure 6.7.   Examples of Wording

Poor Wording	**Simple Wording**
Why were you admitted to the hospital?	Why were you in the hospital?
To deliver a child ____	____ To have my baby
To have a C-section ____	____ To have an operation
To have a surgical procedure____	____ To obtain medical treatment
For medical reasons ____	____ Other (specify): _____
Other (specify): _____	

Figure 6.8.   Example of an Open-Ended Question

Open-Ended Question	**More Structured Question**
How has your coronary artery disease affected your lifestyle?	Now that you have heart disease, are you exercising:
	____ More than before
	____ Same as before
	____ Less than before
	____ Not at all, before or now

Another important issue in survey design involves the use of terms, phrases, and words that are known to both the PI team and the respondent. Careful word choice reduces ambiguity. This clarity of terminology is called an operational definition (Jagger 1982). (See figure 6.10.)

A survey may be personally administered or mailed to the respondents. Either method is effective, but the response rate decreases and turnaround time is greater when the survey is mailed.

Interview Design

Whether conducted face-to-face or over the telephone, an interview can provide important insight into quality issues in healthcare. Interviews may be unstructured or structured. In the unstructured type, the sequence of questions is not planned in advance. Instead, the interview is conducted in a friendly, conversational manner. This type of interview is helpful when the interviewer is trying to uncover preliminary problems that may need in-depth analysis and investigation.

By contrast, a predetermined list of questions is used in a structured interview. The team knows exactly what information is needed, and the interviewer must know and understand the purpose and goal of each question so that a meaningful response can be recognized.

During the interview process, rapport must be established between the interviewer and the respondent. Without this trust, the interviewee may not reveal his or her true opinions. Some techniques to keep in mind during the interview process include funneling,

---

Figure 6.9.   Examples of Closed-Ended Questions

Poor Question Construction	Good Question Construction
What is your present age?	What is your present age?
0–17 ___	___ 20 or younger
17–35 ___	___ 21–30
35–45 ___	___ 31–40
45–60 ___	___ 41–50
60–75 ___	___ 51–60
	___ 61–70
	___ 71 or older

Figure 6.10.   Examples of Terminology

Unclear Terminology	Clear Terminology
Have you received treatment in the ambulatory surgery unit?	Have you had surgery at this hospital for which you came to the hospital in the morning and left after surgery in the afternoon or evening?
Yes ___	___ Yes
No ___	___ No
Don't know ___	

using unbiased questions and clarifying responses. *Funneling* is the process of moving questions from a broad theme to a narrow theme in an unstructured interview. This technique helps the interviewer establish trust with the respondent as well as to address the pertinent quality issues. For example, if a supervisor in health materials management wanted to discuss service to the emergency department with the nursing coordinator from that unit, he might use the following series of questions, which get more specific as the interview proceeds:

Question: Overall, how do you rate our service to the department?
Response: Overall, I think your department is doing a better-than-average job.

Question: A better-than-average job? Where have you experienced problems?
Response: There have been problems at times with getting cath packs when we need them.

Question: Cath packs. Any particular shift?
Response: Definitely in the late afternoons of the day shift.

The interviewer must state each question in a clear and somewhat benign manner so that bias is not introduced. If a specific word or phrase were overemphasized, it might elicit a different response than if all the words were spoken in the same tone. The interviewer should restate each response for clarification. This ensures the correct interpretation of each response.

Sometimes the interviewer may choose to use closed-ended questions, with all of the possible responses specified. In such cases, the interviewer must examine each question carefully to ensure that the wording and delivery of the questions do not bias the response. Questions should begin with broader issues and work toward more specific areas of concern.

Case Study

As previously discussed, one common method of measuring customer satisfaction in healthcare involves conducting a survey. Figure 6.11 shows a survey that was used in a healthcare organization. Students should critique this survey against the criteria listed in the preceding discussion of survey design. Careful consideration should be given to format, wording, appearance, and so on.

Project Application

In the project application for chapter 2, students identified a process needing improvement. They should now identify the expectations of internal and external customers regarding the process being examined. Then, the students should design a survey, interview, or a combination of the two in order to collect data on the process. A minimum of thirty responses should be collected for a survey, or five interviews should be conducted. Student surveys and/or interviews should be critiqued by the instructor before they are administered. Be sure that a consent clause appears in the introduction to the surveys.

Figure 6.11.   Sample Survey Instrument for the Case Study

SSS Questionnaire

Date of Short-Stay Surgery _____
    Type of Procedure _____

Your general impression of the hospital:
    ___ Excellent   ___ Good   ___ Average   ___ Poor

When you spoke with the staff prior to surgery, were they courteous?   ___Yes ___ No

Did they answer your questions about SSS satisfactorily?
___Yes   ___ No   Comments: _____
_____

Were your accommodations in the SSS room:
a.   Clean   Yes ___   No ___
b.   Comfortable   Yes ___   No ___

Treatment by other hospital personnel:

Recovery	exc	good	needs improvement	poor
Concern	☐	☐	☐	☐
Efficiency	☐	☐	☐	☐
Courtesy	☐	☐	☐	☐
Adequate Explanation	☐	☐	☐	☐

Surgery

	exc	good	needs improvement	poor
Concern	☐	☐	☐	☐
Efficiency	☐	☐	☐	☐
Courtesy	☐	☐	☐	☐
Adequate Explanation	☐	☐	☐	☐

X-ray

	exc	good	needs improvement	poor
Concern	☐	☐	☐	☐
Efficiency	☐	☐	☐	☐
Courtesy	☐	☐	☐	☐
Adequate Explanation	☐	☐	☐	☐

When you were discharged, did you receive adequate information and instructions?   ___Yes ___ No

If your surgery was delayed, was an explanation given?
___Yes   ___ No

What determined your selection of Community Hospital of the West as a hospital?
☐ Physician   ☐ Convenience   ☐ Insurance
☐ Friend   ☐ Previous experience
☐ Other:_____

How did you choose the physician who provided your care? _____

Given a choice of hospitals, would you return to Community Hospital of the West?
___Yes   ___ No

In your opinion, how could we improve or add to our services?
_____
_____
_____
_____
_____
_____
_____
_____
_____
_____
_____
_____
_____
_____
_____

We appreciate your confidential opinion of our services. It provides us with the valuable feedback we need in order to continually improve our patient care.

Summary

In healthcare performance improvement, decisions about process or product improvements must be made on the basis of meaningful customer satisfaction data. Customers may be either internal or external to the organization. Many aspects of quality must be considered for each process. Effective ways to collect customers' opinions include surveys and interviews. These tools must be constructed carefully to collect relevant and unbiased data.

References

Atlantic Information Services. 1995. *A Guide to Patient Satisfaction Survey Instruments,* 2nd ed. Washington, D.C.: Atlantic Information Services.

Bowling, Ann. 2002. *Research Methods in Health,* 2nd ed. Philadelphia: Open University Press.

Jagger, Janine. 1982. Data collection instruments: side-stepping the pitfalls. *Nurse Educator,* May–June: 25–28.

Joint Commission on Accreditation of Healthcare Organizations. 2000. *Improving the Care Experience.* Oakbrook Terrace, Ill.: JCAHO.

Sekaran, Uma. 2002. *Research Methods for Business,* 4th ed. New York City: John Wiley & Sons.

Vavra, Terry G. 1997. *Improving Your Measurement of Customer Satisfaction.* Milwaukee, Wis.: ASQ Quality Press.

Chapter 7
Optimizing the Continuum of Care

Learning Objectives

- To explain the reasons why processes are being developed to optimize the continuum of care

- To identify and discuss the steps in the case management function

- To describe how criteria sets contribute to the management of care in the U.S. healthcare system

Background and Significance

Today, American consumers are demanding more extensive and complete healthcare services in the hope of improving the quality and longevity of their lives. Physicians are working to satisfy their entrepreneurial objectives or to realize a professional career with upper-middle-class living standards. Third-party payers, both private and governmental, are trying to maximize profits, minimize costs, and address healthcare fraud. Public and private purchasers, such as businesses buying health insurance for their employees, are looking for comprehensive coverage at affordable premium rates.

The mission of healthcare organizations is to make a positive contribution to the health of their communities and, if possible, to be profitable. The increasing nature of litigation in healthcare, and the increased cost and scarcity of malpractice insurance coverage for providers, adds an additional financial burden to healthcare organizations and physicians. In the United States, this collision of objectives and values has led to a variety of attempts to control the healthcare market, none of which has been entirely successful. It is not the intent of the authors to review the history, successes, and failures of the U.S. healthcare system, but a basic understanding of the system's history, successes, and failures is key to understanding the issues surrounding the concept of the continuum of care. The **continuum of care** can be defined as the totality of healthcare services provided in all settings, from the least extensive to the most extensive. The emphasis is on treating individual patients at the level of care required by their course of treatment. Figure 7.1 shows various types of treatment settings available along the continuum of care.

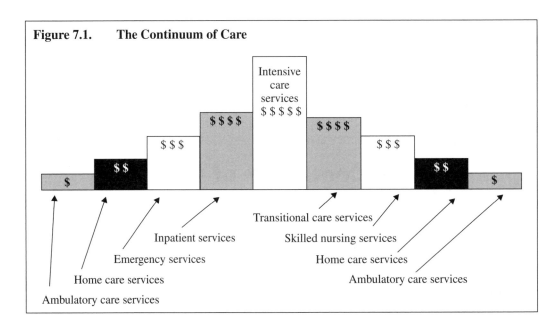

Figure 7.1.   The Continuum of Care

Intensive care services $ $ $ $ $

Transitional care services

Inpatient services

Skilled nursing services

Emergency services

Home care services

Home care services

Ambulatory care services

Ambulatory care services

Understanding that the patient-centered continuum of care (figure 7.3) has evolved from a very authoritarian system of care (figure 7.2) is an important starting point. The practice of medicine has best been described as an art. In the past, each physician was seen as a separate and individual provider who treated only one aspect of a patient's illness or the old family practitioner who treated everything. That physician may order tests and X rays based on the need for information specific to that treatment. Patients did not question physicians' diagnoses, nor whether a test was required. There was scant review of the necessity for tests and care. Medicine was practiced as an art, with most information for treatments driven by the actual skill and knowledge base of the individual practitioner. A surgeon practiced only surgery, and if the patient had other health issues, he or she was referred to another physician for care of those issues. Physicians ordered what they felt was necessary for treatment.

As treatments and technologies developed, the costs of research and knowledge-based care began to rise. As the cost of treatment rose, insurance companies began to experience higher payouts for treatments and interventions that were sometimes duplicated and not always best for the patient. For example, it was routine practice at one point in time for OB/GYN physicians to remove the ovaries and uterus during a hysterectomy as a routine procedure to prevent future problems. It was later discovered that women seemed to fare better hormonally when healthy ovaries were allowed to remain after surgery. Insurance companies then stepped in to state that they would no longer reimburse physicians for this additional aspect of surgery unless there was clear indication of a need to remove the ovaries.

With increased awareness in duplication of services, the drive to determine the medical necessity for treatments and care increased as well. Purchasers and payers of health-care services began to demand a more comprehensive approach to care, one that decreased costs and improved the quality of care being provided. Along with this demand came standards intended to ensure that services being provided were timely, cost-efficient, and

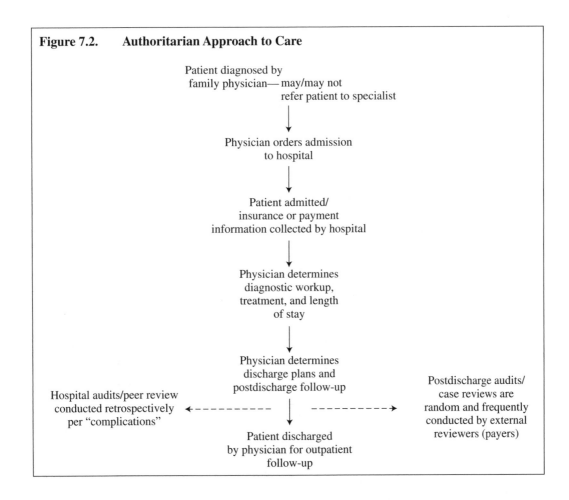

Figure 7.2.   Authoritarian Approach to Care

Patient diagnosed by
family physician— may/may not
refer patient to specialist

Physician orders admission
to hospital

Patient admitted/
insurance or payment
information collected by hospital

Physician determines
diagnostic workup,
treatment, and length
of stay

Physician determines
discharge plans and
postdischarge follow-up

Hospital audits/peer review
conducted retrospectively
per "complications"

Postdischarge audits/
case reviews are
random and frequently
conducted by external
reviewers (payers)

Patient discharged
by physician for outpatient
follow-up

appropriate to the patient's medical condition. As patients were stuck with medical bills that insurance companies refused and providers were unwilling to write off because they were not deemed medically necessary, new processes were developed to address these concerns. The continuum of care model, designed to meet the needs of patients, providers, and payers, was thus developed. A review system was put into place to address medical necessity issues and the creep of fraudulent billing practices that began to occur as a result of the treatment restrictions placed on providers. Millions of dollars were lost in Medicaid and Medicare fraud each year because systems and billing practices were not carefully audited.

Procedures to address the flow of information from one practitioner to another (also known as discharge planning) became a part of the continuum of care efforts that developed. Efforts were made to ensure that information between practitioners and providers could be easily shared so that duplication of testing and procedures could be reduced. Performance improvement (PI) monitoring of the utilization of healthcare services began in many facilities when efforts to control cost and ensure quality were put into practice. Figures 7.2 and 7.3 demonstrate the change in access to care before and after implementation of the utilization review function in healthcare.

Figure 7.3.   Patient-Centered Approach to Care

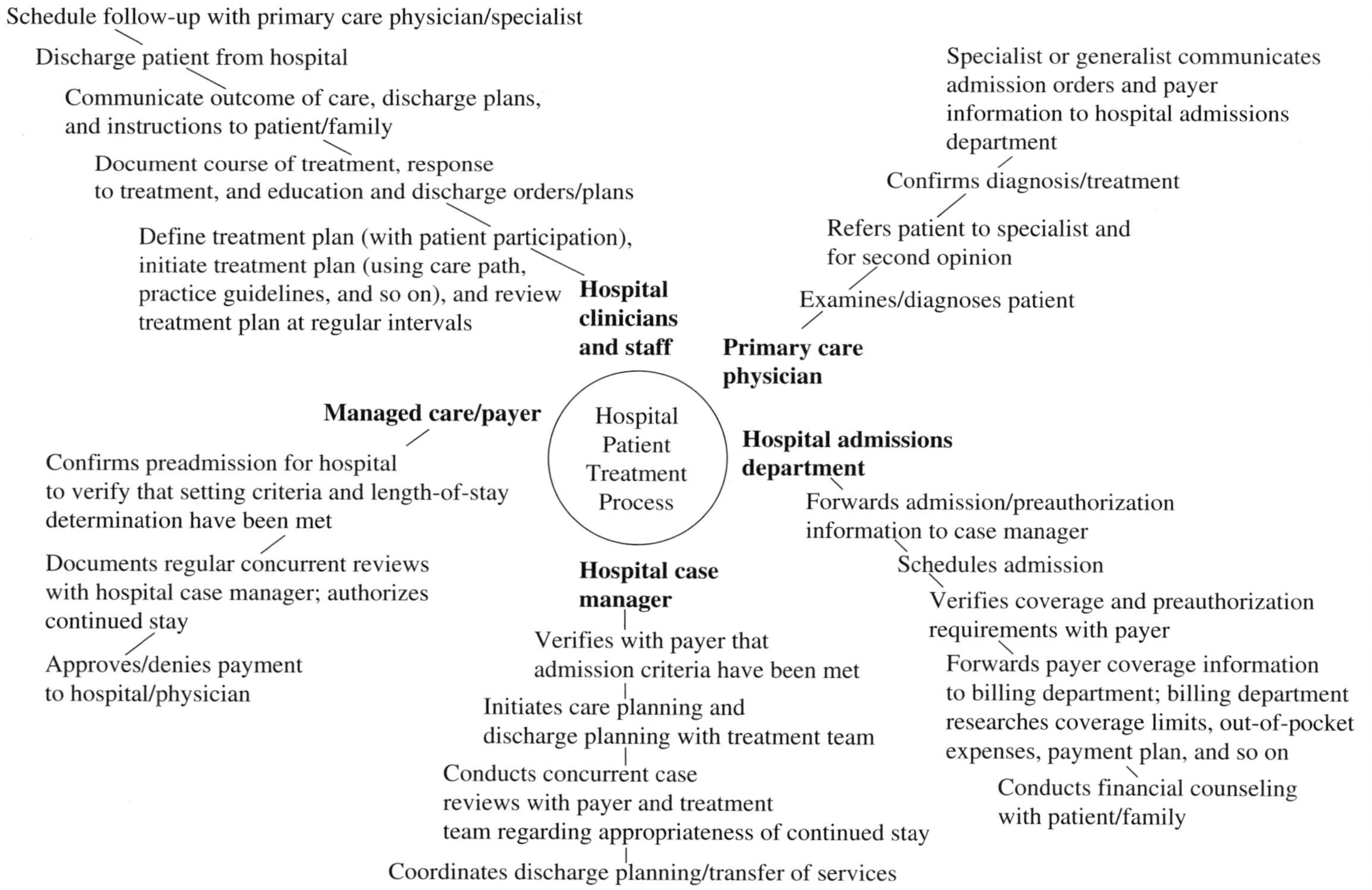

Schedule follow-up with primary care physician/specialist

Discharge patient from hospital

Communicate outcome of care, discharge plans, and instructions to patient/family

Document course of treatment, response to treatment, and education and discharge orders/plans

Define treatment plan (with patient participation), initiate treatment plan (using care path, practice guidelines, and so on), and review treatment plan at regular intervals

Hospital clinicians and staff

Managed care/payer

Confirms preadmission for hospital to verify that setting criteria and length-of-stay determination have been met

Documents regular concurrent reviews with hospital case manager; authorizes continued stay

Approves/denies payment to hospital/physician

Specialist or generalist communicates admission orders and payer information to hospital admissions department

Confirms diagnosis/treatment

Refers patient to specialist and for second opinion

Examines/diagnoses patient

Primary care physician

Hospital admissions department

Forwards admission/preauthorization information to case manager

Schedules admission

Verifies coverage and preauthorization requirements with payer

Forwards payer coverage information to billing department; billing department researches coverage limits, out-of-pocket expenses, payment plan, and so on

Conducts financial counseling with patient/family

Hospital Patient Treatment Process

Hospital case manager

Verifies with payer that admission criteria have been met

Initiates care planning and discharge planning with treatment team

Conducts concurrent case reviews with payer and treatment team regarding appropriateness of continued stay

Coordinates discharge planning/transfer of services

Healthcare in the United States

In discussions of healthcare in the United States, point of view is everything. Common viewpoints include public and private regulation, healthcare economics, and the human desire to benefit personally and collectively.

Throughout the past century, attempts have been made to balance the competing needs and expectations of consumers, providers, and payers. At different times, each group has dominated the marketplace, although none of the three has been on top for long. The products of this competition have included the voluntary hospital system, the finest healthcare technological infrastructure in the world, private health insurance plans, and the broadest range of the most effective pharmaceuticals available, Medicare, Medicaid, preferred provider arrangements, and health maintenance organizations, to name a few. The culmination has come in the most recent experimental solution: managed care. Students of healthcare quality and PI should review the economic and policy issues inherent in U.S. approaches to delivering healthcare.

The overall goal of the U.S. healthcare system is to achieve an equilibrium between health and spending, as illustrated in figure 7.4. As the figure shows, a finite level of optimal collective health can be realized in U.S. society. Although many factors influence collective health, the system today seeks to identify the optimal level of spending that will achieve the optimal level of collective health. Expenditures are funded by a combination of public and private resources: public health, Medicare, Medicaid, insurance, private pay, public sanitation, and others. But there comes a point where more spending does not mean more collective health.

In figure 7.4, the shaded triangle that represents total spending extends above the black line that represents optimal health. This representation acknowledges the realization that all of the healthcare spending a society could possibly do would not necessarily achieve the goal of optimal collective health. At some point for every patient within the healthcare system, no additional health benefit would be achieved by further spending. Any additional expenditure would, in effect, be wasted. The money could have been used for another patient who might still have benefited. Because millions of Americans never approach the optimal health condition, that waste is considered intolerable. The continuum of care system seeks to provide benefits up to the point of optimal health for each individual, thereby realizing optimal health collectively.

For example, Mr. Abraham Smith is a 90-year-old, Caucasian male who has, in recent years, developed a heart condition known as chronic ventricular fibrillation. The condition is manageable with medication, but patients often continue to experience occasional periods of the arrhythmia even when they are taking their medication as prescribed. Mr. Smith experiences such arrhythmias. Each time the fibrillation begins, Mr. Smith feels weak and unwell, his face flushes, and he can feel his heart fluttering in his chest. He is afraid that he is going to die. Immediately, he calls to his wife to take him to the emergency department of his local hospital. By the time the couple drives to the hospital and Mr. Smith is checked by a physician, the fibrillation has subsided, and his heart has returned to a normal sinus rhythm. The emergency department physician on duty examines him, performs an EKG, and draws blood studies, all of which are negative. Each time, the symptoms could be those of a heart attack, but they are not.

Mr. Smith does not want to accept the fact that his heart condition cannot be managed any better than it currently is managed. His demand for services at the emergency department accomplishes nothing, but no one is prepared to tell him not to go when he experiences arrhythmias, especially in light of the litigious nature of U.S. society. Yet emergency visits are among the most expensive types of ambulatory care. Every time Mr. Smith visits the emergency department, he wastes services that could have been used by someone for whom the emergency visit would have been more helpful. Mr. Smith's use of healthcare services is not optimal, neither for him nor for society.

Regulatory approaches seek to control expenditures on individuals such as Mr. Smith in the hope that more resources will be available for those who can still benefit. Medicare regulations mandated utilization review in the mid-1960s. The system required physician committees to review the practice patterns of their colleagues at institutions receiving Medicare dollars. In the 1970s, Medicaid programs in most of the states trained their own reviewers to visit hospitals and make sure that Medicaid patients were staying in the hospital only as long as absolutely necessary. In the mid-1980s, the prospective payment system was implemented for Medicare and Medicaid patients. Under this system, standardized payments are made to hospitals according to the diagnosis-related group (DRG) into which a patient falls. In the 1990s, the private and public sectors wrestled with

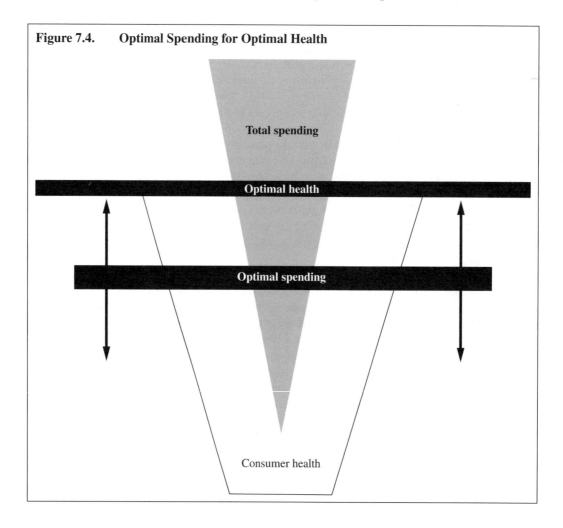

Figure 7.4.   Optimal Spending for Optimal Health

Total spending

Optimal health

Optimal spending

Consumer health

the sometimes unfortunate decisions of managed care officials. All of these efforts have attempted to accomplish a balance between health benefit and health spending.

The issue remains important in the administration of healthcare organizations today. Services are to be accorded to, and expenditures made for, individuals who can still benefit. Organizations must be able to demonstrate that rational decisions were made about a patient's care and that those decisions were in the patient's and society's collective best interest. Organizations must be able to demonstrate as well that the services provided to the patient were appropriate to the patient's physical and quality-of-life needs across the continuum of care. The continuum of care consists of all of the possible settings in which patients may receive care. (See figure 7.1.) Today, care is delivered in homes, physicians' offices, ambulatory care centers, hospitals, long-term care facilities, and residential care facilities. Each of these care settings provides a more or less complex set of services, all of which depend on the identified needs of the patient. In addition, each patient's needs may change, depending on the time that the needs are identified: prior to hospital admission, during the admitting process, during the hospital stay, during the discharge process, and during any immediate subsequent care episode. The expectation of the public and private regulatory agencies is that those needs will be identified and that patients will be cared for in the setting most appropriate to their needs.

Optimizing the Continuum of Care: Steps to Success

The principal process by which organizations optimize the continuum of care for their patients is **case management.** Case managers review the condition of patients to identify each patient's care needs and to integrate patient data with the patient's course of treatment. The case manager in many organizations matches the patient's course with a predetermined optimal course (also known as care map, critical path, or practice guideline) for the patient's condition. He or she identifies the actions to be taken when the patient's care is not proceeding optimally. The same concept under the name managed care is used by many payers to clearly define when a patient may have a procedure or to stipulate a particular course of treatment that the payer believes will be equally effective, but less costly.

Step 1: Perform preadmission care planning

Preadmission care planning is initiated when the patient's physician contacts a healthcare organization to schedule an episode of care service. The case manager reviews the patient's projected needs with the physician. The manager may also contact the patient directly to obtain further information.

In addition, the manager may contact the patient's payer to confirm that all of the necessary preadmission authorizations have been obtained and that the payer will pay for the patient's services. This process is called preauthorization. As part of preauthorization, the payer's representative will have compared the planned services with the payer's criteria of care for the patient's diagnosis. Some payers will pay 80 to 100 percent for care provided in their preferred provider list of agencies. If a patient opts to go to a different "out-of-contract" facility, he or she may have to pay a higher reimbursement rate for services. Some insurance plans allow the patient to have a choice of treating facility and/or physician, but these plans usually have a lower reimbursement rate.

When a patient is to be transferred from one facility to another, the case manager will contact the case manager at the original facility to coordinate the transfer of services.

Step 2: Perform care planning at the time of admission

When the patient is admitted to the hospital, the case manager will review all of the information that has been gathered by the clinicians assigned to the case to confirm that the patient meets the admission criteria for the patient's admitting diagnosis. The manager will confirm that the patient requires services that can be performed in the facility. If it were determined that the facility could not perform the services needed, the case manager would arrange for the patient to be transferred to another facility.

If the facility utilizes care-mapping methodology, the case manager will at this time assign the case to the appropriate critical path. (See figure 7.5.) He or she would verify that all of the services stipulated in the critical path have been initiated.

Step 3: Review the progress of care

The case manager periodically reviews the patient's progress throughout the entire episode of care. When a critical path is being used, the manager will reintegrate care data each time the case is reviewed and compare the patient's progress to the path. When variations from expected progress occur, the case manager coordinates interventions among the clinicians and therapists assigned to the case to move the patient along the path.

From the beginning of the episode of care, the case manager continuously monitors the patient's acuity level and requirements for services. At the same time, he or she plans for the services the patient will need after discharge. The manager will make arrangements for the patient to be transferred to another facility or nursing unit to meet the patient's care needs. Ultimately, the goal is to maintain the patient at the least costly level of care possible.

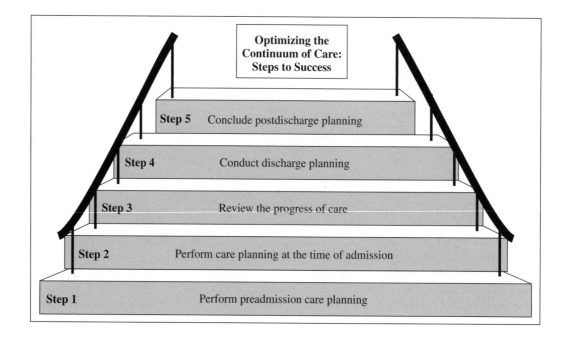

Figure 7.5.   Sample Gantt Chart Format in an Excerpt of a Critical Pathway

Western University Regional Medical Center
Department of Nursing
Case Management Plan

Diagnosis: Idiopathic pediatric scoliosis, with surgery, without complications   **Unit:** 9 East   **DRG:** 215

Average Length of Stay: 7 days   **Usual OR Day (admission day = 1):** 2

Clinical Milestones:

	Prior to Admission	Day 1	Day 2	Day 3	Day 4	Day 5	Day 6	Day 7	Day 8	Day 9	Day 10	Day 11	Day 12
Self-donation of blood	X												
Chest X ray, chem panel	X												
Labs, EKG, blood type & crossmatch	X												
PM admission		X											
H&P		X											
Care planning		X											
Surgery			X										
Surgical ICU			X										
Catheter removed				X									
Patient sits up in bed					X								
Transfer to surgical floor					X								
Physical therapy					X	X	X						
Patient walks down corridor						X							
Patient education for self-care							X						
Patient receives meds and instructions for follow-up care								X					
Outpatient physical therapy scheduled						X							
Patient discharged to home								X					

Health Outcomes:

Diagnosis	Outcome (The patient . . .)	Day–Visit	Intermediate Goal (The patient . . .)	Day–Visit	Process (The nurse . . .)	Day–Visit	Process (The physician . . .)
Fluid–electrolye imbalance: third space shifting secondary to large volume loss and replacement	Has stable vital signs consistent with base-line at admission	5–6 4–14	Is afebrile	4	Takes vital signs every 2 hours	PTA	Arranges for self-donation of blood before surgery
		4–6	Maintains urine out-put over 1 cc/kg/hr while catheterized	4	Measures and records urine output every hour	1	Assesses patient's cardiac status on admission
	Has a baseline nor-mal voiding pattern	6–8	Voids 8 hours after Foley is discontinued			1	Orders lab workup
		4–14	Maintains specific gravity under .1020	4	Monitors specific gravity		
	Has no edema	8–9	Returns to baseline skin turgor	4–8	Balances IV and oral intake to achieve maintenance fluid requirements		

Step 4: Conduct discharge planning

As the patient's requirements for care decrease and the patient moves toward discharge, the case manager undertakes final discharge planning. In this step, the patient's continued care after discharge is planned. Many times family members and significant others are an active part of this process because they will often assist with care after discharge. Postdischarge medications are prescribed and therapies are scheduled. Arrangements to transfer the patient to a subacute facility are made when necessary. Effective discharge planning often begins at the time of admission to ensure that the patient will be prepared to leave the facility as scheduled.

Step 5: Conclude postdischarge planning

Once the patient has been discharged, the case manager conveys the information about the patient's course of treatment to the clinicians who will continue to care for the patient after discharge. At this point, the case management function is returned to the patient's physician and office staff. Some healthcare organizations, however, follow up on patients after discharge to ensure that the transition has gone smoothly and that the patient is receiving all of the services required.

Real-Life Example

Information collected in the form of valid and reliable data is the starting point for the management of the continuum of care. Established criteria for care and clinical paths are crucial to the quality of patient care. Such guidelines, coupled with the clinical expertise of the case manager, can make the process of case management more effective.

At one hospital, care guidelines used internal criteria as well as external criteria for the process of utilization review and case management. It also used the feedback information provided by the third-party payers. Unfortunately, the organization lacked the ability to share that information.

The quality leadership had noted through data collection that the process of admission, treatment, and discharge planning was not well coordinated among the associated caregivers and the business operations of the organization. Key issues included the following:

- Hospital stays were continued after symptoms and treatment had reached a point where the patient could have received appropriate care in a less intensive care setting.

- Patients were admitted to the hospital for diagnosis and treatment in cases where the patients could have received appropriate care in a less intensive care setting.

- Families were not involved in the decisions regarding discharge placement options or care plans. Home health agencies, skilled nursing care facilities, and hospices were not being contacted early enough in the patient's stay to facilitate postdischarge transition to other levels of care.

- Insurance carriers and Medicare were denying payment because services had not been preauthorized. Insurance carriers and Medicare were denying payment because services had been rendered after discharge criteria had been met and documented.

- Medical records contained inadequate documentation of medical conditions, interventions, and outcomes.

At first, hospital utilization/quality department personnel organized themselves to better communicate the needs, problems, and obstacles to an effective case management system. It quickly became apparent, however, that they alone could not make the kind of differences necessary to improve the ailing systems.

A PI team was formed. The team included a representative from the admissions and registration department, business office and financial counselors, the admissions nurse, the operating room scheduler, the operating room manager, a representative from outpatient services, representatives from the nursing units, discharge planners, the utilization review/ case managers, and the director of the utilization/quality department.

The team implemented Shewhart and Deming's model of performance improvement. The team discovered, however, that it did not include all of the individuals and departments who were closely associated with the process. The team realized that it also needed representatives and participants from physicians' offices. It invited key office personnel to participate on the team.

The newly composed team spent several weeks working on team building. The result of their work was a team of individuals who had a common mission and goals. Departmental barriers were ignored. Leading this group of people to form a cohesive team was the most difficult task. Accusations and feelings of failure ensued. But eventually, an important breakthrough occurred. After repeated reinforcement of the concept that improvement of this process would benefit all departments, the team was able to get down to work.

Team members found more purpose and pleasure in their work and felt more powerful and less frustrated. The team collected, analyzed, and reported the information to those persons who could directly help in making a difference. The medical staff was educated regarding the problems and proposed solutions that could help develop an effective case management model. Other office personnel and service departments were included to complete the circle of participants.

The team's achievements included the following:

- Critical information was shared about postdischarge planning, such as treatment plans, goals of care, and financial information, including benefits, limits, out-of-pocket expenses, and deductibles.

- Providers of primary care were included in the decision-making process, and they were given information that benefited not only patients and their families, but also the physicians' practices and the hospital.

- Coordination of services was improved and access to specialists and special services was provided in a timely manner, including options for alternative placement for care and the wise use of financial resources. Often, with completed care plans, the payers were willing to discontinue the patient's insurance contract and pay for services that would complete the healing process and avoid readmission or duplicate extended care and testing.

- The numbers and dollar amounts of denials of payment decreased. The organization's fiscal situation improved, thus allowing the purchase of equipment and expansion of services.

- The quality of documentation in medical records was improved, as was access to records and reports. Some forms and the flow of some medical information were changed.

- Having a proven method of case management served as an asset in subsequent contracting with new providers and payers.

- Departments, personnel within the organization, and provider offices became more unified. The team approach served as a catalyst for change within the organization. Individuals realized that they could make a difference and that they had the ideas and power to make change happen. The PI activity strengthened the commitment of employees to the mission and vision of the organization.

QI Toolbox Techniques

Indicators, or criteria, and Gantt charts are often used in assessments of continuum-of-care issues.

Indicators

An **indicator,** or criterion, is a performance measure that enables healthcare organizations to monitor a process to determine whether it is meeting process requirements. The criteria may be established and implemented internally, externally, or generically.

Internal criteria are usually developed by an interdisciplinary team made up of physicians, nurses, and other clinical staff from the healthcare organization. This type of criterion is developed to monitor specific processes within the organization.

External criteria are created by some organization outside the healthcare facility. Organizations such as insurance companies, peer review organizations, the Centers for Medicare and Medicaid Services, and other regulatory agencies develop healthcare criteria.

Generic criteria have been developed by many of the same agencies for use across the continuum of care and in various regions of the country. The term *generic* implies that the criteria are applicable across many organizations and with many different kinds of patients.

One of the most common applications of generic criteria measures is in the area of admission certification. Admission criteria are used to establish the fact that each patient actually requires care at the level to which he or she has been admitted. Usually, admission criteria in acute care settings have two categories: intensity of service and severity of illness. For a patient to meet the admission criteria, he or she must meet a clinical measure in each of the categories. Intensity of service refers to the type of services or care the patient will require. Severity of illness refers to how sick the patient is or what level of care the patient will require, such as intensive care unit or general medical unit. (See the list of admission criteria in the case study that follows.)

Each indicator, written in the form of a ratio, is used as a tool for monitoring care and service. An indicator for the number of admissions that meet set admission criteria might be the following ratio:

$$\frac{\text{Number of admissions meeting criteria}}{\text{Total number of admissions}}$$

A target or goal set by a healthcare facility might be, "Admission criteria met 99 percent of the time." Meeting such high levels of expectation is very important to healthcare organizations today because care rendered to patients who do not meet admission criteria is generally not reimbursed.

These kinds of indicators can also be used to monitor other important processes in the organization. For example, the Joint Commission on Accreditation of Healthcare Organizations (JCAHO) has developed sets of core performance measures through its ORYX initiatives. (See www.jcaho.org for more information.) Core measurement areas currently include measurement sets for acute myocardial infarction, heart failure, community-acquired pneumonia, and pregnancy-related conditions. Hospitals, for example, are required to select two core measures based upon the healthcare services the organization provides and routinely submit data to a vendor for processing. The vendor then provides comparative information to the organization and the JCAHO on all applicable measures within the selected sets. An example of a data element collected on the core measurement area for acute myocardial infarction can be seen in figure 7.6. There are nine possible measures within the acute myocardial infarction area for which hospitals are required to submit data.

Monitoring these indicators allows the organization's leadership to identify the cases in which the best care may not have been provided to a patient. The indicators also identify excessive numbers of cases in which this was true, thus providing an opportunity for improvement in organizational processes. A set of hospital standard measures (or indicators) utilized by one community hospital is provided in table 7.1. These indicators were used to provide standardized reporting. (See the discussion on standardized reporting in chapter 15.) A corresponding JCAHO standard is cited for each criterion, and the formula is given. Each criterion indicates the benchmark that the organization wants to meet, how often the indicator is measured, which organizational unit is responsible for the measure, from where data are pulled to compute the measure, and where the results are reported.

Gantt Charts

A **Gantt chart** is a project management tool used to schedule important activities. Gantt charts divide a horizontal scale into days, weeks, or months, and a vertical scale into the project activities or tasks.

Gantt charts are used in clinical process improvement to depict clinical guidelines or critical paths in the treatment of common medical conditions. The tool provides a graphic method for showing the simultaneous and interdependent treatments for a clinical condition that are most likely to result in the best possible outcome. Figure 7.5 is an example of a Gantt chart. The chart depicts the clinical guidelines used for cases of idiopathic pediatric scoliosis at a large medical center.

Figure 7.6.   JCAHO Core Measure Set: Acute Myocardial Infarction

Performance Measure Name: Aspirin at arrival

$$\frac{\text{AMI patients who received aspirin within 24 hours before or after hospital arrival}}{\text{AMI patients without aspirin contraindications}}$$

Source: Joint Commission on Accreditation of Healthcare Organizations. 2003. ORYX/Selection Change Form for Hospitals. Accessed online at http://www.jcaho.org/accredited+organizations/hospitals/oryx/selection+change+ forms/index.htm.

Table 7.1. Hospital Standard Measures

Standard	Measure of Process	Type Out/Proc	Measured Y/N	Indication/Formula	Benchmark	How often measured	Owner	Source of Data	Results Given to
				Operative & Invasive Procedures					
PI.3.2.1	Discrepancies--Pre-op/Post-op/Path	O	Y	# cases with discrepancy / # of pathology cases	0%	Quarterly	OR	OR data sheet	QC, Med-Exec, & the Board
PI.3.2.1	Procedure appropriateness: Surgical criteria not met	P	Y	# of appropriate cases / # of cases	100%	Quarterly	OR	OR data sheet	QC, Med-Exec, & the Board
PI.3.2.1	Patient preparation for procedure	P	Y	# of pts w/incomplete prep / # of pts requiring prep	0%	Quarterly	OR	OR data sheet	QC, Med-Exec, & the Board
PI.3.2.1	Procedure performance & patient monitoring: Intraoperative complications	P	Y	# patients w/intraoperative complications / # patient procedures	0%	Quarterly	OR	OR data sheet	QC, Med-Exec, & the Board
PI.3.2.1	Procedure performance & patient monitoring: Unplanned returns to OR	P	Y	# pts w/unplanned return to the operating room / # patients operated on	0%	Quarterly	OR	OR data sheet	QC, Med-Exec, & the Board
PI.3.2.1	Complications of post-procedure care	P	Y	# pts w/complications / # patients operated on	0%	Quarterly	OR	OR data sheet	QC, Med-Exec, & the Board
PI.3.2.1	Complication of post-procedure patient education	P	Y	# pts w/pt ed completed / # patients operated on	100%	Quarterly	OR	OR data sheet	QC, Med-Exec, & the Board
				Medication Use					
PI.3.2.2	Drug selection: DUE--Primaxin appropriateness	O	Y	# meeting criteria / # patients on Primaxin	100%	Quarterly	Pharmacy	DUE	P&T Med-Exec
PI.3.2.2	Prescribing or ordering: DUE--Toradol dosing appropriateness	O	Y	# meeting criteria / # patients on Toradol	100%	Quarterly	Pharmacy	DUE	P&T Med-Exec
PI.3.2.2	Preparing and dispensing: Dispensing Errors	O	Y	# dispensing errors / # inpatient days	0%	Quarterly	Pharmacy	Pharmacy tracking system	P&T Med-Exec
PI.3.2.2	Preparing and dispensing: Med. Delivery Time--Antibiotics	P	Y	# meeting criteria / # patients on med	<120 min	Quarterly	Pharmacy	Meditech documentation	P&T Med-Exec
PI.3.2.2	Administering: Medication Delivery Time--Thrombolytics	P	Y	# meeting criteria / # patients on med	<52 min	Quarterly	Pharmacy	Meditech documentation	P&T Med-Exec
PI.3.2.2	Administering: DUE--Gentamycin appropriateness (peak/trough)	P	Y	# meeting criteria / # patients on med	100%	Quarterly	Pharmacy	DUE	P&T Med-Exec
PI.3.2.2	Monitoring the effects on patients: Adverse drug reactions	O	Y	# ADR's / # of admissions	0%	Quarterly	Pharmacy	Incident report	P&T Med-Exec

Table 7.1. *(Continued)*

		Blood and Blood Components							
PI.3.2.3	Ordering: Blood usage appropriateness	# episodes appropriate / # transfusion episodes	P	Y	100%	Quarterly	Lab	Patient's records	QC & Med-Exec
PI.3.2.3	Ordering: Written order for blood/components	# episodes with orders / # transfusion episodes	P	Y	100%	Quarterly	Lab	Physician's orders	QC & Med-Exec
PI.3.2.3	Distributing, handling, & dispensing: Protocol for checking unit out of bloodbank	# units correctly dispensed / # units dispensed	P	Y	100%	Quarterly	Lab	Bloodbank log	QC & Med-Exec
PI.3.2.3	Administration: Blood hung within 20 minutes of dispensing	# units hung w/in 20 mins. / # units dispensed	P	Y	100%	Quarterly	Lab	Blood slips	QC & Med-Exec
PI.3.2.4	Administration: Blood slips posted in the chart	# blood slips posted / # units transfused	P	Y	100%	Quarterly	Lab	Blood slips	QC & Med-Exec
PI.3.2.5	Administration: Blood slips completed	# slips completed / # units transfused	P	Y	100%	Quarterly	Lab	Blood slips	QC & Med-Exec
PI.3.2.5	Administration: Blood transfusion forms completed	# forms completed / # units transfused	P	Y	100%	Quarterly	Lab	Blood slips	QC & Med-Exec
PI.3.2.3	Administration: Crossmatch: Transfusion ratio	# crossmatches / # transfusions	O	Y	<=2 (I)	Quarterly	Lab	Bloodbank log	QC & Med-Exec
PI.3.2.3	Monitoring effects on patients: Potential transfusion reactions	# patients with potential transfusion reactions / # pts receiving transfusions	O	Y	<=1 (I)	Quarterly	Lab	Bloodbank log or phone notification	QC & Med-Exec
PI.4.5.3	Monitoring effects on patients: Confirmed transfusion reactions	# of true reactions / # transfusion episodes	O	Y	0% (I)	Quarterly	Lab	transfusion reaction investigations	QC & Med-Exec
PI.3.2.3	Transfusion report form completed	# forms completed / # pts receiving transfusions	P	Y	100%	Quarterly	Lab	report log	QC & Med-Exec
		Miscellaneous Measures							
PI.3.2.4	Utilization Management: Patients admitted to observation who should have been inpatients	# pts not admitted as IP's who met inpatient criteria / # pts admitted	P	Y	0%	Quarterly	UR	Utilization mgmt. form	QC & Med-Exec
PI.3.2.4	Utilization Management: Pts admitted who meet criteria on initial review	# appropriate admissions / # of admissions	P	Y	100%	Quarterly	UR	Utilization mgmt. form	QC & Med-Exec
PI.3.2.4	Utilization Management: Pts admitted meeting continued stay criteria	# pts no longer meeting continued stay criteria / # of patients discharged	P	Y	100%	Quarterly	UR	Utilization mgmt. form	QC & Med-Exec
PI.3.2.4	Utilization Management: Pts remaining inpatients after discharge criteria met	# pts remaining after discharge criteria met / # pts discharged	P	Y	0%	Quarterly	UR	Utilization mgmt. form	QC & Med-Exec

(Continued on next page)

Table 7.1. *(Continued)*

				Miscellaneous Measures (cont'd)					
PI.3.2.4	Utilization Management: # of pts for whom discharge planning was done	P	Y	# of pts for whom discharge planning done / # pts discharged	100%	Quarterly	UR	Utilization mgmt. form	QC & Med-Exec
PI.4.5.2	Adverse events during anesthesia	O	Y	# adverse events / # pts given anesthesia	0% (I)	Quarterly	OR	or OR data sheet	QC & Med-Exec
TX.7.1.3.2.3	Restraint & Seclusion: Evidence of less restrictive measures used	O	Y	# pts w/evidence of less restrictive measures used / # patients in restraints	100% documnt.	Quarterly	Nursing	Patient restraint record	QC & Med-Exec
PI.3.3	Mortality Rate	O	Y	# deaths / # IP & OB discharges	1.4% (L)	Quarterly	HIM	Patient record	QC & Med-Exec
PI.3.3.1	Autopsy Results	P	Y	# performed / # met criteria		Quarterly	OR	Patient record & Autopsy criteria	QC & Med-Exec
PI.3.3	C-Section Rate	O	Y	# C-Sections performed / # of deliveries	17% (L)	Quarterly	Labor & Delivery	Obstetric report	QC & Med-Exec
PI.3.3	Vaginal deliveries with complications (ORYX)	O	Y	# vaginal deliveries w/comp. / # vaginal deliveries		Quarterly	OR	Patient report	QC & Med-Exec
PI.3.3	VBAC Rate	O	Y	# VBAC / # repeat C-sections	36% (S)	Quarterly	Labor & Delivery	Obstetric report	QC & Med-Exec
PI.3.3	Attempted VBAC Rate	O	Y	# pts w/previous C-sections who receive trial of labor / # pts w/previous C-sections	36% (L)	Quarterly	Labor & Delivery	Monthly obstetric	QC & Med-Exec
PI.3.3	Joint Replacements with complications (ORYX)	O	Y	# joint replacements with complications / # joint replacements	0%	Quarterly	OR	Patient record & Surgical stats	QC & Med-Exec
				Risk Management Activities					
PI.3.3.2	Notice of Intent	O	Y	# for current quarter	1	Quarterly	Risk Mgmt.	Receipt of atty letter or notice	QC & Med-Exec
PI.3.3.2	Summons and Complaint	O	Y	# for current quarter	1	Quarterly	Risk Mgmt.	Receipt of atty letter or notice	QC & Med-Exec
PI.3.3.2	Potentially Compensable Events	O	Y	# for current quarter	1	Quarterly	Risk Mgmt.	Complaint system	QC & Med-Exec
PI.3.3.2	Small Claims	O	Y	# for current quarter	1	Quarterly	Risk Mgmt.	Complaint system	QC & Med-Exec
PI.3.3.2	Patient/Family Complaints--Care	O	Y	# complaints / # patient days	<=2%	Quarterly	Risk Mgmt.	Complaint system	QC & Med-Exec
PI.3.3.2	Patient/Family Complaints--Billing	O	Y	# complaints / # patient days	<=2%	Quarterly	Risk Mgmt.	Complaint system	QC & Med-Exec
PI.3.3.2	Medical Device Reporting (Manufacturer)	O	Y	# complaints / # patient days	<.1%	Quarterly	Risk Mgmt.	Incident Report	QC, Med-Exec & Safety Comm

Table 7.1. *(Continued)*

				Risk Management Activities (cont'd)					
PI.3.3.2	Medical Device Reporting (FDA)	O	Y	# of reportings / # patient days	<.1%	Quarterly	Risk Mgmt.	Incident Report	QC, Med-Exec & Safety Comm
PI.3.3.2	Tracking Requirements - Operating Room	O	Y	# pts successfully tracked / # pts required to track	100% (I)	Quarterly	Risk Mgmt.		QC, Med-Exec & Safety Comm
PI.3.3.2	Total Product Recalls	O	Y	# recalls / # patient days	<.2%	Quarterly	Risk Mgmt.	Notification from Manufacturer	QC, Med-Exec & Safety Comm
PI.3.3.2	Pharmacy	O	Y	# recalls / # patient days	<.1%	Quarterly	Risk Mgmt.	Notification from Manufacturer	QC, Med-Exec & Safety Comm
PI.3.3.2	Nutrition	O	Y	# recalls / # patient days	<.1%	Quarterly	Risk Mgmt.	Notification from Manufacturer	QC, Med-Exec & Safety Comm
PI.3.3.2	Materials Management	O	Y	# recalls / # patient days	<.1%	Quarterly	Risk Mgmt.	Notification from Manufacturer	QC, Med-Exec & Safety Comm
PI.3.3.2	Bio Medical	O	Y	# recalls / # patient days	<.1%	Quarterly	Risk Mgmt.	Notification from Manufacturer	QC, Med-Exec & Safety Comm
				Quality Control Activities					
PI.3.3.3	Clinical Lab	P	Y	# of QC measures at 100% / # QC measures in dept.	100% (I)	Quarterly	Lab	QC Logs	QC, Med-Exec & the Board
PI.3.3.3	Dietary	P	Y	# of QC measures at 100% / # QC measures in dept.	100% (I)	Quarterly	Dietary	QC Logs	QC, Med-Exec & the Board
PI.3.3.3	Diagnostic Radiology	P	Y	# of QC measures at 100% / # QC measures in dept.	100% (I)	Quarterly	Imaging	QC Logs	QC, Med-Exec & the Board
PI.3.3.3	Nuclear Medicine	P	Y	# of QC measures at 100% / # QC measures in dept.	100% (I)	Quarterly	Imaging	QC Logs	QC, Med-Exec & the Board
PI.3.3.3	Equipment used to administer meds	P	Y	# of QC measures at 100% / # QC measures in dept.	100% (I)	Quarterly	Pharmacy	QC Logs	QC, Med-Exec & the Board
PI.3.3.3	Pharmaceutical equipment used to prepare medications	P	Y	# of QC measures at 100% / # QC measures in dept.	100% (I)	Quarterly	Pharmacy	QC Logs	QC, Med-Exec & the Board
				Patient Rights					
PI.3.2.5	Patient satisfaction: Inpatient	O	Y	percentage reported only	93% (C)	Quarterly	QRS	Gallup results	QC, Med-Exec & the Board
PI.3.2.5	Patient satisfaction: Outpatient tests & treatment	O	Y	percentage reported only	93% (C)	Quarterly	QRS	Gallup results	QC, Med-Exec & the Board
PI.3.2.5	Patient satisfaction: Outpatient surgery	O	Y	percentage reported only	93% (C)	Quarterly	QRS	Gallup results	QC, Med-Exec & the Board
PI.3.2.5	Patient satisfaction: Emergency Room	O	Y	percentage reported only	93% (C)	Quarterly	QRS	Gallup results	QC, Med-Exec & the Board
				Human Resources					
PI.3.2.6 & HR.4.3	Staff views regarding performance and improvement opportunities	O	Y	# employees completing self appraisal forms / # of employees	100% (I)	Annually	Human Resources	Performance Appraisal form	QC, Med-Exec & the Board

(Continued on next page)

Table 7.1. *(Continued)*

	Human Resources (cont'd)	O	Y	Measure		Frequency	Dept.	Record	Reporting
HR.4.3	Collection of data on patterns & trends: Employee Turnover Rate	O	Y	# employees leaving / # of employees		Annually	Human Resources	Resource records	QC, Med-Exec & the Board
HR.4.3	Collection of data on patters & trends: Perf. Review Outcomes	O	Y	# employees scoring <=2 / # of employees	0% (I)	Annually	Human Resources	Performance Appraisal results	QC, Med-Exec & the Board
HR.4.3	Completion of competency testing of employees	O	Y	# employees completing competency testing / # of employees	100% (I)	Annually	Human Resources	Results of tests	QC, Med-Exec & the Board
Environment of Care									
	Safety Plan Performance Measure								
EC.1.3	Patient Incidents--Total	O	Y	# of incidents / patient days	<15% (I)	Quarterly	Risk Mgmt.	Incident Report	Safety, QC & the Board
EC.1.3	Medication	O	Y	# of incidents / patient days	<2%	Quarterly	Risk Mgmt.	Incident Report	Safety, QC & the Board
EC.1.3	IV	O	Y	# of incidents / patient days	<2%	Quarterly	Risk Mgmt.	Incident Report	Safety, QC & the Board
EC.1.3	Adverse effect of medication	O	Y	# of incidents / patient days	<=5%	Quarterly	Risk Mgmt.	Incident Report	Safety, QC & the Board
EC.1.3	Falls	O	Y	# inpatient falls / patient days	<1%	Quarterly	Risk Mgmt.	Incident Report	Safety, QC & the Board
EC.1.3	Diagnostic/Procedure	O	Y	# of incidents / patient days	<5%	Quarterly	Risk Mgmt.	Incident Report	Safety, QC & the Board
EC.1.3	Other	O	Y	# of incidents / patient days	<5%	Quarterly	Risk Mgmt.	Incident Report	Safety, QC & the Board
EC.1.3	Home Health Patient Falls	O	Y	# of falls / # of patients	<10% (I)	Quarterly	Risk Mgmt.	Incident Report	Safety, QC & the Board
EC.1.3	Home Health Medication Errors	O	Y	# of errors / # of patients	<5% (I)	Quarterly	Risk Mgmt.	Incident Report	Safety, QC & the Board
EC.1.3	Visitor Incidents--Total	O	Y	# of incidents / patient days	<.2% (I)	Quarterly	Risk Mgmt.	Incident Report	Safety, QC & the Board
EC.1.3	Falls	O	Y	# of falls / patient days	<.1% (I)	Quarterly	Risk Mgmt.	Incident Report	Safety, QC & the Board
EC.1.3	Other	O	Y	# of other incidents / patient days	<.1% (I)	Quarterly	Risk Mgmt.	Incident Report	Safety, QC & the Board
EC.1.3	Worker's Comp Claims--Total	O	Y	# of claims x 100 / productive man hours	<2%	Quarterly	Risk Mgmt.	Incident Report	Safety, QC & the Board
EC.1.3	Hospital Employees	O	Y	# of claims x 100 / productive man hours	<2%	Quarterly	Risk Mgmt.	Incident Report	Safety, QC & the Board
EC.1.3	Home Health Employees	O	Y	# of claims x 100 / productive man hours	2%	Quarterly	Risk Mgmt.	Incident Report	Safety, QC & the Board
EC.1 3	Employee Infection/Illness	O	Y	# of claims x 100 / productive man hours	2%	Quarterly	Risk Mgmt.	Incident Report	Safety, QC & the Board

Table 7.1. *(Continued)*

EC	Measure			Formula	Threshold	Frequency	Responsible	Source	Reported to
	Safety Plan Performance Measure (cont'd)								
EC.1.3	OSHA Reportables	O	Y	# of reportables x 100 / productive man hours	<1%	Quarterly	Risk Mgmt.	Incident Report	Envir of Care, QC & the Board
EC.1.3	Occupational Injuries	O	Y	# of reportables x 100 / productive man hours	<2%	Quarterly	Risk Mgmt.	Incident Report	Envir of Care, QC & the Board
EC.1.3	Occupational Illness	O	Y	# of reportables x 100 / productive man hours	<1%	Quarterly	Risk Mgmt.	Incident Report	Envir of Care, QC & the Board
EC.1.3	Lost Work Days	O	Y	# of reportables x 100 / productive man hours	<1%	Quarterly	Risk Mgmt.	Incident Report	Envir of Care, QC & the Board
EC.1.3	Restricted Work Days	O	Y	# of reportables x 100 / productive man hours	<2%	Quarterly	Risk Mgmt.	Incident Report	Envir of Care, QC & the Board
EC.1.3	Back Injuries	O	Y	# of reportables x 100 / productive man hours	<1%	Quarterly	Risk Mgmt.	Incident Report	Envir of Care, QC & the Board
EC.1.3	Needlesticks	O	Y	# of reportables x 100 / productive man hours	<1%	Quarterly	Risk Mgmt.	Incident Report	Envir of Care, QC & the Board
EC.1.3	Blood Body Fluid Exposures	O	Y	# of reportables x 100 / productive man hours	<1%	Quarterly	Risk Mgmt.	Incident Report	Envir of Care, QC & the Board
EC.1.8	Chemical Exposures	O	Y	# of reportables x 100 / productive man hours	<1%	Quarterly	Risk Mgmt.	Incident Report	Envir of Care, QC & the Board
	Security Plan Performance Measures								
EC.1.4	Total Security Incidents	O	Y	# of incidents / 91 days or 1/4 yr	<5%	Quarterly	Risk Mgmt.	Incident Report	Envir of Care, QC & the Board
EC.1.4	Violence in Workplace	O	Y	# of incidents / 91 days or 1/4 yr	<1%	Quarterly	Risk Mgmt.	Incident Report	Envir of Care, QC & the Board
EC.1.4	Other Suspicious Circumstances	O	Y	# of incidents / 91 days or 1/4 yr	<5%	Quarterly	Risk Mgmt.	Incident Report	Envir of Care, QC & the Board
	Control of Hazardous Materials and Waste Plan Performance Measures								
EC.1.5	Radiation Monitoring (Over-exposures)	O	Y	# of overexposures / Patient days	0	Monthly / Quarterly	Risk Mgmt.	Monitor tags	QC, Med-Exec & the Board
EC.1.5	Environmental Hazards Inspections	O	Y	# comp / 22 areas / # comp / 10 pt care areas	100%	Annually	Risk Mgmt.	Envir of Care Meeting	QC, Med-Exec & the Board
	Emergency Preparedness Plan Performance Measures								
EC.1.6	Disaster/Disaster Drills	P	Y	# of drills / 1	100% 2/yr	Quarterly	Risk Mgmt.	Envir of Care Meeting	QC, Med-Exec & the Board
	Life Safety Plan Performance Measures								
EC.1.7	Fire Drills	P	Y	# of drills / 3	100% or 3/quarter	Quarterly	Risk Mgmt.	Envir of Care Meeting	QC, Med-Exec & the Board
EC.1.7	Safety Inspections	P	Y	# comp / 22 / # comp / 10	100%	Quarterly	Risk Mgmt.	Envir of Care Meeting	QC, Med-Exec & the Board
EC.1.7	Orientation within 30 days	P	Y	# attended orientation w/in 30 days / # new hires	95%	Quarterly	Risk Mgmt.	Human Resources	QC, Med-Exec & the Board

(Continued on next page)

Table 7.1. (Continued)

	**Life Safety Plan Performance Measures (cont'd)**								
EC.1.7	Yearly Safety Education Compliance	P	Y	# completed education / # of employees	100%	Quarterly	Risk Mgmt.	Human Resources	QC, Med-Exec & the Board
EC.1.8	Biomedical Equipment P.M.'s	P	Y	# pieces of equip done / # pieces of equip due	95%	Quarterly	Risk Mgmt.	National MD	QC, Med-Exec & the Board
EC.1.8	Repairs	P	Y	# pieces of equip repaired / # pieces of failed equipmt	<3%	Quarterly	Risk Mgmt.	National MD	QC, Med-Exec & the Board
EC.1.8	User Errors	P	Y	# user errors identified / # pieces of failed equipment	<3%	Quarterly	Risk Mgmt.	National MD	QC, Med-Exec & the Board
EC.1.8	Laser Safety Procedures	P	Y	# procedures done / # op encounters	no benchmark	Quarterly	Risk Mgmt.	Operating Room	QC, Med-Exec & the Board
EC.1.8	Number of Related Incidents	P	Y	# incidents / # procedures done	<1%	Quarterly	Risk Mgmt.	Operating Room	QC, Med-Exec & the Board
	Utility Systems Plan Performance Measures								
EC.1.9	Plant Ops P.M.'s	P	Y	# PM's done / # PM's due	90%	Quarterly	Risk Mgmt.	Plant Operations	QC, Med-Exec & the Board
EC.1.9	Equipment/Utility Incidents	P	Y	# of incidents / patient days	<.5%	Quarterly	Risk Mgmt.	Plant Operations	QC, Med-Exec & the Board
	Management of Information								
IM.3.2.1	Data Quality Monitoring: Documentation appropriateness	P	Y	# cases w/appropriate doc / # cases reviewed	80% (I)	Quarterly	HIM	DQM form	QC, Med-Exec & the Board
IM.3.2.1	Medical Record Delinquency: Overall	O	Y	# charts 21 days delinq. / # discharges for month	<50% (J)	Quarterly	HIM	Record report	QC, Med-Exec & the Board
IM.3.2.1	Suspensions	O	Y	# suspensions / # physicians	0%	Quarterly	HIM		QC, Med-Exec & the Board
	Infection Control								
IC.2	Overall Nosocomial Infection Rate	O	Y	# nosocomial infections / # patients	<1% (I)	Monthly / Quarterly	Infection Control	Culture reports & Pt records	QR, IC Committee
IC.2	Surgical Wound Infection Rate	O	Y	# post op infections / # surgeries	0.8% (L)	Monthly / Quarterly	Infection Control	Culture reports & Pt records	QR, IC Committee
	Withholding Services								
RI.1.2.5	Advance Directives: Patients asked	P	Y	# pts asked about AD / # patients 18 or older	100%	Quarterly	Admitting	Patient record	QC, Med-Exec & the Board
RI.1.2.5	Advance Directives: Patients receiving info about Advance directives	P	Y	# pts receiving AD info / # pts 18 or older	100%	Quarterly	Admitting	Patient record	QC, Med-Exec & the Board
RI.1.2.5	Advance Directives: Advance directives received by the hospital	P	Y	# AD's hospital obtains / # pts claiming to have AD	100%	Quarterly	Admitting	Tracking form	QC, Med-Exec & the Board
	New Programs								
PI.3	New program effectiveness						CNO		

Copyright ® 2000 by Polly Isaacson. All rights reserved.

Case Study

Table 7.2 is a set of admission criteria for medical/surgical admissions. Figure 7.7 shows an example of a report of one patient's history and physical. Students should compare the patient's history to the admission criteria and determine whether the patient meets or does not meet the criteria for admission to the hospital. The patient must meet at least one criterion in severity of illness and at least one criterion in intensity of service.

Project Application

Students should refer back to chapter 6 and the data they collected for their student projects with surveys and interviews. Students then should identify the performance measure that could capture data about those customer satisfaction issues. For example, if the students were looking at bookstore services, some of the criteria for bookstore performance might be book pricing, availability of books, and buy-back percentage.

Summary

Appropriate utilization of healthcare services has been a major issue in the United States. Utilization management strategies led to the managed care approach, which was common

Table 7.2.	Admission Criteria
Severity of Illness	**Intensity of Service**
Sudden onset of unconsciousness or disorientation	Intravenous medications and/or fluid replacement
Pulse rate: <50/min or >140/min and not typical for patient	Inpatient-approved surgery or procedure within 24 hours of admission
Blood pressure: systolic <90 or >200 mm Hg or diastolic <60 or >120 mm Hg *and* not typical for patient	Vital signs every 2 hours or more often
Acute loss of sight or hearing	Chemotherapeutic agents requiring continuous observation
Acute loss of ability to move body part	Treatment in an ICU, if indicated
Persistent fever	Intramuscular injection every 8 hours
Active bleeding	Respiratory care at least every 8 hours
Severe electrolyte/blood gas abnormality	Glucose monitoring at least 4 times daily
EKG evidence of acute ischemia	
Wound dehiscence or evisceration	
Widely fluctuating blood glucose levels	
Hemoglobin levels 1.4 times upper limit of normal	

Figure 7.7.   Example of a History and Physical Report

Reason for Admission: Severe, short-distance, lifestyle-limiting right lower extremity claudication

History of Present Illness: This is a 32-year-old woman who developed new-onset right lower extremity claudication following right transfemoral cardiac catheterization for routine follow-up 10 years after cardiac transplantation. The catheterization was approximately 10 days ago. Since that time, she describes symptoms of pain in her calf after walking approximately 20 yards or less. If she walks too far, she develops paresthesias and complete numbness in the right foot. The pain is relieved by rest. She does not have rest pain at night. She has never had any symptoms similar to this or any symptoms in the contralateral leg.

She underwent cardiac transplantation 10 years ago. Since that time, she has had annual routine evaluation by transfemoral cardiac catheterization. Dr. Smith, who reviewed the films from the catheterization, reports that there is evidence of mild narrowing in the common femoral artery, possibly due to prior catheterizations. There is also some concern regarding the possibility of arterial dissection more proximally, although this may be an artifact on the angiogram.

Allergies: No known drug allergies

Past Medical History: (1) History of hypertrophic cardiomyopathy, now status post cardiac transplantation. (2) Intermittent episodes of rejection. (3) History of herpes zoster.

Past Surgical History: Cardiac transplantation

Medications: Pepcid 20 mg po bid, Vasotec 5 mg po bid, magnesium oxide 400 mg po bid, aspirin 81 mg po bid, CellCept 1 gm po bid, Neoral 100 mg qam and 75 mg qpm

Social History: The patient is a schoolteacher.

Habits: She drinks alcohol occasionally and does not smoke cigarettes.

Review of Systems: The patient has no active cardiopulmonary symptoms of which she is aware and no history of hepatorenal dysfunction. She has had no other episodes of bleeding or thrombotic disorders.

Physical Examination:

HEENT:	Unremarkable
CHEST:	Clear throughout to auscultation
CARDIOVASCULAR:	Regular rhythm without murmur, gallop, or rub
ABDOMEN:	Soft, nontender with no obvious masses or organomegaly
GENITALIA/RECTAL:	Deferred
EXTREMITIES:	No clubbing, cyanosis, or edema. There is no dependent rub or pallor on elevation. The patient has normal sensation and motor function in the lower extremities. Pulses are 3/3 except in the right lower extremity, where no palpable pulses are present.
LABORATORY DATA:	Potassium 3.8; hematocrit 45; sodium 142
TEST RESULTS:	Angiography demonstrated occlusion of the external iliac artery from near the bifurcation to the distal common femoral artery, which reconstitutes just above its own bifurcation. A guide wire passed easily through this, suggesting soft thrombus. There is excellent collateralization and no evidence of distal abnormalities.
	Duplex ultrasonography performed earlier demonstrated no evidence of deep or superficial thrombophlebitis. Noninvasive vascular studies also suggested aortoiliac/femoral occlusive disease with good collateralization distally.

Impression:
1. Occluded right external iliac and common femoral artery following transfemoral cardiac catheterization
2. Status post cardiac transplantation for hypertrophic cardiomyopathy
3. History of herpes zoster

Plan: The patient will be admitted to the hospital to undergo operative intervention to repair the femoral artery injury. Several possibilities exist, including possible dissection of the artery and injury to the artery during the catheterization or development of a collagen plug post angiography. I have discussed these possibilities with the patient, and I plan to perform an exploration of the right femoral area and, if necessary, a right lower quadrant, retroperitoneal incision to expose the proximal bifurcation and a bypass if necessary. Discussed the possibility of vein patch angioplasty as well. We also discussed the risks of the operation including MI, CVA, death, infection, bleeding, nerve injury, embolization and tissue loss, bowel injury, etc. She understands all these things as well as the indications for operative intervention. We plan operation tonight as soon as an operating room is available.

in the last decade of the twentieth century. Various approaches, including admission criteria and critical pathways, have been developed to assist reviewers in determining the nature and extent of required care. Management and analysis of these issues remains a major component of PI activities in every healthcare organization in the nation.

References

Abdelhak, Mervat. 2000. *Health Information: Management of a Strategic Resource,* 2nd ed. Philadelphia: W. B. Saunders Company.

Groopman, Jerome, M.D. 2000. *Second Opinions.* New York City: Viking Penguin Publishing; Member of Penguin Putman, Inc.

Joint Commission on Accreditation of Healthcare Organizations. 2003. *Comprehensive Accreditation Manual for Hospitals.* Oakbrook Terrace, Ill.: Joint Commission on Accreditation of Healthcare Organizations.

Joint Commission on Accreditation of Healthcare Organizations. 2003. ORYX/Selection Change Form for Hospitals. Accessed online at http://www.jcaho.org/accredited+organizations/hospitals/oryx/selection+change+forms/index.htm.

Meisenheimer, Claire G. 1997. *Improving Quality: A Guide to Effective Programs.* Gaithersburg, Md.: Aspen Publishers, pp. 207–24.

O'Leary, Margaret. 1996. *Clinical Performance Data: A Guide to Interpretation.* Oakbrook Terrace, Ill.: Joint Commission on Accreditation of Healthcare Organizations.

Shewhart, Walter A., and W. Edwards Deming. 1986. *Statistical Method from the Viewpoint of Quality Control.* New York: Dover Publications.

Zander, K. 1997. Use of variance from clinical paths: coming of age. *Clinical Performance and Quality Health Care* 5(1):20–30.

Chapter 8
Preventing and Controlling Infectious Disease

Learning Objectives

- To describe why the control of infection is so important in healthcare organizations

- To differentiate nosocomial infections from community-acquired infections

- To explain the various approaches that healthcare organizations use to manage the occurrence of infection

- To identify the governmental organizations that develop regulations in this area and explain the regulatory approaches often taken

Background and Significance

Infectious disease is an ancient problem for humankind. Epidemics of infectious disease have resulted in major loss of life throughout the centuries. New diseases still erupt onto the world landscape: hanta virus, pneumonia, acquired immune deficiency syndrome (AIDS), and Legionnaire's disease have all been recognized and investigated within the past thirty years. Furthermore, the threat of germ warfare, such as anthrax and smallpox viruses, is becoming more and more prevalent.

The discovery of the germ, or the pathologic organisms bacteria, viruses, and parasites, as the causative agent of infection provided the human species with its most valuable tool in preventing human disease. The most effective means of stopping disease in its tracks is through sanitation procedures of all kinds. Understanding how various germs function has allowed humans the critical opportunity to control and cure disease. As a result of this understanding, vaccines were developed to limit the spread of infectious disease within a population, and antibiotics help limit the growth and spread of pathologic organisms within the human body.

Today, healthcare researchers and providers utilize vaccines, medications, and other interventions to identify germs, limit their spread, and eliminate disease. As a result, a much smaller portion of the population is affected today by infectious disease than in historic times.

In developed countries, government agencies, such as the Centers for Disease Control (CDC) in the United States, maintain national standards for disease prevention and treatment. Professional groups such as the Association of Professionals for Infection and Epidemiology and the American Public Health Association are involved in training communicable disease specialists and establishing standards of care for infectious diseases. In addition, most state licensing agencies and healthcare accrediting agencies publish standards of care for the management of infection and disease prevention.

Infections and infectious disease processes play a prominent role in the management of quality and performance in every healthcare facility, whether it is an acute care hospital, a residential care facility, or a community day-care program for senior citizens. Consumers avoid healthcare facilities that have high rates of nosocomial infection. Patients who acquire an infection while they are hospitalized spread the word about their experience to their families, friends, neighbors, and coworkers. Clearly, the infectious disease experience in healthcare facilities is of paramount importance, in terms of its effect on both patients' lives and a healthcare facility's reputation and accreditation status. There is an increased cost financially and emotionally to patients and facilities when care is extended because of unexpected infection.

Managing the Infectious Disease Experience: Steps to Success

Infection control, surveillance, and management can be performed in a variety of ways. In some facilities, an infection control committee composed of physicians, nurses, and clinical laboratory staff performs weekly reviews of the facilitywide incidence of infectious disease. In others, one individual, with training and expertise, is specifically assigned to infection surveillance and control and granted the authority by the governing board of the organization to institute any measures needed to prevent the spread of infection.

Infection management should be based on an infection control plan developed by the facility's clinical staff. The plan should be specific to that facility's case mix, service lines, and care resources. It should include methods of surveillance and tracking for infections in every area of the facility, from infectious disease units to dietary services, from newborn nurseries to clinical laboratories. The plan should address major communicable diseases that affect the facility's client population, such as tuberculosis, community-acquired pneumonia, hepatitis B and C, and HIV. In addition, the plan should outline the types of routine surveillance and procedures the organization is going to undertake to limit the transmission of infectious agents among patients and staff.

Step 1: Control infection through the use of universal precautions

The mandate for applying universal precautions in healthcare services has been the cornerstone of infection control since the 1980s. **Universal precautions** can be defined as the application of a set of procedures specifically designed to minimize or eliminate the passage of infectious disease agents from one individual to another during the provision of healthcare services. The precautions are described as universal because all caregivers must follow them.

Universal precautions require that caregivers wash their hands between patients and wear gloves when they either examine patients or administer therapies. Caregivers must also wear gloves, gown, mask, and eye protection whenever they perform procedures that disrupt the patient's skin or mucous membranes. Universal precautions assume that all patients carry infectious disease agents. Therefore, barriers are erected between patients, and between patients and caregivers, when blood or bodily fluids are involved in any way.

Step 2: Conduct ongoing infection surveillance procedures

Within the healthcare facility, any occurrence of infection should be evaluated to determine whether the infection was nosocomial or community acquired. A **nosocomial infection** is defined as an infection that was acquired as a result of an exposure that occurred in the healthcare facility after the patient was admitted. A **community-acquired infection** is an infection that was present in the patient before he or she was admitted to the facility.

Specific guidelines have been developed to determine whether an infection is nosocomial or community acquired. For example, if a child were admitted to a hospital with a fever and within 24 hours developed measles, this disease would be considered community acquired. The incubation period for measles is at least 14 days; therefore, it can be determined that the patient's exposure to the disease occurred prior to admission. If, however, a patient was admitted for treatment of an intervertebral disk injury and then developed a urinary tract infection with fever after undergoing surgery, the infection was probably acquired while the patient was hospitalized. This infection would be classified as a nosocomial infection.

Both of these instances of infectious disease would be tracked and reported to the required state agency, but for different reasons. The measles would be tracked to document the fact that the facility's staff took appropriate action to prevent the exposure of other patients. The urinary tract infection would be tracked to document initiation of appropriate treatment interventions and to identify any measures that might be taken to prevent such infections in the future.

It is important to track nosocomial infection rates for each area of the facility that works with patients. Rates of nosocomial infection vary greatly depending upon the nursing unit and the types of patients for whom care is provided. For example, looking at central vascular catheter-related bloodstream infections (CR-BSIs) in the intensive care setting, one finds rates varying from 2.1 per 1,000 central catheter days in the respiratory intensive care unit (ICU) to 30.2 per 1,000 central catheter days in the burn ICU. Looking at noncentral vascular catheter-related bloodstream infections in other intensive care settings (such as the cardiac ICU and postsurgical ICU), one finds rates varying from 0 to 2.0 per 1,000 noncentral catheter days (CDC 2000). Other types of procedures commonly associated with nosocomial infection that should be tracked include the use of indwelling urinary catheters and mechanical ventilation devices.

Decubitus ulcers and surgical site infections are often associated with hospital care. Other possible sources of nosocomial infection include exposure to clinical or nonclinical staff who carry infectious diseases, substandard surgical or postsurgical care, and noncompliance with universal precautions. For a more in-depth discussion of the complexities of tracking and managing nosocomial infection in healthcare facilities, visit the Centers for Disease Control at www.cdc.gov/ncidod. In most facilities, a multidisciplinary approach to infection management involves the patient's physician, a pharmacist, an epidemiologist, a clinical laboratorian, the patient's nurse or case manager, and the patient. The goal is to initiate appropriate treatment, which usually involves the administration of an appropriate antibiotic medication prescribed in an effective dosage. A certified professional trained to evaluate the appropriateness of infection management measures may be assigned to review unusual cases of infection.

The appropriate use of antibiotics has been subject to debate in recent years. Several strains of bacteria are becoming resistant to commonly prescribed antibiotics. Some medical experts implicate the indiscriminate use of antibiotics in cases without clear indication through laboratory culture and sensitivity processes. Specific antibiotics are approved for the treatment of specific conditions, but not for all conditions. Antibiotics are not interchangeable, and an antibiotic approved for the treatment of cellulitis might not be appropriate for the treatment of a respiratory infection. In general, antibiotics are not effective against viral infections, but patients often request and receive antibiotics for common viral infections such as colds.

Another aspect of infection surveillance involves employee health and illness tracking. Policies related to the tracking of employee absences exist for the specific purpose of preventing infection via healthcare workers. Reports of absences are tabulated and examined for any possible connection to cases of nosocomial infection.

Monitoring the care environment is another aspect of infection surveillance. In many facilities, specimens from patient care areas are cultured monthly to identify any pathological bacteria growing in the care environment. The cultures are reported and tracked. When the presence of a significant infectious agent is identified, another specimen is cultured after the area has been cleaned to determine whether the area was sufficiently disinfected.

The cleanliness of food preparation and service areas is also tracked, and records are kept as documentation that all local and state regulations regarding safe food-handling procedures are being followed. The training and safety performance of food service staff are also documented.

Food temperatures and refrigerator temperatures are tracked daily, as are records of the cleaning of food preparation and service areas. Refrigerators containing medications are kept separate from refrigerators containing food, and temperature logs must be kept for medication refrigerators as well.

The infection control committee approves all substances used for cleaning in the facility. In many facilities, the infection control committee conducts monthly environmental rounds to look for areas of noncompliance with standards. The governing body that grants authority for infection control measures in an agency may also review the results of the monthly rounds along with any actions for improvement suggested by the committee.

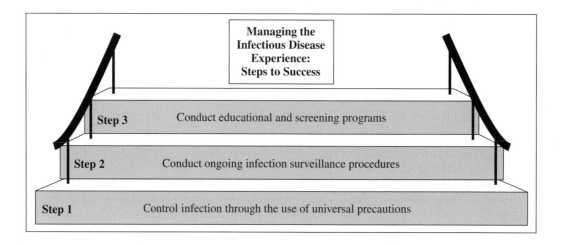

Managing the Infectious Disease Experience: Steps to Success

Step 3   Conduct educational and screening programs

Step 2   Conduct ongoing infection surveillance procedures

Step 1   Control infection through the use of universal precautions

Step 3: Conduct educational and screening programs

An effective infection control program routinely evaluates all the means of transmission of infection throughout the facility and develops educational programs that promote disease prevention. These programs include tuberculosis testing, hand-washing campaigns, influenza vaccination programs, hepatitis B vaccinations for high-risk employees, sterile technique in-services, and needlestick prevention programs. Education on universal precautions is mandatory for all employees in most facilities.

Tracking exposures to blood-borne pathogens such as the human immunodeficiency virus (HIV) is required by most states and is outlined in Occupational Safety and Health Administration (OSHA) regulations. OSHA regulations require the monitoring of employees after exposure for at least one year, with frequent HIV testing. HIV testing and screening for employees who are known to have been exposed to HIV require the informed consent of individual employees. OSHA also mandates that specific education on the risks and outcomes of testing be provided to employees by a certified HIV instructor or physician.

When state law requires the facility to report positive test results to the state department of health, the employee involved must be notified ahead of time. Many state departments of health provide programs to test, treat, and educate employees and clients regarding the HIV disease process. With the advent of new Health Insurance Portability and Accountability Act (HIPAA) regulations, client rights of protection and notification related to infectious diseases have increased. Employers will need to notify and get signed releases in order to send private client information to state agencies that track and treat disease processes.

Most states require facilities to outline a tuberculosis prevention plan that involves the careful screening of employees and/or clients to identify those who have active tuberculosis, as well as those who may have been exposed but do yet not have the active disease process. Annual mandatory testing of workers is usually required by the states as an occupational safety measure for employees, but testing of patients and clients varies considerably from state to state. Often the testing of patients and clients is required in long-term, rehabilitative settings. The tuberculosis plan for most facilities outlines the testing process, staff and client training and education, as well as required treatment interventions. Monthly reporting of the number of employees tested and the number of employees who convert to positive is required in many states. A positive tuberculin test result must be reported to the department of health in all states.

Some states that have populations at high risk for tuberculosis are now requiring a two-step testing process. This involves a skin test for tuberculosis followed by a second test seven days later to confirm a negative result. Positive results are followed up with a chest X ray. The state department of health can help any facility determine whether it serves clients who are at high risk for tuberculosis. Such facilities should institute two-step testing. Facilities that treat patients with active tuberculosis should also provide negative-pressure rooms specially designed for the treatment of tuberculosis to prevent the airborne transmission of the disease.

In summary, the management of infectious disease is an organizationwide performance issue that involves every area of the facility and affects every employee, patient, and visitor. The ability of individuals to perform their jobs often depends upon how carefully infection control is managed in the facility. The concept of wellness centers has been a positive outcome of the focus on infection control and prevention in many organizations. The effort focuses on creating centers that promote activities for health and wellness and limit exposures to debilitating infections.

Real-Life Example

At one time, Marilyn Nelson, RN, RHIA, worked as the nurse manager of the outpatient clinics at the Western States University Medical Center. She was responsible for managing performance improvement (PI) activities in the clinics, facilitating PI teams, and supporting PI activities with the various resources available to her. At that time, the medical center had recently embarked on a benchmarking program. **Benchmarking,** as discussed in chapter 2, is the systematic comparison of one organization's outcomes or processes with the outcomes or processes of similar organizations.

Marilyn and some of her colleagues began the benchmarking process by comparing the outcomes of the clinic's patients with published information on the outcomes of other organizations and with clinical trends apparent in the literature. In the course of examining pharmacy and therapeutics data, Marilyn and her clinical colleagues recognized that their organization appeared to be treating a significantly higher number of urinary tract infections (UTIs) than other, similar organizations. To identify the reason for the apparent discrepancy, the team first examined the published literature on the frequency of UTIs in ambulatory care practice. Next, they contacted clinicians at other ambulatory care centers in the country to gather information on their experiences. The team's investigations confirmed that while the sex-specific distribution of the incidence of UTI was similar to other institutions (mostly occurring in females), the overall incidence was significantly higher at Western States University Medical Center than at any of the other medical centers contacted. The situation appeared to provide an excellent PI opportunity, the kind of improvement that clinicians would see as important and valuable to patients as well as one that had important cost implications for the organization.

Marilyn convened a PI team, which included the director of the outpatient pharmacy, the director of the outpatient clinic laboratories, and leaders of the nursing teams from each of the ambulatory clinics in which patients with UTIs were commonly treated: internal medicine, family practice, obstetrics/gynecology, urology, general surgery, and pediatrics.

The team's initial discussion of the situation revealed the complexity of the processes in place to evaluate urinary tract function. Because the clinics are affiliated with a major university medical center, a number of caregivers might be involved in any one patient's care. First, there were the attending physicians from a variety of specialties and areas of expertise. In addition, there were the nurses and medical assistants in each clinic. Because the medical center was a teaching facility, there were also any number of house staff rotating through the clinics on a monthly basis. Finally, there were the technicians who performed urinalysis procedures in the clinical laboratory. Where, when, and by whom a patient's urinalysis was performed could take any one of a number of paths in the organization.

Marilyn and her colleagues decided to develop flowcharts for the various care paths. (See figures 8.1 through 8.5.) They collected data on the outcomes of each of the paths to see whether they could identify the organization's true experience with UTIs.

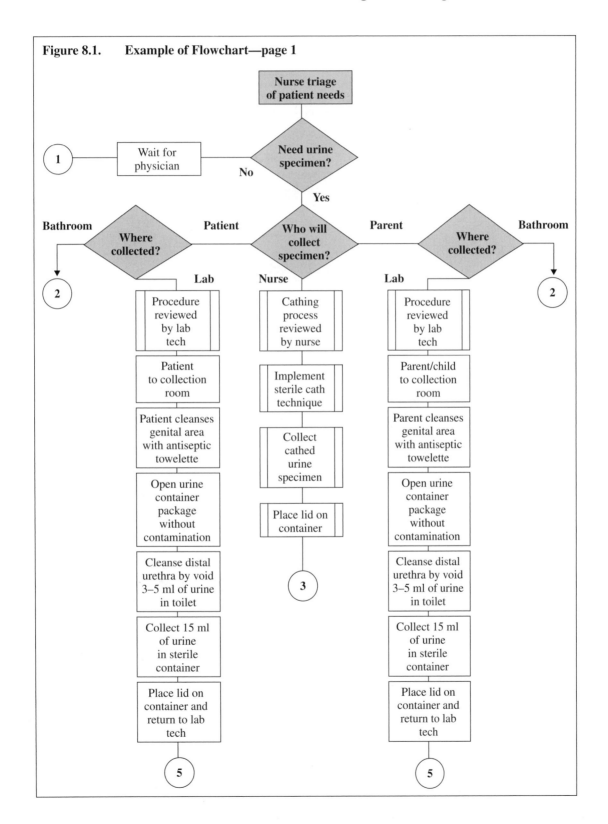

Figure 8.1.   Example of Flowchart—page 1

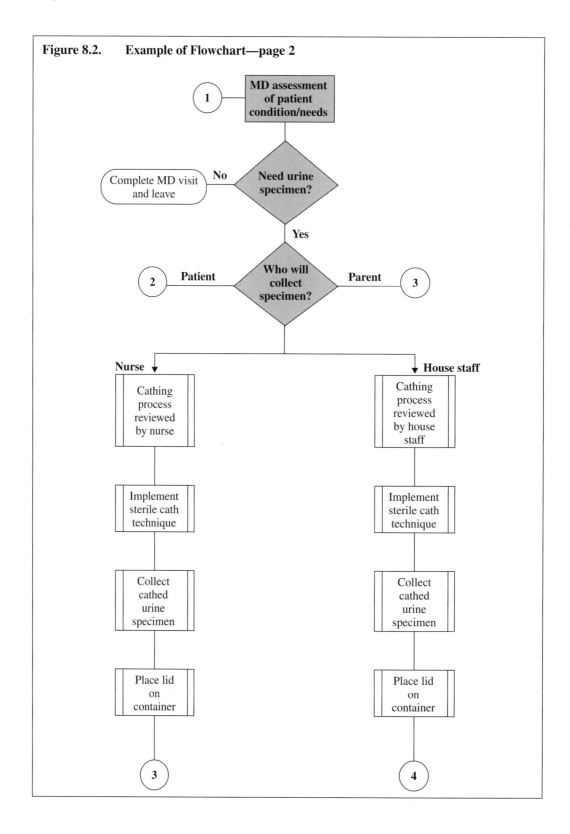

Figure 8.2.   Example of Flowchart—page 2

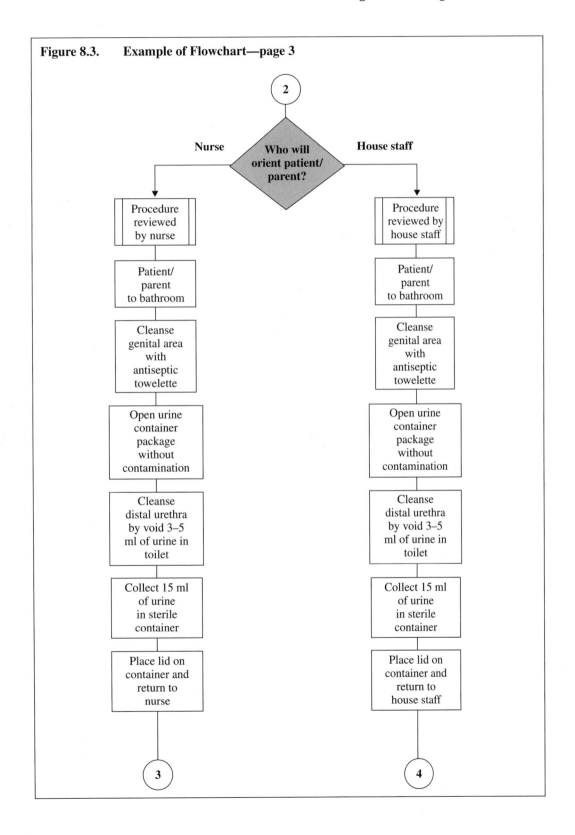

Figure 8.3.   Example of Flowchart—page 3

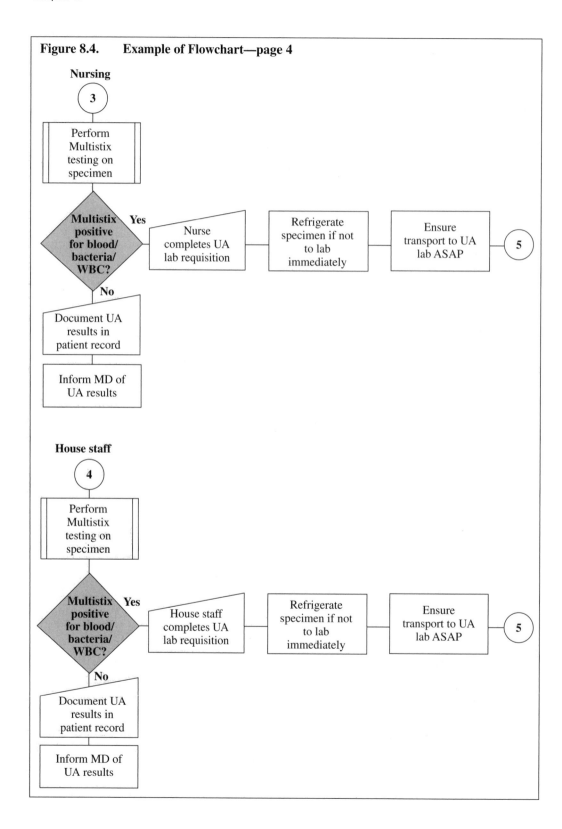

Figure 8.4.   Example of Flowchart—page 4

Figure 8.5.   Example of Flowchart—page 5

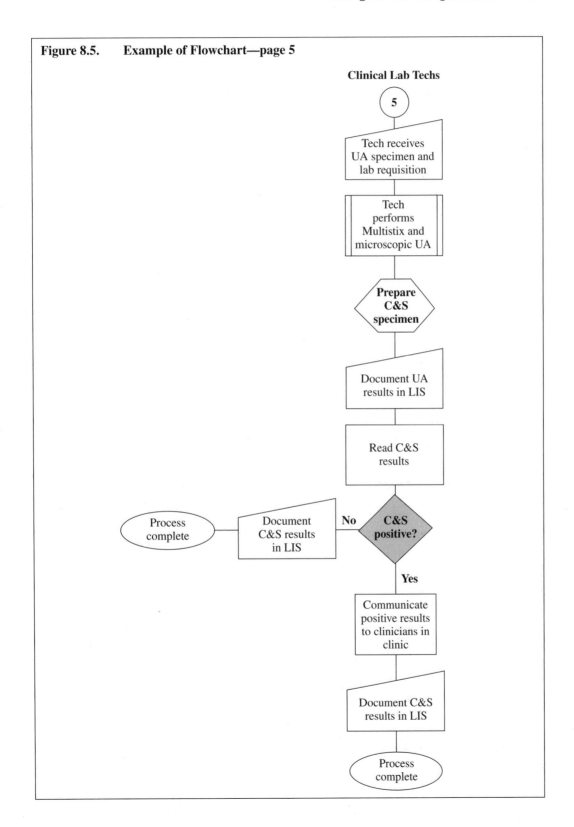

QI Toolbox Technique

The use of **flowcharts** allows a PI team to examine the process under investigation from all directions. The technique makes it possible for the team to gather the most important details so that everyone on the team can understand the process and its contributing subprocesses in the same way. When a flowchart is well designed, few misconceptions can survive.

Flowcharts are used to represent standard functions within processes. Representative and commonly used **icons** are discussed below:

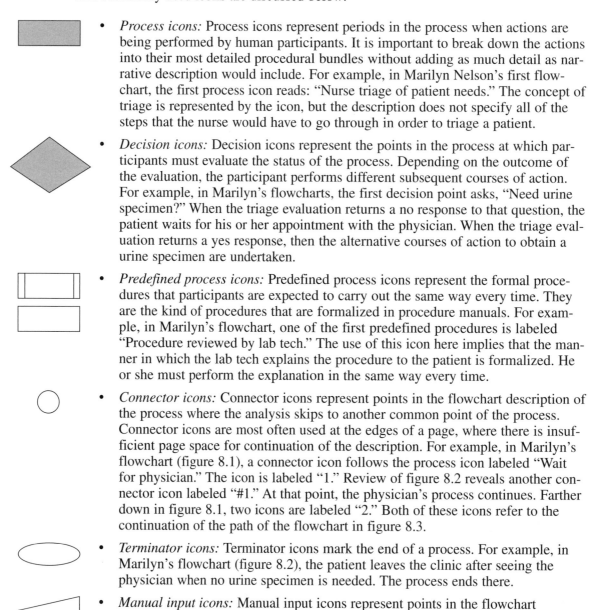

- *Process icons:* Process icons represent periods in the process when actions are being performed by human participants. It is important to break down the actions into their most detailed procedural bundles without adding as much detail as narrative description would include. For example, in Marilyn Nelson's first flowchart, the first process icon reads: "Nurse triage of patient needs." The concept of triage is represented by the icon, but the description does not specify all of the steps that the nurse would have to go through in order to triage a patient.

- *Decision icons:* Decision icons represent the points in the process at which participants must evaluate the status of the process. Depending on the outcome of the evaluation, the participant performs different subsequent courses of action. For example, in Marilyn's flowcharts, the first decision point asks, "Need urine specimen?" When the triage evaluation returns a no response to that question, the patient waits for his or her appointment with the physician. When the triage evaluation returns a yes response, then the alternative courses of action to obtain a urine specimen are undertaken.

- *Predefined process icons:* Predefined process icons represent the formal procedures that participants are expected to carry out the same way every time. They are the kind of procedures that are formalized in procedure manuals. For example, in Marilyn's flowchart, one of the first predefined procedures is labeled "Procedure reviewed by lab tech." The use of this icon here implies that the manner in which the lab tech explains the procedure to the patient is formalized. He or she must perform the explanation in the same way every time.

- *Connector icons:* Connector icons represent points in the flowchart description of the process where the analysis skips to another common point of the process. Connector icons are most often used at the edges of a page, where there is insufficient page space for continuation of the description. For example, in Marilyn's flowchart (figure 8.1), a connector icon follows the process icon labeled "Wait for physician." The icon is labeled "1." Review of figure 8.2 reveals another connector icon labeled "#1." At that point, the physician's process continues. Farther down in figure 8.1, two icons are labeled "2." Both of these icons refer to the continuation of the path of the flowchart in figure 8.3.

- *Terminator icons:* Terminator icons mark the end of a process. For example, in Marilyn's flowchart (figure 8.2), the patient leaves the clinic after seeing the physician when no urine specimen is needed. The process ends there.

- *Manual input icons:* Manual input icons represent points in the flowchart description of the process where the participants must record data in paper-based or computer-based formats. For example, Marilyn's flowchart (figure 8.4) includes a manual input icon labeled "Document UA results in patient record." The subprocess represented by the manual input icon follows the Dipstix analysis performed in the clinic by either a nurse or a member of the house staff. When the clinician performing the Dipstix does not find blood, bacteria, or white blood cells in the specimen, the analysis is complete and only the documentation and communication of findings to the physician remain to be done.

- *Line connector icon:* Line icons direct the flow of the processes from one step to another from decision points to subprocesses.

Case Study

The case study for this chapter is a continuation of Marilyn Nelson's investigation of the UTI issue in the ambulatory clinics of the Western States University Medical Center.

After flowcharting the process for collecting urine specimens (depicted in figures 8.1 through 8.5), Marilyn and her colleagues recognized how complex the issue was within their organization. They decided to collect data from all process paths evident in the flowchart. Because so many people were involved in the processes and because significant delays could be involved, they also began to wonder what part contaminated specimens played in the situation.

Although it was an expensive project, the team designed an investigative study to collect data. Urine specimens were routinely tested by nursing personnel or house staff in the clinic to determine each specimen's pH and specific gravity and to classify each specimen according to its color, clarity, and presence of gross hematuria. Each specimen was then tested with Multistix to determine whether microscopic bacteria, red blood cells, or white blood cells were present. When a specimen failed any of the Multistix screens, it was referred to Clinical Laboratory Services for microscopic analysis, culture, and sensitivity analysis by a laboratory technician.

First, the team collected data regarding time elapsed between the collection of the specimen, the point-of-care testing with Multistix, and receipt of the specimen in Clinical Laboratory Services. In addition, the team investigated the sequence of events that occurred in the interim. A summary of the data collected is provided in table 8.1.

Table 8.1.	Average Time to Point-of-Care Screening with Confidence Intervals (in minutes)						
Point-of-Care Training & Processing	Internal Medicine Clinic	Pediatrics Clinic	General Surgery Clinic	Orthopedic Surgery Clinic	Obstetrics Clinic	Gynecology Clinic	Specialty Clinics
Nurse	5.0 (4.0, 6.0)	4.25 (3.0, 5.5)	6.25 (5.0, 7.5)	6.0 (5.0, 7.0)	3.25 (2.75, 3.75)	3.0 (2.0, 4.0)	3.0 (2.0, 4.0)
House staff	10.0 (7.0, 13.0)	8.0 (6.0, 10.0)	12.25 (10.0, 14.25)	11.0 (9.0, 13.0)	6.5 (4.0, 9.0)	6.0 (4.0, 8.0)	7.5 (5.5, 9.5)

Second, on a temporary and random basis, the team obtained urine specimens from each clinic immediately following collection and had a complete analysis performed stat in the clinic labs. This analysis identified pH, specific gravity, color, clarity, cell counts, and bacterial counts almost immediately after the specimen was delivered by the patient or collecting clinician. All specimens that showed microscopic bacteria either on Multistix or on microscopic analysis were cultured. A summary of the data collected is provided in table 8.2.

Third, the team compared the incidence of UTI identified in the randomly collected specimens to incidence identified in specimens going through the usual process. A summary of the data collected is provided in table 8.3.

Questions for Case Study

1. Upon examination of the data sets, Marilyn and her colleagues identified several areas where the analysis revealed situations that were probably contributing

Table 8.2.   Random STAT Processing Clean Catch/Cath Urine Specimens (percentage of positive specimens for culture)

Collector	Internal Medicine Clinic	Pediatrics Clinic	General Surgery Clinic	Orthopedic Surgery Clinic	Obstetrics Clinic	Gynecology Clinic	Specialty Clinics
Nurse or patient	5.6	13.2	4.2	3.4	7.0	3.9	5.7
House staff or patient	4.2	12.1	3.4	5.2	6.8	4.2	6.2
Parent	—	27.5	11.7	9.0	—	25.2	—

Table 8.3.   Routine Processing Clean Catch/Cath Specimens (percentage of positive specimens for culture)

Collector	Internal Medicine Clinic	Pediatrics Clinic	General Surgery Clinic	Orthopedic Surgery Clinic	Obstetrics Clinic	Gynecology Clinic	Specialty Clinics
Nurse or patient	5.7	12.4	5.1	4.1	7.2	4.1	5.2
House staff or patient	36.8	25.6	10.2	8.7	10.0	9.2	14.2
Parent	—	25.6	12.2	8.0	—	23.0	—

to the clinic's high UTI rate. Look at the UTI rate for children whose parents had collected the specimen versus the rate for children who had been catheterized by nursing personnel to collect the specimen. What do you see? What might be the reason for the higher rate in children whose parents had collected the specimen?

2. Are there any other areas in the data that reveal important aspects that may be contributing to the high UTI rates? What might be the reasons for these higher rates?

3. Upon discussion of the findings and examination of the flowcharted processes, the laboratory manager noted a subtle change in clinic processes that probably was contributing to the problem. Clinic staff had begun at some point to Multistix the specimens in the original collection containers. Thus, what had been a clean-catch specimen could become a contaminated one when staff opened the container to perform the Multistix. What should really have been done was to pour off a small amount or "aliquot" of the specimen into another container, reseal the original for the laboratory, and perform the Multistix on the aliquot container. This process would minimize the possibility of contaminating the original specimen. How would this change in process be represented in the flowcharts presented in the case study?

Project Application

Students should consider using flowchart techniques in their projects.

Summary

The management of infectious disease is an organizationwide performance issue that involves every area of the facility and affects every employee, patient, and visitor. The ability of individuals to perform their jobs depends on how carefully infection control is managed in the facility. So does the ability of patients to recover as rapidly as possible without complications. Means by which the impact of infection is limited in healthcare organizations include use of universal precautions, infection surveillance procedures, appropriate treatment regimens, and staff and patient screening.

References

Arias, Kathleen Meecham, ed. 2000. *Quick Reference to Outbreak Investigation and Control in Health Care Facilities.* Sudbury, Mass.: Jones & Bartlett.

Benneyan, J. C. 1998. Statistical quality control methods in infection control and hospital epidemiology, part I: introduction and basic theory. *Infectious Control and Hospital Epidemiology* 19(3):194–214, March (review).

Boex, J. R., J. Cooksey, and T. Inui. 1998. Hospital participation in community partnerships to improve health. *Joint Commission Journal on Quality Improvement* 24(10):541–48, October.

Centers for Disease Control. 2000. www.cdc.gov.

Friedman, C., C. A. Baker, J. L. Mowry-Hanley, K. Vander Hyde, M. S. Stites, and R. J. Hanson. 1993. Use of the total quality process in an infection control program: a surprising customer-needs assessment. *American Journal of Infection Control* 21(3):155–59, June.

Howland, R., and M. D. Decker. 1992. Continuous quality improvement and hospital epidemiology: common themes. *Quality Management in Health Care* 1(1):9–12, fall.

Hurt, N. 1993. The role of the infection control nurse in quality management in the ambulatory care setting. *Journal of Healthcare Quality* 15(3):43–44, May–June.

Jhaveri, S. 1996. A total quality management approach to infection control. *Executive Housekeeping Today* 17(3):12, 14, March.

Lathrop, C. B. 1997. Using graphs to consolidate reports to the board. *Journal of Healthcare Quality* 19(1):26–33, January–February.

Seto, W. H. 1995. Training the work force: models for effective education in infection control. *Journal of Hospital Infections* 30 (suppl.):241–47, June.

Simmons, B. P., and S. B. Kritchevsky. 1995. Epidemiologic approaches to quality assessment. *Infection Control and Hospital Epidemiology* 16(2):101–4, February.

Welch, L., A. C. Teague, B. A. Knight, A. Kenney, and J. E. Hernandez. 1998. A quality management approach to optimizing delivery and administration of preoperative antibiotics. *Clinical Performance and Quality Health Care* 6(4):168–71, October–December.

Woomer, N., C. O. Long, C. Anderson, and E. A. Greenberg. 1999. Benchmarking in home health care: a collaborative approach. *Caring* 18(11):22–28, November.

Chapter 9
Decreasing Risk Exposure

Learning Objectives

- To describe the importance of managing risk exposure in the contemporary healthcare organization

- To explain the importance of the use of occurrence reporting in decreasing risk exposure

- To define the concept of a sentinel event

- To understand how sentinel events can point to important safety opportunities for improvement in healthcare organizations

- To explain how risk managers use their skills in patient advocacy to lessen the impact that potentially compensable events can have on healthcare organizations

- To introduce national patient safety goals for healthcare organizations and strategies for proactive risk reduction activities

Background and Significance

By nature of the business, healthcare organizations are susceptible to claims of liability. Claims result for incidents that adversely affect patients, visitors, and employees. Every day, employees in healthcare organizations work with equipment and substances that are dangerous even when used appropriately. Clinical laboratories perform analyses that use chemicals that can cause burns. Radiological instruments deliver high doses of radiation. Hospital pharmacies package and deliver medications that are poisonous to human beings when administered imprecisely.

Patients contribute health risks as well. Some refuse to wait for assistance before getting out of bed to use the bathroom. Some patients do not like sleeping with side rails on their beds and risk dangerous falls. Belongings disappear due to theft or negligence. Patients pull out their own intravenous lines and expose nursing staff to blood-borne pathogens. There are also the unintended consequences of treatment to consider. Response to therapy does not always go as planned. The surgeon's knife sometimes nicks contiguous structures. The possibility of unintended injury is everywhere in healthcare organizations.

In 1999, the Institute of Medicine (IOM) published a report on medical errors in healthcare entitled "To Error Is Human: Building a Safer Health System" that sent shock waves through the American healthcare industry. The IOM reported that "at least 44,000 people, and perhaps as many as 98,000 people, die in hospitals each year as a result of medical errors that could have been prevented." Since the IOM's report, there has been both government and private sector attention placed on safety issues in healthcare. The Clinton administration issued an executive order instructing government agencies that conduct or oversee healthcare programs to create a task force to find new strategies for reducing errors, and Congress appropriated $50 million to support efforts targeted at reducing medical errors. In response to these initiatives, the Patient Safety Improvement Act of 2003 was introduced to create a new system for voluntary reporting of medical errors. In addition, mandatory reporting systems such as state licensing agencies require healthcare organizations to report adverse medical events that result in death and serious harm, and regulatory agencies such as the Joint Commission on Accreditation of Healthcare Organizations (JCAHO) have established standards that require the monitoring of national patient safety goals as a condition of accreditation. On the private sector side, businesses buying insurance coverage for their employees are being encouraged to make safety a prime concern in their contracting decisions, and patients and their families are being urged to take a proactive role in reducing medical errors by participating in safety initiatives established by healthcare providers.

In this atmosphere of complexity, danger, chance, and emotion, risk managers work to accomplish their professional objectives. Risk managers seek to manage the organization's risk exposure and improve its processes so that the threat of liability is minimized. **Risk** in this context is a formal insurance term denoting liability to compensate individuals for injuries sustained in the healthcare facility. Occurrences involving injury or property loss are called **potentially compensable events** (PCEs).

Patients and clients come to healthcare organizations expecting physical and emotional benefits for themselves and their family members. Employees come to work intending to provide the best-quality healthcare they can deliver. Because everyone comes to the healthcare setting with the best intentions, negative occurrences can result in anger and guilt. Generally, errors in healthcare are caused by faulty systems, processes, and conditions that lead people to make mistakes or fail to prevent them. Such occurrences present healthcare organizations with opportunities for reducing errors and improving patient safety. The cornerstone of patient safety for healthcare organizations is to foster an environment that acknowledges the unintentional nature of human error and how it can learn from its mistakes.

Decreasing Risk Exposure: Steps to Success

Most people think of risk management as the process of working through a malpractice suit. Although that is sometimes the case, risk managers are more often trying to identify organizational conditions that increase risk exposure before occurrences involving injury happen. Identification before injury then allows the organization to be proactive in improving care processes prior to incurring the exposure.

Most recently, any healthcare organization's patient safety program defined by the JCAHO must include proactive error reduction activities. In other words, healthcare organizations are to identify and address underlying system problems that may result in adverse patient incidents. Most healthcare organizations are integrating these proactive error reduction activities into their performance improvement (PI) plan and goals. According to the JCAHO, beginning in 2004 all healthcare settings will be required to identify high-risk processes in which a failure of some type could jeopardize the safety of individuals to which they provide services. The standards further require that at any given time at least one high-risk process be the subject of an ongoing intense analysis by utilizing methods such as failure mode, effects and criticality analysis (FMECA), a technique that promotes systems thinking. The FMECA includes defining high-risk processes using flowcharts, identifying potential failure points in current processes, and scoring each potential failure by considering factors such as the frequency of failure, potential harm, and the likelihood that the failure will be detected before it reaches the patient. Potential failures with the highest criticality score become the focus of process redesign.

Other initiatives related to patient safety and proactive error reduction include the JCAHO's national patient safety goals. Initiated in 2003, the JCAHO established six goals to help accredited organizations address specific areas of concern in relation to patient safety. No more than six goals, with a maximum of two recommendations per goal, are established for any given year. Each year, the goals are reevaluated. Some of the patient safety goals may continue, while others may be replaced because of emerging new priorities and areas of focus. All accredited organizations must implement the goals and associated recommendations that are relevant to the services provided by their organization. First-year goals can include recommendations related to the following topics:

- Patient identification

- Communication between caregivers

- High-alert medications

- Wrong site, patient, or procedure surgery

- Infusion pumps

- Clinical alarm systems

These goals and recommendations are the result of lessons learned from organizations that have experienced a **sentinel event.** Lessons learned are the findings from a **root-cause analysis** and why the sentinel event may have occurred. Additional information on sentinel events and associated root-cause analysis is addressed in step 5 below.

Step 1: Develop risk management policies and procedures

The risk manager leads the development of risk management policies and procedures for the organization. Policies and procedures in the risk management area include a policy describing the organization's insurance strategy, procedures outlining its claims-tracking and negotiation system, and procedures outlining how its databases are maintained and risk management reports developed. Risk managers also routinely review the operational policies and procedures of all departments to ensure that they have been designed in a way that reduces, rather than increases, the possibility of risk exposure. Risk management procedures also define requirements for occurrence reporting and reporting to insurers, licensing agencies, public health departments, governing bodies, the National Practitioner Data Bank, and other accrediting and regulatory agencies. In addition, risk managers review all policies and procedures approved in the organization to ensure that none of the activities would expose the organization to risk.

Step 2: Develop risk management educational programs

In conjunction with staff development coordinators, risk managers develop educational activities for employees. Examples of such activities include hand-washing technique, needle safety protocol, and incident documentation and reporting. Educational topics are based on data derived from risk management reports. Program attendance is documented

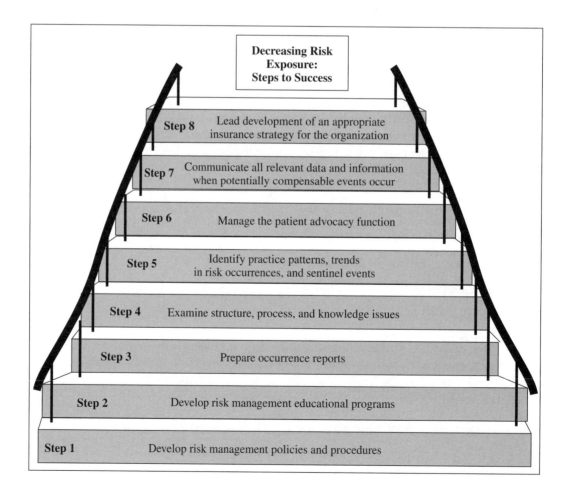

in each employee's personnel file, and competency testing for comprehension and retention is performed.

Step 3: Prepare occurrence reports

The risk manager's principal tool for capturing the facts about PCEs is the **occurrence report,** sometimes called the **incident report.** Effective occurrence reports carefully structure the collection of data, information, and facts in a relatively simple format. An excellent example of an occurrence report is provided in figure 9.1. Note that this example captures information about the persons involved. It records information about the time of day, the day of the week, and the type of shift that the employees involved were working. Page 2 of the report gathers information about the witnesses to the incident and the results of contact with and examination by a physician or nurse practitioner. Next, pages 3, 4, and 5 of this example gather information regarding specific aspects of the most common incidents that occur in healthcare. For example, in section 7, the report collects data on the type of burn. All data collected in these incident-specific sections are coded for easy checking and entry into a database management system for incident tracking.

Following completion by individuals on the unit where the incident occurred, the occurrence report is sent to the risk manager. The risk manager reviews policy, procedure, and other aspects of the occurrence, and involves management and administrative staff as necessary. After the cause analysis and revision of policy, the long-term follow-up of the incident is documented on page 6 of the report. For example, this would include actions regarding policies and procedures, the conclusion for the injured individual, contacts with manufacturers, and retraining of personnel.

The second important document for managing risk exposure is the patient's medical record. Every response to an incident that relates to a patient must be documented in the patient's medical record. If a case of injury should become a malpractice suit or other legal action, this documentation in the medical record would become crucial. The clinicians involved in an incident must be certain that all appropriate clinical documentation regarding the occurrence is added to the medical record as well as being documented on the occurrence report.

Occurrence reports are generally not open to view by the plaintiff's attorney. Thus, to communicate the care given the patient in its entirety, and especially regarding an occurrence, the record must be documented carefully because it will be used to portray the events to the public. The risk manager must develop a careful balance while overseeing the documentation of a potentially compensable event. Complete details on all persons involved, actions taken, condition and responses of the patient should be recorded on the occurrence record. However, in the patient's medical record, the details should be limited to those documenting care given to the patient. No details about how or what contributed to the occurrence should be recorded. No mention should be made in the patient's medical record of an incident or occurrence report having been completed. The documentation of a medication error in the sample occurrence report (figure 9.2) is a good example. The JCAHO has implemented new patient right standards related to disclosing outcomes. The standard does not require that the organization inform the patient about an error, but it does require that all patients be informed of all outcomes, unanticipated or otherwise.

Figure 9.1.   Sample Occurrence Report

Med Rec # *00–05–45*
Name: *Jackson, Julia*
Date of Birth: *06–22–23*
Street: *6401 Fremont Ave*
City: *Western City, CA*

Risk Management use only_____

Patient ID/name of individual involved.
Use addressograph for patient.

INSTRUCTIONS: (1) Fill out the first page of the Incident Report Form. (2) Select the type of incident from the bottom of page 2. (3) Fill out all appropriate sections as directed. The report must be dated and filled out by the end of the shift in which the incident occurred or was discovered. **DO NOT COPY THIS FORM.** Please print, this report must be legible. **Please fill out all applicable parts of this form.** Upon completion of this form route it to your Nurse Manager or Supervisor. Do not leave this form in the patient's chart.

Date of incident: *05 / 29 / 03*   Time (2400 Clock): *1645*   Hospital Unit: *Med/Surg*

What day of the week
did it occur on?

				Did the incident occur during:		Employee involved worked a(n)	
Sun	[√]	Thurs	[ ]	Day 0701–1500	[ ]	8 Hour shift	[√]
Mon	[ ]	Fri	[ ]	Evening 1501–2300	[√]	10 Hour shift	[ ]
Tues	[ ]	Sat	[ ]	Night 2301–0700	[ ]	12 Hour shift	[ ]
Wed	[ ]					Double shift	[ ]
						Other_____	

Where did the incident occur? *patient room*

Description of incident. Include: follow-up care given, i.e., vital signs, X-Ray, laboratory tests, etc.
Pt. developed a macular rash over trunk and extremities after 10 mg

dose of Compazine given for postop nausea. Compazine stopped

and Benadryl given IM.

IMMEDIATE EFFECT OF THE INCIDENT: *Severe macular rash over trunk*

and extremities

Involved Person Data

Date of Admission *05 / 29 / 03*

What sex is the person?
Male   [ ]
Female   [√]

What is the person's age? *76*

Inpatient	[√]
Outpatient	[ ]
Student	[ ]
Employee	[ ]
Visitor	[ ]
Volunteer	[ ]
Other_____	

Current Diagnosis/Reason for visit: *Bowel Obstruction*

Is the involved person aware of the incident?	Yes	[√]	No	[ ]
Is the family aware of incident?	Yes	[ ]	No	[√]
Did the incident involve equipment?	Yes	[ ]	No	[√]
If Yes, was Bioengineering notified?	Yes	[ ]	No	[ ] N/A [ ]

CONFIDENTIAL: This material is prepared pursuant to Code Annotated, §26-25-1, et seq., and 58-12-43(7, 8, and 9), for the purpose of evaluating health care rendered by hospitals or physicians and is NOT PART of the medical record.

Figure 9.1.   *(Continued)*

DO NOT COPY

****   PLEASE PRINT   ****

Person preparing report (Signature): *Gwen Nelson, R.N.* Print *Gwen Nelson, R.N.*

Name of individual witnessing incident (Print): *Bob Patterson, R.N.*

Dept/Address: *Med/Surg Team Leader*

Name of employee involved in incident: *Gwen Nelson, R.N.* Dept/Address *Med/Surg*

Name of employee discovering incident: *Gwen Nelson, R.N.* Dept/Address *Med/Surg*

**** STAFF TO NOTIFY ATTENDING PHYSICIAN AND/OR DESIGNATED RESIDENT/NURSE PRACTITIONER OF INCIDENT ****

I notified Dr./NP *Jeff Cook* at *1650* (time).

M.D./NP responded ☐ in person ☑ by phone at *1705* (time).

Was the attending physician notified?

Yes [√] Date: *05* / *29* / *03* Time: *1650*

No [ ] Why not? _____

Examining Physician/Nurse Practitioner statement regarding condition/outcome of person involved:

Pt. was examined by me at 1700 hours. Trunk and extremities show a macular rash on them. One dose of Benadryl given IM to pt. and rash began to subside. Compazine stopped.

Examining MD/NP signature: *Tom Lander, M.D. House Staff*

Examining MD/NP name (print): *Tom Lander, M.D.*

Date: *05* / *29* / *03* Time: *1700* Clinical Service: *Medicine*

CHOOSE THE TYPE OF INCIDENT YOU ARE REPORTING. Use the index below to locate the type of incident you are reporting, go to that section and mark the appropriate box(es). THERE MAY BE MORE THAN ONE ITEM APPLICABLE IN A SECTION. CHECK BOX(ES) IN APPROPRIATE SECTIONS.

Medication/IV Incident	Page 3, Section 1	Patient Behavioral Incident	Page 5, Section 6
Blood/Blood incident	Page 3, Section 2	Safety Incident	Page 5, Section 9
Burn	Page 5, Section 7	Security Incident	Page 5, Section 8
Equipment Incident	Page 5, Section 10	Surgery Incident	Page 5, Section 4
Fall	Page 4, Section 3	Treatment/Procedure Incident	Page 5, Section 5
Fire Incident	Page 5, Section 11		

CONFIDENTIAL: This material is prepared pursuant to Code Annotated, §26-25-1, et seq., and 58-12-43(7, 8, and 9), for the purpose of evaluating health care rendered by hospitals or physicians and is NOT PART of the medical record.

(Continued on next page)

Figure 9.1. *(Continued)*

DO NOT COPY

SECTION 1 MEDICATION/IV INCIDENT

1A. TYPE OF MEDICATION

Fill in specific medication/solution on the adjacent line.

Analgesic _____
Anesthetic agent _____
Antibiotic _____
Anticoagulant _____
Anticonvulsant _____
Antidepressant _____
Antiemetic___*Compazine*_____
Antihistamine _____
Antineoplastic _____
Bronchodilator _____
Cardiovascular _____
Contrast media _____
Diuretic _____
Immunizations _____
Immunosuppressive _____
Insulin _____
Intralipids _____
Investigational drug _____
IV solution _____
Laxative_____
Narcotic_____
Oxytocics _____
Psychotherapeutic _____
Radionuclides _____
Sedative/tranquilizer _____
TPN _____
Vasodilator _____
Vasopressor _____
Vitamin _____
Other _____

1B. TYPE OF MEDICATION OR IV INCIDENT

Adverse reaction	[√]	1B01
Allergic/contraindication	[ ]	1B02
Delayed stat order	[ ]	1B03
Improper order (MD/NP)	[ ]	1B04
Incompatible additive	[ ]	1B05
Incorrect additive	[ ]	1B06
Incorrect dosage	[ ]	1B07
Incorrect drug	[ ]	1B08
Incorrect narcotic count	[ ]	1B09
Incorrect patient	[ ]	1B10
Incorrect rate of flow	[ ]	1B11
Incorrect route	[ ]	1B12
Incorrect schedule	[ ]	1B13
Incorrect solution/type	[ ]	1B14
Incorrect time	[ ]	1B15
Incorrect volume	[ ]	1B16
Infiltration	[ ]	1B17
Given before culture taken	[ ]	1B18
Medication given before lab results returned	[ ]	1B19
Medication missing from cart	[ ]	1B20

Not documented	[ ]	1B21
Not prescribed	[ ]	1B22
Omitted	[ ]	1B23
Outdated	[ ]	1B24
Out-of-sequence	[ ]	1B25
Patient took unprescribed medication	[ ]	1B26
Repeat administration	[ ]	1B27
Transcription error	[ ]	1B28
Other_____		1B29

1C. ROUTE OF MEDICATION ORDERED:

IM	[ ]	1C01
IV	[ ]	1C02
PO	[ ]	1C03
Other___*Suppository*_____	√	1C04

1D. MEDICATION DISPENSING INCIDENT

Meds not sent/delayed from pharmacy	[ ]	1C01
Incorrectly labeled	[ ]	1C02
Incorrect dose	[ ]	1C03
Incorrect drug sent	[ ]	1C04
Incorrect IV additive	[ ]	1C05
Incorrect IV fluid	[ ]	1C06
Incorrect route (IV, PO, IM, PR)	[ ]	1C07
Mislabeled	[ ]	1C08
Other_____		1C09

SECTION 2
BLOOD/BLOOD COMPONENT INCIDENT

2A. BLOOD/BLOOD COMPONENT TYPE

Albumin	[ ]	2A01
Cryoprecipitate	[ ]	2A02
Factor VIII (AHF)	[ ]	2A03
Factor IX (Konyne)	[ ]	2A04
Fresh frozen plasma	[ ]	2A05
Packed red blood cells (PRBC)	[ ]	2A06
Plasmanate	[ ]	2A07
Platelets	[ ]	2A08
Rhogam	[ ]	2A09
Washed red blood cells (WRBC)	[ ]	2A10
Whole blood	[ ]	2A11
Other_____		2A12

2B. TYPE OF BLOOD/BLOOD COMPONENT INCIDENT

Crossmatch problem	[ ]	2B01
Improper unit verification	[ ]	2B02
Inappropriate IV fluids administered with blood components	[ ]	2B03
Inappropriate documentation	[ ]	2B04
Inappropriate storage	[ ]	2B05
Incomplete patient ID	[ ]	2B06
Incorrect patient	[ ]	2B07
Incorrect rate	[ ]	2B08
Incorrect type	[ ]	2B09
Incorrect volume	[ ]	2B10
Patient refused	[ ]	2B11
Other_____		2B12

CONFIDENTIAL: This material is prepared pursuant to Code Annotated, §26-25-1, et seq., and 58-12-43(7, 8, and 9), for the purpose of evaluating health care rendered by hospitals or physicians and is NOT PART of the medical record.

Figure 9.1. *(Continued)*

DO NOT COPY

SECTION 3
FALLS

3A. FALL CODE STATUS OF PATIENT

Attended	[ ] 3A01
Unattended	[ ] 3A02

3B. LOCATION OF FALL

Bathroom in patient's room	[ ] 3B01
Bathroom (other location)	[ ] 3B02
Elevator	[ ] 3B03
Examining/treatment room	[ ] 3B04
Hallway/corridor	[ ] 3B05
Nursing station	[ ] 3B06
Parking lot	[ ] 3B07
Patient's room	[ ] 3B08
Recreation area	[ ] 3B09
Shower/tub room	[ ] 3B10
Stairs	[ ] 3B11
Waiting room	[ ] 3B12
Walkway/sidewalk	[ ] 3B13
Other_____	3B14

3C. FALL OCCURRED IN CONJUNCTION WITH:

Bedside commode	[ ] 3C01
Chair	[ ] 3C02
Due to toy	[ ] 3C03
During transfer	[ ] 3C04
Exam table	[ ] 3C05
Fainting/dizzy	[ ] 3C06
Fall/slip	[ ] 3C07
From bed	[ ] 3C08
Improperly locked device	[ ] 3C09
Recreational activity	[ ] 3C10
Scales	[ ] 3C11
Stretcher	[ ] 3C12
Table	[ ] 3C13
Tripped	[ ] 3C14
While ambulating unattended	[ ] 3C15
While ambulating with assist.	[ ] 3C16
While entering or leaving bed	[ ] 3C17
While using ambulatory device	[ ] 3C18
Other_____	3C19

3D. PATIENT ACTIVITY PRIVILEGES
(As per medical order)

Ambulate with assistance	[ ] 3D01
Ambulate with walker	[ ] 3D02
Ambulate without assistance	[ ] 3D03
Bathroom privileges with assistance	[ ] 3D04
Bathroom privileges without assistance	[ ] 3D05
Bedrest	[ ] 3D06
Up Ad lib	[ ] 3D07
Up in chair/wheelchair	[ ] 3D08
Other_____	3D09

3E. PATIENT MENTAL CONDITION AT THE
TIME OF THE FALL

Confused/poor judgment	[ ] 3E01
Language barrier	[ ] 3E02
Oriented	[ ] 3E03
Unconscious	[ ] 3E04
Uncooperative	[ ] 3E05
Unresponsive/medicated	[ ] 3E06
Other_____	

3F. PATIENT'S CALL LIGHT WAS:

On	[ ] 3F01
Off	[ ] 3F02
Not within reach	[ ] 3F03
Patient unable to use	[ ] 3F04
Not applicable	[ ] 3F05

3G. POSITION OF BED

High	[ ] 3G01
Low	[ ] 3G02
Intermediate	[ ] 3G03
Not applicable	[ ] 3G04

3H. BED ALARM

On	[ ] 3H01
Off	[ ] 3H02
Not applicable	[ ] 3H03

3I. POSITION OF SIDE RAILS
(At the time of the fall)

Half Rails	[ ] 3E01	Full Rails	[ ] 3E06
1 Up	[ ] 3E02	1 Up	[ ] 3E07
2 Up	[ ] 3E03	2 Up	[ ] 3E08
3 Up	[ ] 3E04		
4 Up	[ ] 3E05		

Not applicable [ ] 3E09

3J. PATIENT RESTRAINTS

Removed by patient	[ ] 3J01
Restraints intact	[ ] 3J02
Not applicable	[ ] 3J03
Other_____	3J04

3K. CONDITION OF AREA WHERE FALL
OCCURRED

Normal/dry	[ ] 3K01
Wet floor	[ ] 3K02
Ice condition	[ ] 3K03
Other_____	3K04

3L. FALLS IN CONJUNCTION
WITH MEDICATION

Narcotic or sedative received by patient in the past 12 hours?	[ ] 3L01
When was the last dose? _____	3L02
What was the drug? _____	3L03
What was the route of administration?_____	3L04

CONFIDENTIAL: This material is prepared pursuant to Code Annotated, §26-25-1, et seq., and 58-12-43(7, 8, and 9), for the purpose of evaluating health care rendered by hospitals or physicians and is NOT PART of the medical record.

(Continued on next page)

Figure 9.1.   *(Continued)*

DO NOT COPY

SECTION 4
SURGERY INCIDENT
Anesthesia occurrence   [ ] 0401
Contamination. . . . . [ ] 0402
Incorrect needle count   [ ] 0403
Incorrect sponge count. . . . . [ ] 0404
Informed consent absent   [ ] 0405
Informed consent incorrect . . . . [ ] 0406
Instrument lost/broken   [ ] 0407
Retained foreign body . . . . [ ] 0408
Other_____   0409

SECTION 5
TREATMENT/PROCEDURE INCIDENT
Adverse reaction   [ ] 0501
Allergic response. . . . . [ ] 0502
Application/removal of cast/splint   [ ] 0503
Cancellation of procedures . . . . [ ] 0504
Catheter or tube related   [ ] 0505
Delay. . . . . [ ] 0506
Dietary problem   [ ] 0507
Dressing/wound occurrence. . . . . [ ] 0508
Informed consent absent   [ ] 0509
Informed consent incorrect . . . . [ ] 0510
Injection site   [ ] 0511
Invasive procedure/placement . . . . [ ] 0512
Mislabeled specimen   [ ] 0513
Missing specimen . . . . [ ] 0514
Not documented   [ ] 0515
Omitted. . . . . [ ] 0516
Patient/site identification   [ ] 0517
Positioning . . . . [ ] 0518
Prep problem   [ ] 0519
Repeat procedure. [ ] 0520
Reporting of test results   [ ] 0521
Thermoregulation problem . . . . [ ] 0522
Transcription error   [ ] 0523
Transfer/moving of patient. . . . . [ ] 0524
Other_____   0525

SECTION 6
PATIENT BEHAVIORAL INCIDENT
Attempted AWOL   [ ] 0601
AWOL. . . . . [ ] 0602
Inappropriate sexual behavior   [ ] 0603
Injured by other patient . . . . [ ] 0604
Patient altercation   [ ] 0605
Self-inflicted injury . . . . [ ] 0606
Suicide gesture   [ ] 0607
Other_____   0608

SECTION 7
BURNS
Chemical   [ ] 0701
Electrical. . . . . [ ] 0702
Inhalation   [ ] 0703
Radioactive . . . . [ ] 0704
Thermal   [ ] 0705

SECTION 8
SECURITY INCIDENTS
Bomb threat   [ ] 0801
Breaking and entering . . . . [ ] 0802
Drug theft   [ ] 0803
Secure area key loss/missing . . . . [ ] 0804
Major theft (over $250)   [ ] 0805
    Amount:_____
Minor theft . . . . [ ] 0806
    Amount:_____
Personal property damage/loss. . . . . [ ] 0807
    Amount:_____
Hospital property damage . . . . [ ] 0809
    Amount:_____
Other_____

SECTION 9
SAFETY INCIDENTS (patients and visitors only)
Body fluid exposure   [ ] 0901
Chemical exposure . . . . [ ] 0902
Chemotherapy spill   [ ] 0903
Drug exposure. . . . . [ ] 0904
Hazardous material spill   [ ] 0905
Needlestick . . . . [ ] 0906
Other_____   0907

SECTION 10
EQUIPMENT INCIDENT
Disconnected   [ ] 1001
Electrical problem. [ ] 1002
Improper use   [ ] 1003
Malfunction/defect . . . . [ ] 1004
Mechanical problem   [ ] 1005
Not available. . . . . [ ] 1006
Electrical shock   [ ] 1007
Electrical spark . . . . [ ] 1008
Struck by   [ ] 1009
Wrong equipment . . . . [ ] 1010
Tampered with
    By patient   [ ] 1011
    Non-patient   [ ] 1012
Other_____   1013

SECTION 11
FIRE INCIDENT
Equipment caused   [ ] 1101
Cigarette caused . . . . [ ] 1102
Laser caused   [ ] 1103
Other_____   1104

CONFIDENTIAL: This material is prepared pursuant to Code Annotated, §26-25-1, et seq., and 58-12-43(7, 8, and 9), for the purpose of evaluating health care rendered by hospitals or physicians and is NOT PART of the medical record.

Figure 9.1. *(Continued)*

DO NOT COPY

**EMPLOYEES DO NOT COMPLETE BELOW,
FOR NURSE MANAGER/SUPERVISOR USE ONLY.**

Recommendations and/or corrective actions based on review of report and discussion with employee:

NURSE MANAGER/SUPERVISOR Follow-Up [Check appropriate box(es)]/Corrective action]

Policy/Procedure:

Evaluate	[ ] 1201	**Discussed with:**	
Recommend change	[ ] 1202	Physician	[ ] 1209
Changed	[ ] 1203	Staff . . . . [ ] 1210	
No action taken	[ ] 1204	Patient	[ ] 1211
Non-compliance	[ ] 1205	Other . . . . [ ] 1212	
Inadequate	[ ] 1206		
Needs enforcement	[ ] 1207	Date: _____	
Review with involved individual(s). . . .	[ ] 1208	Time: _____	

Describe specific follow-up actions taken (if applicable include names of depts) _____
_____
_____

SIGN AND DATE: (Indicates review of report)

1. Quality Management/Risk Management_____ ___/___/___

2. Nurse Manager/Supervisor (as applicable) _____ ___/___/___

3. Department Head/DON (As applicable) _____ ___/___/___

4. QM Coordinator (As applicable)_____ ___/___/___

5. Other: Title_____ Name _____ ___/___/___

BIOENGINEERING USE ONLY

Manufacturer contacted	[ ] 1301
Manufacturer instructions followed	[ ] 1302
Needs enforcement of policy/procedure	[ ] 1303
Include instructions in staff education and training	[ ] 1304
Preventative maintenance or biomedical evaluation of equipment ordered	[ ] 1305
Recommend repair or replacement	[ ] 1306
Removed from service	[ ] 1307
Other _____	1308

RISK MANAGEMENT USE ONLY

IMMEDIATE EFFECT OF THE INCIDENT

Alteration in skin integrity	[ ] 1401	Patient discomfort/inconvenience	[ ] 1411
Birth related injury . . . . [ ] 1402		Psycho/social trauma. . . . . [ ] 1412	
Breach of confidentiality	[ ] 1403	Reproductive injury or loss	[ ] 1413
Death . . . . [ ] 1404		Sensory impairment. . . . . [ ] 1414	
Disability	[ ] 1405	Severe internal injuries	[ ] 1415
Disfigurement. . . . . [ ] 1406		Substantial disability . . . . [ ] 1416	
Drug/blood reaction	[ ] 1407	Unanticipated neuro deficit	[ ] 1417
Fluid imbalance . . . . [ ] 1408		Unanticipated systemic deficit. . . . . . . [ ] 1418	
Neuro deficit	[ ] 1409	Indeterminate	[ ] 1419
Orthopedic injury . . . . [ ] 1410		None. . . . . [ ] 1420	
		Other_____	1421

Description_____

CONFIDENTIAL: This material is prepared pursuant to Code Annotated, §26-25-1, et seq., and 58-12-43(7, 8, and 9), for the purpose of evaluating health care rendered by hospitals or physicians and is NOT PART of the medical record.

Figure 9.2.   **Sample Progress Note in Patient's Record**

PROGRESS NOTES	Med Rec # 00-05-45 Jackson, Julia

DATE & TIME	NOTES MUST BE DATED AND TIMED
5/29/03 1650	Patient developed a macular rash over entire trunk and extremities after 10 mg of Compazine given for nausea. Dr. Cook and house staff notified. *Gwen Nelson, RN*
5/29/03 1700	Called to pt. for rash on trunk & extremities. Pt. examined, adverse reaction to Compazine most likely. Patient to receive 20 mg of Benadryl IM now. If nausea continues, Dramamine 50 mg IV prn. *J. Lander, MD*
5/29/03 1700	Dr. Lander examined patient and ordered Benadryl 10 mg IM. Patient injected IM 10 mg of Benadryl. *Gwen Nelson, RN*
5/29/03 1810	Rash is subsiding and nausea less. *Gwen Nelson, RN*

PROGRESS NOTES

Step 4: Examine structure, process, and knowledge issues

As department members and process improvement teams begin to identify customers and review performance, issues in organizational structures, processes, outcomes, and knowledge may become apparent. Commonly, these issues are documented in department or team communications to a PI council. (See the discussion of documentation recommendations in chapter 5.) Risk managers are usually members of the council and routinely review communications from the departments and teams for this purpose. Issues in turn should be documented in risk management databases so that corrective action may be initiated if and when they meet a performance threshold (the level above which the occurrence is not occurring by chance and cannot be tolerated). If it becomes apparent that members of the organization do not have appropriate background in procedure or policy, the risk manager may have to initiate educational sessions to develop staff background regarding the issue.

Step 5: Identify practice patterns, trends in risk occurrences, and sentinel events

Using aggregate data summarized from the occurrence report discussed in step 1, the risk manager attempts to identify trends in risk occurrences within the organization. For instance, if there were an increase in blood transfusion reactions, the manager might request a focused review by blood bank personnel to be sure that typing and grouping procedures are being followed appropriately. If there were an increase in occurrence reports documenting patients' falls on a particular nursing unit, the risk manager might ask the nurse manager on that unit to be sure that staff understand how to identify a patient at risk for falls and that they are taking appropriate preventive measures with such patients.

Occurrences that are classified as sentinel events require a more intense investigation. The JCAHO defines a sentinel event as "an unexpected occurrence involving death or serious injury, or the risk thereof. Serious injury specifically includes loss of limb or function. The phrase 'or risk thereof' includes any process variation for which a recurrence would carry a significant chance of serious adverse outcome. Such events are called 'sentinel' because they signal the need for immediate investigation and response" (JCAHO 2003). An organization should also include near misses in its definition of events that require intense investigation. Near misses include occurrences that do not necessarily affect an outcome but, if they were to recur, carry significant chance of being a serious adverse event. Near misses fall under the definition of a sentinel event but are not reviewable by the Joint Commission under its current sentinel event policy.

The JCAHO requires accredited organizations to conduct an in-depth investigation of an occurrence, including root-cause analysis and action plan, which meets the definition of a sentinel event within forty-five calendar days of either the event or awareness of the event. A root-cause analysis is a process used to identify the basic or causal factors that underlie variation in performance. The root-cause analysis and action plan conducted as a result of a sentinel event are reviewable under the JCAHO sentinel event policy.

Currently, sentinel events are voluntarily reportable to the JCAHO. Some state licensing agencies, however, require mandatory reporting of these events. The intent of voluntary reporting to the JCAHO is so that all healthcare organizations can benefit from lessons learned and ideally reduce the risk of a similar error occurring within their organization. A JCAHO publication called "Sentinel Event Alert" features aggregate data related to root causes and risk reduction strategies for sentinel events that occur with significant frequency.

These lessons learned from frequently occurring sentinel events form the basis for error-prevention advice to organizations.

Another important PI activity that contributes to the risk manager's databases is the credentialing of physicians and the validation of nurses' and other clinicians' licenses to practice. (See chapter 12 for additional information on competency and credentialing.) Standardized monitoring of clinicians' practice patterns and outcomes are conducted by committees of the medical staff and other disciplines. Documentation of these reviews is analyzed by these committees, disciplines, and the risk manager to identify clinicians who may be practicing outside their scope of licensure or who may benefit from additional education regarding policy, procedure, or the current standard of practice in the region where the healthcare organization does business. A critical area of liability for a healthcare organization is the processes for verification of staff clinical competency and practitioners working outside their scope of licensure or approved privileges. Malpractice claims against a provider would be in favor of the plaintiff if unfavorable information related to competency or privileging is revealed during the discovery process of a lawsuit.

Step 6: Manage the patient advocacy function

The first major objective of a proactive risk management program is to minimize the organization's exposure to risk or legal action. The second major objective is to minimize the economic impact of indemnity and expense payments related to claims. The economic impact of a quality risk program should be measured by the development of goodwill and the satisfaction of its customers. Historically, this second objective has been a function of risk management and is known as patient advocacy. Some healthcare organizations have removed this responsibility from the risk manager and established a separate patient advocate service that works with the risk manager on complaints specific to potentially compensable events.

The objective of patient advocacy is to support the patient through difficult interactions with the healthcare organization. The JCAHO (2003) defines an advocate as "a person who represents the rights and interests of another individual as though they were the person's own, in order to realize the rights to which the individual is entitled, obtain needed services, and remove barriers to meeting the individual's needs." Most businesses can be somewhat bureaucratic, and healthcare organizations are no exception. Add the confusion, complexity, danger, and chance of potentially compensable events, and the customer may feel shunned and isolated by the organization at a time of great personal strife. The customer needs to express his or her anger to lessen it and needs someone in the organization to validate his or her right to feel it. All licensed healthcare organizations are required to inform patients of their rights to initiate a complaint or grievance and to have their complaints reviewed and, if possible, resolved.

The risk manager as patient advocate must be sure that the customer knows the facts associated with an occurrence. The patient advocate must accept responsibility for the occurrence in the customer's eyes when the organization's employees were responsible for the situation during which the incident occurred. If the organization's employees were not responsible for the situation, but the customer believes that they were, the patient advocate should continue to serve a neutral role in resolving the situation. The challenge for the patient advocate is to investigate the complaint and address the issues specific to the event in such a way as to avoid legal action against the organization. Sometimes just listening to the patient's concerns, offering an apology, or negotiating monetary compensation can resolve the complaint without any further legal action against the organization.

Step 7: Communicate all relevant data and information when potentially compensable events occur

Inevitably, some claims of liability do involve formal legal action. The facility's attorneys, the representatives of the facility's insurance company, and the plaintiff's attorneys then become involved.

The risk manager should continue to function as the representative of the organization in such legal environments. He or she coordinates all requests for information by *subpoena duces tecum* from attorneys or from the courts. He or she explains the organization's perspective to attorneys regarding the completion of interrogatories and coordinates the appearance of employees at depositions or at trial. He or she remains open to negotiation of an appropriate settlement to conclude the action before it goes to trial.

The risk manager is responsible for communicating all relevant data and information to the organization's insurer as soon as possible after a potentially compensable event (PCE) is recognized. The risk management department routinely provides the organization's insurance company with copies of the occurrence reports for all potentially compensable events. The timeliness and efficiency of such communications are extremely important if potentially compensable events are to be resolved rapidly and without litigation. Subsequently, the risk manager must keep the insurer apprised of important communications with the customer or the customer's attorneys. Insurance company representatives must have complete information in order to make appropriate resolution decisions.

Step 8: Lead development of an appropriate insurance strategy for the organization

In addition to the transmission of information regarding PCEs, the risk manager also takes the lead in designing an insurance strategy that meets the needs of the healthcare organization. To be effective in this area, the risk manager must have an in-depth knowledge of the organization's services, facilities, equipment, procedures, and staff capabilities.

The insurance strategy is developed with reference to the healthcare service lines of the organization. For example, the perinatal service is one of the service lines that have the greatest inherent risk of liability. Numerous complications can occur during a woman's pregnancy, labor, and the delivery, and many of these complications cannot be foreseen. The fact that the organization provides perinatal services would be taken into consideration by the organization's management and insurer.

Real-Life Example

At Community Hospital of the West, Dr. Low, an obstetrician, delivered Mrs. Yu's infant with relatively little difficulty. However, when the placenta was delivered, a rush of blood appeared at the patient's cervical os. Dr. Low attempted to explore the patient's uterus to see whether there were still some pieces of the placenta inside that were causing the bleeding, but there was so much blood that she could not explore the uterus adequately. After some minutes of trying to deal with the situation, she realized that the bleeding did not appear to be abating even though the uterus was contracting appropriately. Dr. Low decided to take Mrs. Yu to surgery to perform an exploratory laparotomy and possible emergency hysterectomy. The physician knew that if she could not stop the bleeding, the patient's life would be in danger. She packed the uterus as tightly as possible, instructed nursing staff to find blood for a transfusion, covered the patient with a sheet, placed the patient on oxygen, and began wheeling the patient's gurney to the elevator.

Community Hospital of the West is a major tertiary care facility in a large American city. It had always provided obstetrical delivery and neonatal services in the north wing of the second floor of the facility. Delivery rooms were developed in this wing. Surgical services and the operating rooms were developed on the third floor in the north wing. When patients required cesarean section deliveries or other surgical treatment, they had to be transferred from the delivery rooms on the second floor to the operating rooms on the third floor.

Dr. Low and the obstetrical nurses assisting her waited for approximately one minute before an elevator arrived. Most of the hospital staff found the elevators very slow and had commented on this many times over the years. Dr. Low, the nurse, and the patient on the gurney arrived in about another minute and a half on the third floor and rushed into an operating room. Crash induction of anesthesia was begun. As the operating room staff tried to get a line in to start the blood transfusion and Dr. Low began to remove the packing from the uterus, a massive amount of blood gushed from the organ. The patient's heart went into ventricular fibrillation, and despite emergency resuscitative efforts, Mrs. Yu died.

Because the death occurred during a surgical procedure, it was reportable to the county coroner's office. The coroner accepted the case and performed an autopsy. Mrs. Yu was found to have an anomalous uterine artery that had been opened upon delivery of the placenta, and she bled to death.

QI Toolbox Techniques

The death of a patient as discussed in the real-life example is always classified as a sentinel event. In such cases, the JCAHO requires the organization to do a root-cause analysis of the event to discover what processes in the organization led to the occurrence. Within the human body, there is always the possibility that unusual and unexpected events will occur. No one could really have known that Mrs. Yu's uterus was anomalous in its blood supply. Does that mean that Mrs. Yu's death was truly inadvertent? Could the death of this patient have been averted despite the anomaly in her anatomy?

The toolbox technique that is used most often in root-cause analysis is the **cause-and-effect** or **fishbone diagram.** This technique structures the root-cause inquiry and helps the investigators to be sure that they have examined the situation from all perspectives. As figure 9.3 shows, the fishbones delineate the causes of the situation (the effect is at the head of the fish) as classified in four categories. In the structure shown, the categories all begin with the letter *M.* This design was intended to make it easier to remember the categories as "the 4 Ms": Manpower, Material, Methods, and Machinery. Other approaches use other names for the categories, such as equipment, procedures, people, and policies.

- **Manpower** examines influences of the human worker on the situation. In the case of the death of Mrs. Yu, a human worker influence was the obstetrical nurses' lack of training in surgical procedures. This lack of training meant that surgical procedures could not be performed in the delivery room.

- **Material** examines the influences of supplies and equipment on the situation. In the real-life example, surgical supplies and equipment were not available to Dr. Low in the delivery room, and so she could not perform the necessary exploratory procedure there.

- **Methods** examines influences of policies and procedures on the situation. In Ms. Yu's case, it was the policy of the institution to take all obstetrical cases requiring surgical delivery or other surgical procedures to the operating room on an entirely different floor of the hospital. This policy caused a fairly long period of time to elapse during transport and in this situation led to the patient's death from blood loss.

- **Machinery** examines influences of machines or other major pieces of equipment on the situation. In this case, the slowness of the elevator in the hospital contributed to the delay in effective treatment.

In root-cause analysis, it is important to continue to ask the question *why?* until the absolute root cause has been discovered. Omachonu discusses proximate versus root causes

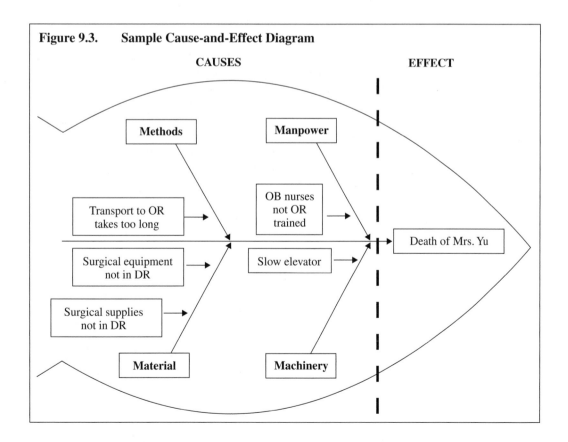

Figure 9.3.   Sample Cause-and-Effect Diagram

(Omachonu 1999). Proximate causes are those that can be pointed to directly. In the real-life example, a proximate cause was the inability of the obstetrical nurses to assist in emergency surgical procedures. However, that proximate cause has, in turn, many underlying causes, some of which may be the root cause of the sentinel event. Those underlying causes must be identified because the existence of those underlying causes is what really allowed the sentinel event to happen.

Investigators of Mrs. Yu's death had to ask why many times to get to the root causes of all the problems contributing to that occurrence. With respect to the lack of expertise in surgical procedures of obstetrical nursing staff, asking why uncovered significant negative attitudes on the part of all involved. Obstetrical nurses were reluctant to take on new responsibilities. Nursing administration was unwilling to commit the funds to develop new expertise in the nurses. Obstetrical physicians were skeptical that acceptable levels of competence could be developed in the obstetrical nurses. The institution's administration recognized that significant remodeling of the delivery suites would be required to provide the surgical services there. If root-cause analysis had not been performed, all of these contributing factors would have remained under the surface and would not have been dealt with.

In any situation, several levels of proximate causes must be identified and worked through to finally uncover the root causes. Asking why repeatedly helps investigators to arrive at root causes. (See figure 9.4.)

Following investigation and root-cause analysis of the factors that contributed to Mrs. Yu's death, administrators at Community Hospital of the West recognized that the current configuration of the delivery suite in the facility had directly contributed to a patient's death. In a situation where her anomalous uterine anatomy presented the care team with dangerous and unexpected consequences, she need not have died. If the exploratory laparotomy and possible emergency hysterectomy procedures could have been performed in the delivery room, Mrs. Yu's life probably could have been saved.

The administrators recognized that every transport of a patient from the delivery room to the operating rooms exposed the patient to unnecessary risk. The time wasted in transport in this case had proved fatal. The administrators undertook reconfiguration of the system to prevent the recurrence of this situation. The delivery rooms were rapidly remodeled and equipped with all the supplies and equipment necessary to support the performance of surgical procedures. Obstetrical nurses were retrained to assist on surgical

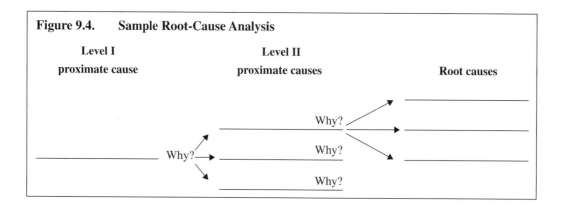

Figure 9.4.   Sample Root-Cause Analysis

procedures performed in the delivery room as they would be performed in the operating rooms. Following that episode, all cesarean sections and other emergency procedures were performed without requiring that the patient be transported to an operating room.

Case Study

Derek Johnson, M.D., had been an anesthesiologist at Community Hospital of the West for fifteen years. He was forty-five years of age. The physician was board certified to perform all kinds of anesthesia procedures, including every type of surgical procedure and obstetrical anesthetic procedure. His colleagues had noticed for some time that he was becoming more and more haggard looking, but they ascribed this to the hectic work schedule that anesthesiologists often must maintain.

One day, Derek Johnson was admitted to the intensive care unit at Community Hospital of the West. The news spread through the organization quickly. Dr. Johnson was suffering from a compromise of his immune system and was close to dying from septicemia. Some of the operating room (OR) nurses discussed the situation on dinner break. They speculated that Dr. Johnson was a drug addict. The anesthesiologists provided morphine to patients from a lockbox to which only the anesthesiologists had keys. They were not required to account for narcotics beyond signing out dosages from the lockbox on a clipboard that hung beside it. Some of the OR nurses had noticed over the past six months to a year that Dr. Johnson's patients invariably received morphine for pain as recorded on the clipboard. In many cases, the patients clearly did not need it. For most, there was no documentation in their medical charts of morphine administration.

One of the nurses reported that three months earlier she had cornered the chief of the anesthesia service and told him of Dr. Johnson's narcotics irregularities. Not wanting to challenge or accuse a fellow physician, the chief said indignantly that there must be some other explanation and terminated the conversation. The nurse went nowhere else to discuss the problem—not to the director of surgical services, nor to the chief of surgery, nor to the director of nursing, nor to an administrator.

Case Study Questions

1. Besides basic human weakness, what other reasons are evident for Dr. Johnson's narcotics problem?

2. Does Community Hospital of the West bear any responsibility for Dr. Johnson's predicament?

3. What are the root causes of this situation? Build a cause-and-effect diagram on the basis of the findings in this case study.

4. What is the likely response of the hospital's governing body to this situation?

Project Application

Students should consider using a cause-and-effect diagram or findings from a root-cause analysis in their storyboard projects.

Summary

Healthcare facilities are dangerous places. Employees in healthcare organizations must be continuously aware of situations that might result in injury to patients, visitors, or staff. To help manage the dangerous work setting, healthcare organizations use occurrence reporting systems that track and document incidents. Healthcare organizations are experiencing greater scrutiny and regulation from external supervisory agencies to proactively identify and monitor potential areas of high risk. Risk managers carefully review the organization's policies and procedures and assess the staff's ability to execute them. Risk managers also maintain open channels of communication with any person injured in the facility until a satisfactory resolution of the claim or litigation has been reached. Finally, risk managers provide information to the organization's insurers and represent the organization at all formal meetings and legal proceedings related to injury claims.

References

Institute of Medicine. 1999. *To Err Is Human: Building a Safer Health System.* Washington, D.C.: National Academy Press.

Joint Commission on Accreditation of Healthcare Organizations. 1998. Sentinel events: approaches to error reduction and prevention. *Joint Commission Journal of Quality Improvement* 24(4):175–86.

Joint Commission on Accreditation of Healthcare Organizations. 2003. www.jcaho.org.

Omachonu, Vincent K. 1999. *Healthcare Performance Improvement.* Norcross, Ga.: Engineering and Management Press.

Spath, Patrice L. 2002. Target: patient safety. *Journal of the American Health Information Management Association* 73(3): 26–33.

Spath, Patrice L., ed. 2000. *Error Reduction in Health Care.* San Francisco: Jossey-Bass.

Troyer, Glenn T., and Steven L. Salman. 1986. *Handbook of Health Care Risk Management.* Gaithersburg, Md.: Aspen.

Wakefield, D. S., et al. 1999. Understanding why medication administration errors may not be reported. *American Journal of Medical Quality* 14(2):81–88.

Youngberg, Barbara J., ed. 1999. *Essentials of Hospital Risk Management.* Gaithersburg, Md.: Aspen.

Chapter 10
Optimizing Patient Care

Learning Objectives

- To recognize the common means by which healthcare organizations monitor and improve the quality of patient care

- To understand the expectation that healthcare will be individualized across the care continuum

- To understand the roles that clinical practice guidelines and evidence-based medicine play in standardizing patient care

- To explain the contribution that the Minimum Data Set for Long-Term Care and the Health Plan Employer Data and Information Set can make to improving the quality of patient care

Background and Significance

The preceding chapters discussed the performance improvement (PI) model, its goals, and some of the factors involved in working with PI processes. This chapter focuses on a systematized approach to process improvement that can ultimately benefit the patient.

Even the individual medical requirements of a single patient can initiate the PI cycle. The critical factor in optimizing the care of patients is the organization's ability to improve patients' understanding of their health, their ability to care for themselves, their independence, and their quality of life. The goal of PI in healthcare is to design and implement systems that provide consistency and quality in all of the patient care processes performed to improve each individual patient's health.

The standards of the Joint Commission on Accreditation of Healthcare Organizations (JCAHO) describe the goal of patient care as providing "individualized, planned, and appropriate interventions" (JCAHO 1999a). These interventions may be in the form of care, services, treatment, or rehabilitation. Performance improvement efforts provide a process for evaluating every service, provider, setting, and outcome. The cornerstone of patient care is the establishment of a treatment plan that is specific, individualized, and based on a thorough assessment of the patient's physical, emotional, social, cognitive, and cultural needs.

Treatment/Care Plans

In today's healthcare settings, the care pathway for an individual patient (figure 10.1) is often completed by a multidisciplinary team using data developed through team assessment processes. The assessment process may be as simple as having a patient complete a questionnaire in a physician's office or as complicated as having physicians, radiologists, and nurses take part in an assessment in a hospice setting with a terminal cancer patient and family.

The goals of the care pathway are broad in scope and set the overall direction for care. The pathway defines the specific treatment, its timing and frequency. Many healthcare facilities use a system of care based on established national clinical standards for treatment interventions.

Risser and his colleagues (2000) advocate the use of team approaches in every care process. They identify the team approaches as significant contributions to improving patient care and to decreasing the number of patient care errors. The teamwork approach:

> teaches team members to actively coordinate and support each other in the course of clinical task execution using the structure of work teams. Teams and teamwork behaviors do not replace clinical skills but rather serve to ensure that clinical activities are properly integrated and executed to deliver effective . . . care. Teamwork is a management tool to expedite care delivery to patients, a mechanism to give caregivers increased control over their constantly changing environments, and a safety net to help protect both patients and caregivers from inevitable human failings and their consequences. . . . The goal of each core team is to deliver high-quality clinical care to the set of patients assigned to it. To achieve this goal, team members coordinate directly and repeatedly with each other to ensure proper and timely clinical task execution and to detect and help overloaded teammates. Each team member works to maintain a clear understanding (a common situation awareness) of the care status and care plan for each patient assigned to the team and the workload status of each team member. Teams hold brief meetings to make team decisions, assign/reassign responsibilities and tasks, establish/reestablish situation awareness, and learn lessons. The team oversees and directly manages the use of all care resources needed by the patient assigned to the team (Risser et al. 2000, pp. 241–42).

Regulatory and licensing agencies usually define the items required in an assessment at any particular healthcare or medical site. Many agencies also define the degree and experience required of team members to perform an adequate assessment. One of the key requirements is that only educated, trained, licensed, and competent caregivers are allowed to perform patient assessment. The team and patient establish priorities for treatment. If this is done at an early time in treatment, it is classified as an initial treatment plan. This plan is usually a blueprint for a more complex plan that will follow and addresses the most emergent need of the client. An initial problem list is also developed to reflect the most critical problems to be managed. Some treatment issues may be delayed or postponed on the basis of urgency or immediacy of care needs. For example, a patient admitted to an emergency department with an actively bleeding injury and diabetes may undergo treatment for the acute injury first and an evaluation of the diabetes later. The diabetes may be listed as a problem on the treatment plan and documented as deferred to a later time. Goals and interventions are developed by the team to address the patient's need for services and to evaluate clinical improvement. Staff are assigned to each goal, and a time frame for completion or frequency of intervention is identified.

Figure 10.1.   Sample Care Pathway
BACK PAIN (Prolapsed Intervertabral Disc) PID
CARE PATHWAY: BACK PAIN (Prolapsed Intervertabral Disc) (PID)
ADMISSION DAY
MEDICAL
Medical assessment
Explain procedure to be carried out
Request investigations as required
Prescribe usual medication
Prescribe medication of consultant's preference, e.g., muscle relaxants, NSAIDs, anticoagulant, and opiates, if required
NURSING
Nursing assessment
TPR and BP record 4 hly. daily*
Urine analysis
Explain team nursing
Discuss discharge requirements
Refer to OT if required   Yes/No*
Refer to SW if required   Yes/No*
Explain how pain will be dealt with
Assess for self-medication
Complete manual handling assessment
Assess pressure sore risk using Waterlow
Explain ward layout, including location of fire exits
Provide ward information booklet
Provide identity bracelets
Give patient time to ask questions about condition
Maintain bed rest—as flat as possible or in a semi-recumbent position
May be allowed up to the toilet if the consultant permits
Assist with hygiene needs as required
Assist with elimination needs as required
Explain importance of movement in reducing the risk of chest infections, DVT, and pressure sores while on bed rest
Check circulation, sensation, and movement (CSM) to lower limbs
Assess patient's pain score; if > 2 give analgesia
Give medications as prescribed
Give normal diet and fluids—assist as required
Provide environment conducive to sleep

(Continued on next page)

Figure 10.1.   *(Continued)*
CARE PATHWAY : BACK PAIN (PID)
OUTCOMES
1. Plan of care
2. Discharge date
3. Pain control
4. Ability to pass urine
5. CSM intact to lower limbs
ACUTE PHASE
MEDICAL
Review patient daily
Review test results
Attend to any problems that may arise
Request further investigations as required
NURSING
Maintain Safe Environment
The patient should be given a firm bed to fully support the bony framework
If possible, the patient should remain as flat as possible or in a semi-recumbent position, if tolerable, to help prevent complications of the chest from occurring
Anti-embolic stockings should be applied as per consultant protocols to help reduce the risk of deep vein thrombosis. The stockings should be changed on alternate days
Medications should be given as prescribed
Assess patients pain score; if > 2 give analgesia
Record TPR daily to aid assessment of general condition
Check CSM to lower limbs
Ensure patient able to pass urine
Breathing
The nurse should check that the patient is able to breathe freely and deeply
Deep breathing exercises should be encouraged to help reduce the risk of chest infection
Communicating
The patient and family should be kept informed of procedures and investigations to be performed and explanations for these procedures should be given to the patient or family
Allow the patient opportunity to express worries or fears
Eating and Drinking
Meals should be served in a way that the patient can easily handle; for example, the food should be cut up into small helpings to aid digestion
A high-fiber diet and adequate fluid intake should be encouraged to reduce risk of constipation
Flexible straws or feeder cups should be provided to help maintain independence

Figure 10.1.   *(Continued)*

CARE PATHWAY : BACK PAIN (PID)

Eliminating

Give assistance as required

Provide urinary bottles/bed pans

Encourage fluids

If allowed by the consultant and the patient is able, he or she may be allowed up to the commode or toilet

Personal Cleansing and Dressing

Encourage patient to be as independent as possible

Give assistance as required, while at the same time ensuring that privacy is maintained

Facilities should be made available for cleaning teeth and hair washing

Ability to Maintain Body Temperature

Measure patient's temperature daily

Observe for signs of abnormal temperature, e.g., feel and color of skin, lethargy, and so on

Mobility

Maintain bed rest during acute phase

Assess patient's skin for abnormal signs, e.g., discoloration, sores, loss of sensation

Assess the patient in regard to pressure sore formation using Waterlow

Encourage movement (as bed rest and pain allow) to reduce the risk of pressure sore formation

Resting and Sleeping

Provide environment conducive to sleep

Give sedation if required and prescribed by doctor

OUTCOMES

1. Plan of care

2. Tolerates diet and fluids

3. *Pain control*

4. *Ability to pass urine*

5. Temperature 38° C or below

6. Patient has bowel movement

7. CSM intact to lower limbs

REHABILITATION PHASE MEDICAL

Review patient daily

Attend to any problems that may arise

Assess patient's condition—fit for discharge

Prescribe TTOs and request outpatient appointment

(Continued on next page)

Figure 10.1.   *(Continued)*

CARE PATHWAY : BACK PAIN (PID)
NURSING
Maintaining Safe Environment
Allow patient to mobilize as pain allows
Advise patient to rest on bed for periods
Advice re: back care should given in readiness for discharge
Medications should be given as prescribed
Assess patient's pain score; if > 2 give analgesia
Record TPR daily
Communicating
The patient and family should be kept informed of plans and explanations of these plans should be provided to patient and family
Give the patient opportunity to express worries or fears
Eating and Drinking
A high-fiber diet and adequate fluid intake should be encouraged to reduce risk of constipation
Eliminating
Encourage independence
Advise patient to mobilize to the toilet
Advise dietary intake
Personal Cleansing and Dressing
Encourage patient to be as independent as possible; allow patient to take shower if patient prefers
Give assistance as required; ensure that privacy is maintained
Facilities should be made available for cleaning teeth and hair washing
Working and Playing
Assess the need for a sick note upon discharge
Discuss with patient whether their work may have been a contributory factor to their back pain
Resting and Sleeping
Provide environment conducive to sleep
PHYSIOTHERAPY
Advise patient on safe practices in order to reduce the risk of another occurrence of back pain
Provide written advice
Organize follow-up physiotherapy appointment, if appropriate

Figure 10.1.   *(Continued)*
CARE PATHWAY : BACK PAIN (PID)
OUTCOMES
Plan of care
Tolerates diet and fluids
Mobilizes within own limitations
Urinates without difficulty
Pain is controlled
No signs of DVT
Patient has bowel movements
Discharge plan
Independent within own limitations
Apyrexial
CSM intact to lower limbs
DISCHARGE DAY
MEDICAL
Review patient; ensure fit for discharge
Request outpatient appointment
NURSING
Assist with hygiene needs as required
Assist with dressing as required
Daily TPR
Complete Discharge Plan
Ensure stairs practiced safely
Give written discharge advice
Explain contents if necessary
Advise re: routine precautions
PHYSIOTHERAPIST
Ensure independent mobility
Encourage home exercises and give written reminders
OUTCOMES
Patient demonstrates understanding of discharge plan
Patient discharges safely home or to appropriate place of care
Adapted from National Institutes of Health. National Electronic Library for Health: Back Pain (Prolapsed Intervertebral Disc) PID. Accessed online at http://www.nelh.shef.ac.uk/nelh/kit/cps/paths.nsf/Lookup/Backpain67/$file/Backpain67.doc.

Once the care pathway has been developed, implementation begins. The patient's status is continuously monitored for signs of stabilization, improvement, or destabilization. Revisions of the care pathway and treatment interventions are developed in response to changes in the patient's status. The frequency of revision in the treatment plan is often determined in facility policy and developed from licensing agency regulations. Revisions in the plan are also completed based upon the path of clinical status for the client.

Most treatment/care teams meet on a regular basis to evaluate patient data, interventions, and improvement. The flow of patient care moves in the following way:

- Assessment to treatment planning

- Treatment planning to care or service

- Care or service to reassessment

- Reassessment to continuation of care when improvement is demonstrated or reassessment to new interventions when improvement is not demonstrated

As the patient's care proceeds, some key care processes are implemented. These key care processes include such services as laboratory, radiology, pharmacy, dietary, nursing, and physical and respiratory therapy.

The eventual outcome of this flow of care is discharge to the patient's home or to a different care setting. (See figure 10.2.) The coordination of patient services and care is often discussed among care team members. Patients should be involved in this process and should be actively encouraged to take part in the planning of their own care. The strengths and limitations of the patient should be considered in the development of the treatment plan and in completion of the goals. Care pathways need to reflect the cultural values that impact a client and family in the care setting as well. For instance, in some Asian countries, certain types of cold and hot foods must be served to a newly delivered mother to ensure her health and well-being. Making dietary accommodations to meet an Asian client's needs can ease the transition and help the client's perception of holistic individualized care. Family members and other significant persons should be encouraged to become involved and should be educated about the patient's treatment process and care needs.

Optimizing Patient Care: Steps to Success

The treatment and care of patients in a healthcare system can be measured in many different ways. The different measures can be used to identify problems, to demonstrate compliance with regulations, or to reflect improvement in patient care processes. Some problems, however, are more easily measured than others. For example, it is easier to measure the number of patients who receive a presurgical dose of antibiotics to prevent postoperative wound infection than it is to measure that particular intervention's effect on the occurrence of infection in an individual. Different people have different responses to medication interventions. Such differences create variations in the overall response to medication. The focus of improvement processes, therefore, must be tied to patient-specific data about the care processes being provided at any given facility.

Several common improvement processes are discussed in the following steps.

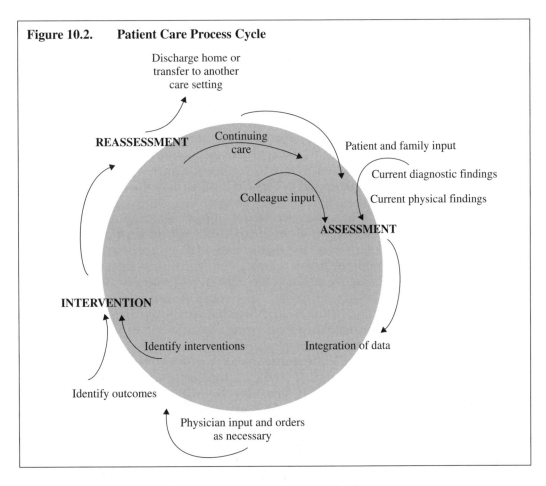

Figure 10.2.   Patient Care Process Cycle

Step 1: Conduct evaluations of the organization's medication systems and processes

Medication systems and processes should be evaluated to ensure the safe and effective use of medications in any facility that provides medications to patients. Methods for prescribing, ordering, and securing medications should be defined in policy and procedure and monitored for compliance. Patient and family education regarding medication use, contraindications, and side effects are other areas for monitoring, as medication errors are the most widely publicized area of medical malpractice. The Institute of Medicine (IOM) reported in a November 1999 report that besides "the cost in human lives, preventable medical errors exact other significant tolls. They have been estimated to result in total costs (including the expense of additional care necessitated by the errors, lost income and household productivity, and disability) of between 17 billion and 29 billion per year in hospitals nationwide" (IOM 1999). The cost of extended stays, corrective care, and litigation are extensive. The wrong medication to a particular patient could be deadly: "Most errors are caused by faulty systems, processes, and conditions that lead people to make mistakes or fail to prevent them" (IOM 1999). The goal of IOM is to reduce medication errors over the next five years by 50 percent. The guidelines for proper medication practice include the five rights of medication management: the right patient, the right medication, the right dose, the right route, and the right time. One of the new safety standards for JCAHO outlines the need for proper patient identification procedures, especially in medication management.

Pharmacy and therapeutics (P and T) monitor the prescribing, ordering, preparation, dispensing, and storage of medications. These pharmacy-related functions may be coordinated through a committee (commonly referred to as the **P and T committee**) or independent of a committee by the pharmacy department. Another important function includes managing the organization's **formulary,** which specifies the drugs approved for use in the organization. The selection of items to be included in the formulary is based on objective evaluation of relative therapeutic merit, safety, and cost, and to minimize duplication of the same basic drug type, entity, or product. The kinds of medication that are monitored depend on the care setting and type of services provided.

The P and T committee or pharmacy department (if no committee structure is defined for this function) plays an advisory role for the medical staff and recommends policies related to medication and other therapies. Its recommendations have far-reaching therapeutic and economic effects for the healthcare organization. This committee is the organizational link between the medical staff and the pharmacy department. In addition, the committee develops educational programs on issues related to medications and medication use to meet the needs of the professional staff. Annual competency testing for medications may be a part of the credentialing process for professionals or part of the annual review process for licensed staff in a facility.

The composition of a P and T committee varies from organization to organization. Membership, however, usually consists of physicians, nurses, administrators, and pharmacists. The committee may meet quarterly, bimonthly, or monthly depending on the size of the healthcare organization and the needs of the medical staff. Its meetings, minutes should be maintained in a permanent record. The functions and scope of this committee have been described by the American Society of Health-System Pharmacists (ASHP). According to the ASHP, the committee's functions include the following:

- Serving in an advisory capacity to the medical staff and administration in all matters pertaining to the use of drugs, including investigational drugs

- Developing and regularly revising a formulary of drugs accepted for use in the organization

- Establishing programs and procedures that help ensure cost-effective drug therapy

- Establishing or planning suitable educational programs for the organization's professional and nonprofessional staff on matters related to drug use such as side effects and adverse drug reactions

- Participating in PI activities related to the distribution, administration, and safe use of medications, including the handling of medication errors and patient or staff complaints

- Reviewing all adverse drug reactions

- Initiating or directing drug use review programs or studies and reviewing the results of such activities

- Advising the pharmacy in the implementation of effective drug distribution and control procedures, especially for schedule I and II narcotics

- Making recommendations to determine which drugs should be stocked in the facility's patient care areas

Developing an effective formulary is a significant responsibility. With the number of drugs available, every organization should have a program for objectively evaluating and selecting the drugs to be included in its formulary. Limiting the number of drugs routinely available from the pharmacy produces substantial financial and patient care benefits. Because of the changing nature of healthcare delivery, the P and T committee can serve as an effective bridge between the clinical factors and economic consequences related to drug therapy.

For example, the pharmacy and therapeutics program of Western States University Hospitals and Clinics has two major components. The first component is the drug usage evaluation (DUE), and the second is its therapeutic interchange program.

The DUE is a systematic, criteria-based, ongoing function of the medical staff. It evaluates how drugs are used at the University Hospital in four key areas:

- Medication ordering, including appropriate indications and dosing

- Medication preparation and dispensing

- Medication administration

- Monitoring of the effects of medications and their therapeutic outcomes

Drug use evaluation projects can be either disease focused or drug specific. Projects are usually selected on the basis of the following criteria:

- The drug is less effective than another drug that can be used to treat the same condition; the physician and pharmacist may identify a more effective drug therapy.

- The drug is considered high risk or problem prone. Drugs may fall into this category because they have an adverse reaction profile or because certain populations have additional risk.

- The drug is frequently prescribed.

- The drug is costly.

The DUE subcommittee at the Western States University Hospitals and Clinics uses several screening mechanisms to monitor drug use and identify issues that may require additional evaluation. These mechanisms include the following:

- Receiving information from other PI activities including the infection committee, critical care committee, and risk management committee

- Reviewing incident reports and medication errors

- Reviewing all nonformulary drug requests

- Reviewing all adverse drug reactions reported
- Reviewing drug purchases
- Maintaining the hospital formulary

The major component of the P and T PI program at Western States is the investigation of **adverse drug reactions** (ADRs) to improve the use of drugs and prevent adverse reactions. An ADR is a detrimental response to a medication that is undesired, unintended, or unexpected in dosages recognized in accepted medical practice. Adverse reactions that should be reported to the P and T committee include those that prolong the length of a patient's hospital stay, cause discontinuation of drug therapy, or provoke a change in drug therapy requiring corrective measures such as antidotes.

The committee should also be concerned with **medication error** rates. The committee monitors the frequency with which clinicians administer medications to the wrong patient. It also monitors food–medication or medication–medication interactions that produce symptoms in patients.

In managing the facility's formulary, the Western States P and T committee implemented a program known as therapeutic interchange. Therapeutic interchange is the authorized exchange of therapeutic alternatives in accordance with previously established and approved guidelines or protocols within a formulary system. The following activities summarize the therapeutic interchange process.

The P and T committee designates a group of drugs as therapeutically equivalent and appropriate for therapeutic interchange after seeking feedback from the medical staff. The committee approves guidelines for interchanging the drugs. The pharmacy then evaluates the cost of using each drug and determines which drug will be the drug of choice. Factors in these analyses include the drug cost, special contract price incentives, and other costs associated with use of the drug.

Drugs included on the interchange program are listed in the computerized formulary, and notations are made to indicate the drug of choice for a given group of drugs. When the pharmacy receives an order for any drug in the group, the pharmacist substitutes the selected formulary agent according to the guidelines established by the P and T committee. Prescribers are notified of the change in drug by interchange. Prescribers may order an alternative agent within the group and document a reason why the patient cannot receive the selected agent through interchange.

Step 2: Conduct evaluations of the organization's use of seclusion, restraints, and protective devices

The use of seclusion, restraints, and protective devices requires intensive performance monitoring. Protective devices include wrist restraints, jacket restraints, chairs with restraining tables, restraints to stabilize a patient's body during surgery, and side rails on hospital beds. Protective devices prevent something from happening. One example is the use of head stabilization devices for dental procedures. The JCAHO refers to the use of any of these devices as **special treatment procedures** (STPs).

The use of special treatment procedures involves an increased risk of client deaths with pursuant legal risks. Monitoring should assess compliance with physician's orders, mandated maximum time limits of two hours for children and two to four hours for adolescents and adults, physical assessment during STPs every two hours, and an appropriateness assessment by a physician within one hour of initiation of the order. Patients must be checked every fifteen minutes during the procedure. Nourishment and bathroom privileges must be offered at least every two hours during the procedure. Clients must remain in the direct line of sight of supervising clinicians.

Independent licensed practitioners should be credentialed for the privilege of assessing and applying STPs. Continued training and documentation of staff competency for special treatment procedures and protective devices must be documented. It is also very important to document leadership education and commitment to the reduction of the use of STPs facilitywide. The primary focus of training for facilities includes de-escalation techniques and safety procedures. The JCAHO wants to reduce the frequency of STPs across all facilities. The JCAHO is particularly concerned about the use of STPs and protective devices in long-term care and rehabilitation settings because the procedures may infringe on patients' rights.

Step 3: Conduct evaluations of laboratory services and the use of blood products

The organization's compliance with established standards related to the use of laboratory equipment and the handling of laboratory specimens must be monitored. Laboratory services are regulated by established protocols from the Clinical Laboratory Improvement Amendments (CLIA) and the Centers for Disease Control. Equipment calibration and other parameters of laboratory values must be monitored daily.

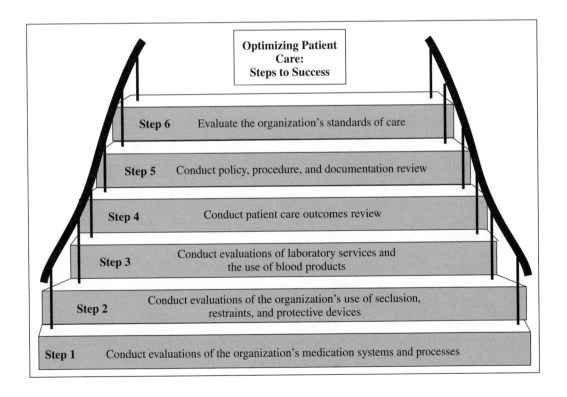

Optimizing Patient Care: Steps to Success

Step 6   Evaluate the organization's standards of care

Step 5   Conduct policy, procedure, and documentation review

Step 4   Conduct patient care outcomes review

Step 3   Conduct evaluations of laboratory services and the use of blood products

Step 2   Conduct evaluations of the organization's use of seclusion, restraints, and protective devices

Step 1   Conduct evaluations of the organization's medication systems and processes

The handling of blood and blood products for transfusions is also regulated and monitored. Measuring, assessing, and improving the ordering, typing, matching, dispensing, and administering of blood and blood products are a standard part of continuous monitoring for most clinic and hospital settings. The review process seeks to validate the need for transfusion, the use of the appropriate type of blood product, and effective procedures for blood product administration.

The cause of every **transfusion reaction** must be investigated. It has been cited that in eight deaths resulting from hemolytic transfusion reactions, all were primarily attributable to incomplete patient/blood verification. Three cases were a result of improper handling and processing of samples for many patients at the same time in the same location. Some risk reduction strategies identified through the root-cause analysis of these events were in-service training on transfusion-related processes, revision of staffing models for these work areas, improved patient identification processes for blood verification, and environmental redesign to accommodate fewer specimens in one location and procedures to restrict simultaneous cross-matching of many patients (JCAHO 1999b). The JCAHO now considers hemolytic transfusion reactions to be sentinel events, and, as such, they require in-depth analysis, data collection, and a written communication to the JCAHO concerning actions taken to prevent such occurrences in the future.

Step 4: Conduct patient care outcomes review

Patient care outcomes are reviewed to improve the safety and quality of care as well as to identify issues related to medical necessity for treatment and appropriateness of care. The most recent evolution of outcomes review initiated by the JCAHO for accreditation purposes is to require healthcare organizations to implement recommendations of findings from root-cause analyses related to recurring sentinel events. For example, medication errors related to a wrong medication administered to the wrong patient have helped identify (through root-cause analyses) the need to improve the accuracy of patient identification by requiring healthcare organizations to use at least two patient identifiers before administering a medication. Another outcomes review process related to high-risk procedures and occurrences (introduced by the JCAHO in 1997) is the ORYX initiative. This initiative has evolved to include core measure sets that are based upon the services the healthcare organization provides. The intent of requiring organizations to collect performance data on their outcomes of care allows the JCAHO—and the healthcare organization—to review and compare data trends and patterns in treatment among like organizations to help improve patient care processes.

For example, one core measure is Community-Acquired Pneumonia (CAP). The set of treatment requirements associated with this measure is as follows (National Quality Management Program 2002):

- Oxygenation assessment within 24 hours of facility arrival

- Inpatients screened for and/or given pneumococcal vaccination

- Blood cultures obtained prior to initial antibiotic administration

- Documentation of time from first hospital arrival to first dose of antibiotic

- Adult CAP smoking-cessation counseling

- Pediatric CAP smoking-cessation counseling

Monitoring of each of these items would be collected and reported through the PI system in the facility as well as to the JCAHO. The JCAHO uses the core measure data as part of its scoring mechanism for accreditation of facilities. There are other core measures for pregnancy and related conditions, heart failure, and acute myocardial infarction. Accredited facilities were asked to select two of four core measure sets to monitor and begin reporting by July 2002. JCAHO is continuing to develop core measures and is currently working on an intensive care unit (ICU) core measure set.

Step 5: Conduct policy, procedure, and documentation review

The development of policies on standard practices in the facility should be multidisciplinary in nature and design. Most facilities operate within a standard set of policies that have been developed by a multidisciplinary team of clinical and administrative professionals who meet regularly. Updates and revisions of policies are undertaken as national standards of care change. The governing board and leadership of an organization hold the ultimate responsibility for the services provided in the facility and generally recommend changes and set time frames for the review of every policy and procedure to be followed. Some facilities operate with a separate policy and procedure committee.

A key quality performance concern is the adequate and reliable documentation of care. Poor documentation leads to the largest number of risk management and legal situations in the industry. Accreditation and licensing agencies have standards on the documentation of patient care and expect that a sample of clinical documentation will be reviewed regularly as part of an organization's PI activities. The expectation is that all records are complete in terms of authentication and necessary reports and that they appropriately document the condition and treatment of the patient. An example of this type of standard would be the time line requirement for signing off on a verbal order from a physician to a nurse in the client record. Each licensing agency will have a directive regarding this standard. Verbal order sign-offs for seclusion and restraint orders must be completed by a physician within 24 hours of the order. Facility policies will direct the time frame for other verbal order sign-offs based on the facility and type of care provided.

Step 6: Evaluate the organization's standards of care

In order for a healthcare organization to define optimal care, it must first establish **standards of care** and care policies. Some healthcare organizations have moved from a policy and procedure format to a **clinical practice standards** model. This model defines practice based on diagnosis. The flow of treatment interventions and the patient's progress are evaluated on the basis of nationally accepted standards of care for the diagnosis. As each standard is developed and approved, a baseline for performance in the healthcare setting develops. Variations from the standards of care, sentinel events, and high-risk, problem-prone activities must be examined. Action plans are then developed to improve care in areas identified through the monitoring process. A decline in performance or a lack of improvement may require further evaluation of the processes and redesign of care processes.

Ongoing Developments

It is important to recognize that the PI processes discussed in the preceding section are part of a continuum of development in the evaluation of patient care. This continuum of development began, effectively, with the initiation of accreditation and standardization programs decades ago, and will continue in the future.

The principal reason for this long continuum of development is that the evaluation of patient care across all settings is extremely difficult. The expectation is that healthcare will be individualized because what works well for one patient may not work well for another. Some healthcare researchers, particularly those working in the federal government, have spent millions of dollars and years of research to develop **clinical guidelines** that attempt to standardize the care of a single condition across the entire country. Many clinical practitioners, however, find the guidelines difficult to implement, or even contraindicated in some cases, because of comorbid conditions or social ramifications in a patient's clinical presentation. (The use of **critical paths** was discussed in more detail in chapter 7.)

Other healthcare researchers have developed the concept of **evidence-based medicine.** Evidence-based medicine attempts to identify the care processes or interventions that achieve the best outcomes in different types of medical practice. Researchers perform large, population-based studies. Such studies, however, are difficult to do without a well-developed information infrastructure to provide data for analysis. The United States does not yet have a well-developed information infrastructure.

Another group of researchers developed the concept of **indicators** to point healthcare organizations toward areas where improvement is necessary. One example of this approach is the JCAHO's ORYX program. In this approach, JCAHO staff attempted to define indicators for all of the care situations commonly encountered in U.S. healthcare. The JCAHO then mandated that healthcare organizations participate in some kind of aggregate indicator comparison program so that data could be drawn from multiple organizations across the country. Ideally, this approach will facilitate the identification of effective patterns of care over the next decade.

The National Committee for Quality Assurance (NCQA) began accrediting managed care organizations in 1991 in response to the need for standardized, objective information about the quality of the services provided by managed care organizations. The NCQA introduced the Health Plan Employer Data and Information Set (HEDIS) in the early 1990s as a means of gathering information about care, outcomes, and member satisfaction with managed care organizations and other health plans. HEDIS gathers a significant amount of information about the ambulatory care experiences of millions of health plan members from across the country. Specifically, HEDIS gathers data in the following areas:

- Measures of quality, such as immunization, cholesterol screening, mammography, and prenatal care

- Measures of access, with at least one visit to a provider within three years used as an indicator of assessment of healthcare need

- Measures of membership, with particular attention to coverage cancellation as an indicator of dissatisfaction

- Measures of utilization, including factors such as high-occurrence, high-cost diagnosis-related groups; frequency of procedures; general hospital acute care; outpatient and emergency visits; cesarean section rates; complicated neonatal care; and outpatient drug utilization

- Measures of financial performance, such as cost per member, cost per member plus dependents, and indicators of financial stability

The NCQA reports these findings to employers, who use the information to make decisions about contracts with health plans. In this way, the NCQA influences the kind of care offered by managed care plans and provides consumers with information about the health-care offerings of different plans.

Another new approach to monitoring care and identifying opportunities for improvement within healthcare organizations has arisen in the long-term care setting. In June 1998, the federal government mandated the use of the **Minimum Data Set (MDS) for Long-Term Care** to plan the care of long-term care residents. This data set structures the assessment of long-term care residents in the following areas:

- Cognitive patterns
- Communication and hearing patterns
- Vision patterns
- Physical functioning and structural problems
- Continence
- Psychosocial well-being
- Mood and behavior patterns
- Activity-pursuit patterns
- Disease diagnoses
- Other health conditions
- Oral/nutritional status
- Oral/dental status
- Skin condition
- Medication use
- Treatments and procedures

The federal government requires that long-term care facilities receiving Medicare or Medicaid funding transmit the patient-specific data to the state departments of health for processing and use in the long-term care certification and survey review process. The certification and survey review process is carried out by the state departments of health on behalf of the federal government to certify that facilities receiving Medicare or Medicaid funds are complying with federal regulations. The departments of health pay special attention to data on the occurrence of decubitus ulcers in low-risk patients (those who can ambulate, can turn over in bed, are not cognitively impaired, and so on), dehydration, and fecal impaction.

On the basis of the data gathered via the MDS, the facility is provided a **facility quality indicator profile** that shows what proportion of the facility's residents have deficits in each area of assessment during the reporting period and, specifically, which residents have which deficits. The profile also provides data comparing the facility's current experience with its preestablished comparison group. Data from the facility quality indicator profile are also forwarded to the Centers for Medicare and Medicaid Services. CMS has developed and maintains data for the Outcomes and Assessment Information Set (OASIS), which is used in home health agencies. The OASIS lists a core of items for the comprehensive assessment of an adult home care patient. The data also help measure patient outcomes for outcome-based quality improvement. Home health agencies are required to electronically submit these data to the state standard system and they will become part of the Conditions of Participation for CMS. An example of a facility quality indicator profile is provided in figure 10.3.

Real-Life Example

A typical example of an improvement effort might begin with an assessment of an over-weight adolescent who is being treated for psychosis with the antipsychotic drug Zyprexa. One side effect of Zyprexa is a decrease in satiety factors in the brain, which creates a sense of hunger even when adequate food is provided. Upon assessment of the patient based on national standards for height and weight of adolescents, the patient might be identified as being in a high-risk category for obesity. An evaluation of the causative factors might reveal a genetic component, a disease component, medication side effects, poor personal habits, or knowledge deficit regarding healthy nutrition. A multilayered action plan would be initiated after an investigation of all aspects of the contributing factors. This plan might include a nutritional consult, a dietary regimen, a medication evaluation, or a psychiatric consult. At some point in the process, an activity therapist might be consulted to direct physical exercise as an intervention to increase and support muscle strengthening as weight is lost. This example of the PI process demonstrates how the process may be individualized for a patient.

A systemwide PI measure could be instituted in a care setting in regard to the problem of obesity in adolescents. If treating physicians noted significant weight increase over a brief period of time in many adolescent patients started on medications with the side effect of promoting weight gain, they might standardize their case management. Baseline evaluations of weight and height could be mandated for all patients and for all patients started on this type of medication. Significant weight gains could be tracked, and interventions could be instituted for the most effective outcomes. The outcomes may lead to a change in treatment that would lead physicians to order dietary consults and weight monitoring for all patients taking this medication. Other preventive measures could be instituted early in treatment to prevent excessive weight gain and alleviate the patient's risk of cardiac disease related to obesity. The potential outcome from this PI process could lead to healthier patients who require fewer medical services in the future.

QI Toolbox Technique

A common toolbox technique applicable to the improvement of patient care is the patient care outcome review criteria set. Criteria sets are used to identify opportunities for improvement with respect to the appropriateness, processes, and outcomes of medical care and surgical procedures in healthcare facilities.

The JCAHO Core Measures are an example of criteria sets that healthcare organizations use for ongoing patient care monitoring. The list of JCAHO Core Measures includes: Acute Myocardial Infarction, Heart Failure, Community-Acquired Pneumonia, and Pregnancy and Related Conditions. In figure 10.4, the JCAHO Core Measure data set for heart failure is listed. Cases are identified using the ICD-9-CM diagnosis codes, and then each data measure is captured in a retrospective review of the patient health record. These data are then monitored and rated, ultimately for improvement in patient care outcomes.

Figure 10.3.   Example of a Long-Term Care Facility Quality Improvement Profile

Facility Characteristics

Run Date: 2/1/03   12:36:15 p.m.   **Report Period:** 8/1/02 to 1/31/03
Facility: Western Gardens   **Date Submitted:** 1/31/03
Comparison Group Used: **Facility Login ID:** AT 4763
    All State Facilities,
    October–December 1998

Resident Population	Number of Residents	Facility Percentage*	Comparison Group Percentage*
Gender			
Male	19	28.8	33.7
Female	47	71.2	66.2
Age			
>25 years	1	1.5	0.5
25–54 years	4	6.1	7.5
55–64 years	2	3.0	6.5
65–74 years	13	19.7	14.8
75–84 years	23	34.8	32.7
84+ years	23	34.8	38.0
Payment source (all that apply)			
Medicaid per diem	48	72.7	45.1
Medicare per diem	11	16.7	22.1
Medicare ancillary part A	18	27.3	19.2
Medicare ancillary part B	6	9.1	5.8
Self-pay/family-pay per diem	4	6.1	16.3
Medicaid resident liability			
or Medicare copayment	1	1.5	6.1
Private insurance per diem	3	4.5	8.1
All other per diem	1	1.5	2.6
Diagnostic characteristics			
Psychiatric diagnosis	8	12.1	9.7
Mental retardation	2	3.0	2.5
Hospice	0	0.0	0.7
Type of assessment			
Admission	12	18.2	34.0
Annual	7	10.6	11.0
Significant change in status	0	0.0	4.6
Significant correction of prior			
full assessment	1	1.5	0.5
Quarterly	42	63.6	49.7
Significant correction of			
prior quarterly	4	6.1	0.2
All other	0	0.0	0.0
Stability of conditions			
Conditions/disease make resident unstable	5	7.6	35.6
Acute episode or chronic flare-up	2	3.0	4.6
End-stage disease, ≤6 months to live	2	3.0	1.3
Discharge potential			
None	44	66.7	56.6
Within 30 days	2	3.0	15.3
Within 31–90 days	5	7.6	3.9
Uncertain	14	21.2	21.6

(Continued on next page)

Figure 10.3.   (*Continued*)

Facility Quality Indicator Profile	
Run Date: 2/1/03 12:36:15 p.m.	**Report Period:** 8/1/02 to 1/31/03
Facility: Western Gardens	**Date Submitted:** 1/31/03
Comparison Group Used:	**Facility Login ID:** AT 4763
All State Facilities, October–December 1999	

Domain/Quality Indicator	Number in Numerator	Number in Denominator	Facility Percentage*	Comparison Group Percentage*	Percentile
Accidents					
1. Incidence of new fractures	0	53	0.0	1.1	0
2. Prevalence of falls	1	54	1.9	14.4	0
Behavioral/emotional patterns					
3. Prevalence of behavioral symptoms affecting others	13	54	24.1	28.2	41
High risk	10	40	25.0	32.7	34
Low risk	3	14	21.4	16.9	70
4. Prevalence of symptoms of depression	8	54	14.8	21.1	42
5. Prevalence of symptoms of depression without antidepressant therapy	1	54	1.9	8.7	18
Clinical management					
6. Use of nine or more medications	21	54	38.9	37.7	56
Cognitive patterns					
7. Incidence of cognitive impairment	1	14	7.1	12.5	41
Elimination/incontinence					
8. Prevalence of bladder or bowel incontinence	27	51	52.9	52.0	51
High risk	13	13	100.0	88.6	100
Low risk	14	38	36.8	39.9	38
9. Prevalence of occasional or frequent bladder or bowel incontinence without a toileting plan	9	18	50.0	55.7	40
10. Prevalence of indwelling catheter	3	54	5.6	4.9	60
11. Prevalence of fecal impaction	0	54	0.0	0.7	0
Infection control					
12. Prevalence of urinary tract infection	0	54	0.0	0.7	0
Nutrition/eating					
13. Prevalence of weight loss	1	54	1.9	2.5	54
14. Prevalence of tube feeding	1	54	1.9	2.5	54
15. Prevalence of dehydration	0	54	0.0	0.9	0
Physical functioning					
16. Prevalence of bedfast residents	1	54	1.9	6.1	29
17. Incidence of decline in late-loss activities of daily living	3	43	7.0	14.7	23
18. Incidence of decline in range of motion	7	47	14.9	9.8	79

Figure 10.3.   (*Continued*)

Domain/Quality Indicator	Number in Numerator	Number in Denominator	Facility Percentage*	Comparison Group Percentage*	Percentile
Psychotropic drug use					
19. Prevalence of antipsychotic drug use in the absence of psychosis or related conditions	8	50	16.0	18.7	46
High risk	4	9	44.4	35.4	75
Low risk	4	41	9.8	13.9	38
20. Prevalence of antianxiety/hypnotic use	9	50	18.0	18.6	61
21. Prevalence of hypnotic use more than two times in past week	1	54	1.9	3.4	50
Quality of life					
22. Prevalence of daily physical restraints	11	54	20.4	9.4	91
23. Prevalence of little or no activity	33	54	61.1	36.1	86
Skin care					
24. Prevalence of stage 1–4 pressure ulcers	4	54	7.4	7.9	65
High risk	4	23	17.4	13.5	81
Low risk	0	31	0.0	2.8	0

*Percentages may not total 100 owing to missing data.

Note: Original form designed by the Center for Health Systems Research and Analysis at the University of Wisconsin–Madison.

Figure 10.4.	JCAHO Core Measure Criteria Set—Heart Failure	
Performance Measure Name	**Numerator**	**Denominator**
Discharge Instructions	Heart failure patients with documentation that they or their caregivers were given written instructions or other educational material addressing all of the following: activity level, diet, discharge medications, follow-up appointment, weight monitoring, what to do if symptoms worsen	Heart failure patients discharged home
LVF assessment	Heart failure patients with documentation in the hospital record that left ventricular function (LVF) was assessed before arrival, during hospitalization, or is planned for after discharge	Heart failure patients
ACEI for LVSD	Heart failure patients who are prescribed an Angiotensin-converting enzyme inhibitor (ACEI) at hospital discharge	Heart failure patients with LVSD and without ACEI contraindications
Adult smoking-cessation advice/ counseling	Heart failure patients who receive smoking-cessation advice or counseling during the hospital stay	Heart failure patients with a history of smoking ciga-rettes anytime during the year prior to hospital arrival

Source: http://www.jcaho.org/pms/core+measures/. © JCAHO 2002. Used with permission.

Case Study

Students should consider a personal healthcare experience in any setting. They should identify the key processes in relation to the areas discussed in this chapter (for example, pharmacy, medication administration, nutritional assessment, or special procedure).

Case Study Questions

1. How might one monitor the processes the student has experienced? What types of data would be useful? Describe variations and possible causes of variations in the care processes.

2. What was working and what was not working in the processes? Consider inter-viewing healthcare professionals to discuss these issues.

Project Application

Students should consider using a standardized criteria set in evaluating customer satisfaction for the student project.

Summary

The application of performance improvement processes to patient care is varied and multi-faceted. Opportunities for improvement can be identified at many levels, and organizations are limited only by constraints on their creativity and resources. Typical activities focusing on improvement of care include assessment of pharmacy and therapeutics usage; blood products usage; policy, procedure, and documentation; surgical case review; and special treatment procedures. There is also important work being done at the national level in the areas of clinical practice guidelines, clinical paths, evidence-based medicine, indicator monitoring, and data set analysis.

References

Allison, J. J., et al. 2000. The art and science of chart review. *Joint Commission Journal of Quality Improvement* 6(3):173–81.

American Society of Health-Systems Pharmacists. 1999–2000. *Best Practices for Health-System Pharmacy: Position and Practice Standards of the ASHP.* Bethesda, Md.: ASHP.

Ashton, C. M., et al. 1999. An empirical assessment of the validity of explicit and implicit process-of-care criteria for quality assessment. *Medical Care* 37(8):798–808.

Bodenheimer, Thomas. 1999. The American health care system: the movement for improved quality in health care. *New England Journal of Medicine* 340(6):488–92.

Brook, Robert H., Elizabeth A. McGlynn, and Paul D. Cleary. 1996. Measuring quality of care. *New England Journal of Medicine* 335(13):966–70.

Chassin, M. R. 1993. Improving quality of care with practice guidelines. *Frontiers of Health Services Management* 10(1):40–44.

Cohen, M. R., J. Senders, and N. M. Davis. 1994. Failure mode and effects analysis: a novel approach to avoiding dangerous medication errors and accidents. *Hospital Pharmacy* 29(4):319–30.

Cook, Deborah J. 1997. The relation between systematic reviews and practice guidelines. *Annals of Internal Medicine* 127(3):210–16.

DeBruin, A. F. 1994. The sickness impact profile: SIP68, a short generic version. *Journal of Clinical Epidemiology* 47(8):863–71.

Dreachslin, J. L., P. L. Hunt, and E. Sprainer. 1999. Communication patterns and group composition: implications for patient-centered care team effectiveness. *Journal of Healthcare Management* 44(4):252–66.

Eddy, David M. 1998. Performance measurement: problems and solutions. *Health Affairs* 17(4).

Feinstein, Alvin R. 1995. Meta-analysis: statistical alchemy for the 21st century. *Journal of Clinical Epidemiology* 48(1):71–79.

Freund, Deborah. 1999. Patient outcomes research teams: contribution to outcomes and effectiveness research. *Annual Review of Public Health* 20:337–59.

Greenfield, Sheldon, and Eugene C. Nelson. 1992. Recent developments and future issues in the use of health status assessment measures in clinical settings. *Medical Care* 30(suppl.).

Grimshaw, Jeremy M., and Ian T. Russell. 1993. Effect of clinical guidelines on medical practice: a systematic review of rigorous evaluations. *Lancet* 342:1317–22.

Guyatt, Gordon, and the Evidence-Based Medicine Working Group. 1992. Evidence-based medicine. *Journal of the American Medical Association* 268(17):2420–25.

Iezzoni, Lisa I. 1995. Risk adjustment for medical effectiveness research: an overview of conceptual and methodological considerations. *Journal of Investigative Medicine* 43(2):136–50.

Iezzoni, Lisa I. 1997. The risks of risk adjustment. *Journal of the American Medical Association* 278(19):1600–1607.

Institute of Medicine. 1999. *To Err Is Human: Building a Safer Health System.* Washington, D.C.: National Academy Press.

Johr, J. J., et al. 1996. Improving health care: clinical benchmarking for best patient care. *Joint Commission Journal of Quality Improvement* 22(9):599–616.

Joint Commission on Accreditation of Healthcare Organizations. www.jcaho.org.

Joint Commission on Accreditation of Healthcare Organizations. 1999a. *1999–2000 Standards for Behavioral Health Care.* Oakbrook Terrace, Ill.: JCAHO.

Joint Commission on Accreditation of Healthcare Organizations. 1999b. *Sentinel Event Alert.* Issue 10, August 30.

Keller, R. B., D. E. Wennberg, and D. N. Soule. 1997. Changing physician behavior: the Maine Medical Assessment Foundation. *Quality Management in Health Care* 5(4):1–11.

Laupacis, Andreas, Nandita Sekar, and Ian G. Stiell. 1997. Clinical prediction rules: a review and suggested modifications of methodological standards. *Journal of the American Medical Association* 277(6):488–94.

Lomas, Jonathon, et al. 1989. Do practice guidelines guide practice? *New England Journal of Medicine* 321(19):1306–11.

Longo, Daniel R. 1993. Patient practice variation. *Medical Care* 30(5, suppl.).

National Quality Management Program Presentation/Tricare Conference, Washington , D.C., February 5, 2002.

McHorney, Colleen. 1999. Health status assessment. *Annual Review of Public Health* 20:309–35.

Mosser, G. 1996. Clinical process improvement: engage first, measure later. *Quality Management in Health Care* 4(4):11–20.

Risser, Daniel T., et al. 2000. A structured teamwork system to reduce clinical errors. In *Error Reduction in Health Care,* P. L. Spath, editor. San Francisco: Jossey-Bass Publishers.

Chapter 11
Improving the Care Environment and Life Safety

Learning Objectives

- To identify the link between the environment of care and patient safety

- To list the seven most important areas in managing the care environment and care safety, with special emphasis on conducting a hazard vulnerability analysis to identify and prioritize potential emergencies

- To describe the safety monitoring process

- To identify the importance of regular posteducation assessment and occurrence reporting in documenting performance improvement activities involving the care environment and safety management

Background and Significance

As discussed in chapter 9, healthcare organizations can be dangerous places. They house complex equipment, chemicals, and hazardous instruments and materials that can injure patients, visitors, and staff when used inappropriately. Because of these issues, healthcare organizations must continuously monitor and evaluate their patient environment and the safety of the individuals receiving or providing care services. With the new emergency management standards put in place by the Joint Commission on Accreditation of Health-care Organizations (JCAHO) after September 11, 2001, this issue extends beyond the boundaries of the facility to the community. Public facilities are at greater-than-ever risk from external threats, and the need to have emergency preparations in place for disasters has come to the fore of national concern. Within its Environment of Care Standard EC. 1.4, the JCAHO has redirected its focus in the area of emergency management and offers new guidance for communitywide emergency preparation using both backup and contingency plans (JCAHO 2003a). This could become critical should a national emergency occur, such as a bioterrorist attack using anthrax or smallpox.

Improving Care Environment and Life Safety: Steps to Success

The monitoring and performance improvement (PI) of the patient care environment and safety involve the continuous examination and evaluation of activities in seven important areas (referred to as plans):

- Security management plan

- Hazardous materials and waste management plan

- Emergency management plan, including the four phases of emergency management planning: mitigation, preparedness, response, and recovery

- Fire prevention plan

- Medical equipment management plan, including infusion pumps and clinical alarm systems (JCAHO 2003c)

- Utility management plan

- Safety plan, including worker safety, no-smoking policies, and National Patient Safety Goals related to patient identification processes, high-alert medication management, and communication improvement between caregivers

This chapter discusses each area (program/plan) in terms of scope, goals, and objectives. Common to each plan is an orientation and education program for personnel with their specific roles and responsibilities defined, as well as an annual evaluation of each plan's objectives, scope, performance, and effectiveness.

Step 1: Monitor and improve the security management program

According to JCAHO's Environment of Care Standard EC.1.2, the security management program accords highest priority to the following activities (JCAHO 2003a):

- Addressing security concerns affecting patients, visitors, employees, medical staff, and volunteers, particularly in the event of a security incident or failure, or as necessary in handling situations involving VIPs or the media

- Reporting and investigating all security incidents

- Designating staff responsible for developing, implementing, and monitoring security

- Providing identification for patients, visitors, employees, medical staff, and volunteers, as appropriate

- Controlling access to security-sensitive areas such as cashier's offices, pharmacy services, nursing medication areas, and laboratories where access to money, monetary instruments, or drugs and paraphernalia must be restricted

- Providing worker safety in parking garages, elevators, and other low-traffic areas

- Controlling human and vehicle traffic in and around the environment during disasters

The primary goal of security management activities is to ensure that employees are able to follow established security procedures. Employees must understand the processes in place for minimizing security risks and handling security breaches. Occurrence reports should be tracked carefully to identify incidents in which security policy and procedures were ineffective.

Step 2: Monitor and improve the hazardous materials and waste management program

An organization's hazardous materials and waste management program should be designed to identify all types of materials and waste (specific to the healthcare organization) that require special handling to minimize the risk of injury to patients, visitors, employees, medical staff, and volunteers. According to the JCAHO's Environment of Care Standard EC.1.3, the program should accord highest priority to the following activities (JCAHO 2003a):

- Selecting, handling, storing, transporting, using, and disposing of hazardous materials appropriately and safely

- Identifying, evaluating, and maintaining an inventory of hazardous materials and waste used or generated by the organization

- Managing chemical, chemotherapeutic, radioactive, body fluid, and infectious waste, including sharps such as needles

- Monitoring and disposing of hazardous gases and vapors

- Providing adequate and appropriate space and protective equipment for the safe handling and storage of hazardous materials and wastes

- Reporting and investigating all hazardous material or waste spills, exposure, and other incidents

The hazardous materials and waste management program must provide an integrated, coordinated effort that complies with state statutes and meets the standards of the JCAHO, the National Fire Protection Association (NFPA), the Life Safety Code, and the Occupational Safety and Health Administration (OSHA). The program must work effectively to eliminate risk exposure and maximize loss prevention related to the numerous hazardous materials used in healthcare organizations.

Hazardous materials and waste management activities address employees' knowledge of and ability to perform appropriate procedures. Because a variety of different materials are inventoried, used, and destroyed in many areas of healthcare organizations, department managers must take a leading role in ensuring employee compliance.

Any equipment used in conjunction with hazardous materials and waste management must be inspected, maintained, and tested at regular intervals to ensure proper operation. Occurrence reports should be tracked carefully to identify incidents in which hazardous material and waste management policy and procedures were not followed.

Most important, all employees must know the procedures to follow in the event of actual or suspected exposure to hazardous materials or waste. They must recognize the health hazards of mishandling such materials and understand procedures for reporting exposure. Employees must understand how to clear spilled material from exposed areas and whether protective equipment or special precautions must be taken in the course of cleaning up and disposing of spilled materials.

As backup for employee training, department managers must maintain a **material safety data sheet** (MSDS) on every hazardous material used in their respective departments. Material safety data sheets should include the following information:

- Identification of the material, including its common and chemical names, family name, and product codes

- Risks associated with the material, including overall health risk, flammability, reactivity with other chemicals, and effects at the site of contact

- Descriptions of the protective equipment and clothing that should be used to handle the material

- Precautionary labeling that must appear on containers or storage units

- Physical characteristics of the material, including its boiling and melting points, specific gravity, solubility, appearance and odor, vapor pressure and density, evaporation rate, and physical state

- Information on potential fire and explosion hazards, including the material's flash point and autoignition temperature as well as unusual explosion hazards or toxic gases associated with the material

- Information on appropriate fire-extinguishing media and related special procedures

- Information on specific health hazards associated with the material, including the toxicity of components, carcinogenicity, reproductive effects, effects of overexposure, target organs, medical conditions aggravated by exposure, and routes of entry

- Information on emergency first-aid procedures

- Reactivity data, including storage, use conditions to avoid, and other incompatible materials

- Instructions on the disposal of spilled material

- Industrial protective equipment needed, such as ventilation hoods and respiratory protective devices

- Storage and handling precautions

- Procedures and precautions to be followed when the material is being transported

See figure 11.1 for an example of an MSDS.

Figure 11.1.   Example of a Material Safety Data Sheet

Eli Lilly and Company
Material Safety Data Sheet

MSDS Index

Heparin Sodium Injection
Effective Date: 18-Oct-2002

Section 1—Chemical Product and Company

Manufacturer:
Eli Lilly and Company
Lilly Corporate Center
Indianapolis, IN 46285

Emergency Phone:
1-317-276-2000
CHEMTREC:
1-800-424-9300 (North America)
1-703-527-3887 (International)

Common Name: Heparin Sodium Injection

Chemical Name: Heparin sodium salt
Synonym(s): 017265 Formulation; Heparin Sodium Formulation; Heparin Injection; Heparina Sodica Formulation
Tradename(s): Ariven
Lilly Item Code(s): AM0405; AM0520; AM0622; AM0642; VL0405; VL7323

See attached glossary for abbreviations.

Section 2—Composition / Information on Ingredients

Ingredient	CAS	Concentration %
Heparin Sodium	9041-08-1	0.7–15
Benzyl Alcohol	100-51-6	1
Water	7732-18-5	84–98

Contains no hazardous components (one percent or greater) or carcinogens (one-tenth percent or greater) not listed above.

Exposure Guidelines:
Benzyl alcohol—WEEL 10 ppm (44.2 mg/m3) TWA.

Section 3—Hazards Identification

Appearance: Clear aqueous solution
Physical State: Liquid
Odor: Odorless

Emergency Overview

Emergency Overview Effective Date: 18-Oct-2002

Lilly Laboratory Labeling Codes:

Health 0   **Fire** 0   **Reactivity** 0

Primary Physical and Health Hazards: Not Considered a Health Hazard.

Caution Statement: Heparin Sodium Injection is not considered to be a health hazard

(Continued on next page)

Figure 11.1.   *(Continued)*

Routes of Entry: Inhalation and skin contact.

Effects of Overexposure: Heparin sodium is not expected to be active orally. Effects of exposure by injection may include delayed clotting of blood. Dilute solutions of benzyl alcohol are not expected to be irritating.

Medical Conditions Aggravated by Exposure: Hemophilia and individuals on coumadin, heparin, or other anticoagulant therapy.

Carcinogenicity: Heparin sodium—No carcinogenicity data found. Not listed by IARC, NTP, ACGIH, or OSHA.
Benzyl alcohol—Not listed by IARC, NTP, ACGIH, or OSHA. Two-year carcinogenicity studies conducted by NTP demonstrated no evidence of carcinogenicity in mice and rats.

Section 4—First Aid Measures

Eyes: Flush eyes with plenty of water. Get medical attention.

Skin: Remove contaminated clothing and clean before reuse. Wash all exposed areas of skin with plenty of soap and water. Get medical attention if irritation develops.

Inhalation: Move individual to fresh air. Get medical attention if breathing difficulty occurs. If not breathing, provide artificial respiration assistance (mouth-to-mouth) and call a physician immediately.

Ingestion: Do not induce vomiting. Call a physician or poison control center. If available, administer activated charcoal (6–8 heaping teaspoons) with two to three glasses of water. Do not give anything by mouth to an unconscious person. Immediately transport to a medical care facility and see a physician.

Section 5—Fire Fighting Measures

Flash Point: No applicable information found
UEL: No applicable information found
LEL: No applicable information found

Extinguishing Media: Use water, carbon dioxide, dry chemical, foam, or Halon.

Unusual Fire and Explosion Hazards: None known.

Hazardous Combustion Products: May emit toxic fumes when exposed to heat or fire.

Section 6—Accidental Release Measures

Spills: Prevent further migration into the environment. Use absorbent/adsorbent material to solidify liquids. Wear protective equipment, including eye protection, to avoid exposure (see Section 8 for specific handling precautions).

Section 7—Handling and Storage

Storage Conditions: Controlled Room Temperature:15 to 30 C (59 to 86 F).

Section 8—Exposure Controls / Personal Protection

See Section 2 for Exposure Guideline information.

Under normal use and handling conditions, no protective equipment is required. The following is recommended for a production setting:

Respiratory Protection: Use an approved respirator.

Eye Protection: Chemical goggles and/or face shield.

Figure 11.1. *(Continued)*

Ventilation: Laboratory fume hood or local exhaust ventilation.

Other Protective Equipment: Chemical-resistant gloves and body covering to minimize skin contact.If handled in a ventilated enclosure, as in a laboratory setting, respirator and goggles or face shield may not be required. Safety glasses are always required.

Section 9—Physical and Chemical Properties

Boiling Point: No applicable information found

Melting Point: Not applicable

Specific Gravity: No applicable information found

pH: 5.0 to 7.5

Evaporation Rate: No applicable information found

Water Solubility: Soluble

Vapor Density: No applicable information found

Vapor Pressure: No applicable information found

Section 10—Stability and Reactivity

Stability: Stable at normal temperatures and pressures.

Incompatibility: May react with strong oxidizing agents (e.g., peroxides, permanganates, nitric acid, etc.).

Hazardous Decomposition: May emit toxic fumes when heated to decomposition.

Hazardous Polymerization: Will not occur.

Section 11—Toxicological Information

Acute Exposure
No data available for mixture or formulation. Data for ingredient(s) or related material(s) are presented.

Oral: Heparin sodium—Mouse, median lethal dose greater than 5000 mg/kg.

Skin: No applicable information found.

Inhalation: No applicable information found.

Intravenous: Heparin sodium—Mouse, median lethal dose 2800 mg/kg.

Skin Contact: Benzyl alcohol—Rabbit, irritant

Eye Contact: 30% Heparin solution—Rabbit, slight irritant
5% Heparin sodium solution—Rabbit, nonirritant
Benzyl alcohol—Rabbit, corrosive

Chronic Exposure
No data available for mixture or formulation. Data for ingredient(s) or related material(s) are presented.

Target Organ Effects: Heparin—Blood effects (decreased red blood cell count, decreased hemoglobin).
Benzyl alcohol—Nervous system effects (nerve tissue change, staggered gait, drowsiness)

Reproduction: Heparin—Rats administered subcutaneous doses up to 10 mg/kg/day demonstrated no effects on conception or pregnancy or on teratogenicity, implantation sites, or fetal weight (when administered during organogenesis).
Benzyl alcohol—One study reported decreased fetal weight at maternally high doses. No developmental effects have been reported in other animal studies.

(Continued on next page)

Figure 11.1.   *(Continued)*

Sensitization:
Benzyl alcohol—Guinea pig, not a contact sensitizer.

Mutagenicity: Heparin—Negative in Ames assay. No increase in chromosome aberrations in human lymphocytes in vitro. Negative in rate bone marrow micronucleus test in vivo.
Benzyl alcohol—Not mutagenic in bacterial cells, inconclusive results in mammalian cells.

Section 12—Ecological Information

No environmental data for the mixture or formulation. The environmental data for ingredient(s) or related material(s) are presented.

Ecotoxicity Data: Benzyl alcohol
Fathead minnow 96-hour median lethal concentration: 460 mg/L
Bluegill sunfish 96-hour median lethal concentration: 10 mg/L
Tidewater silverfish 96-hour median lethal concentration: 15 mg/L
Daphnia magna 48-hour median lethal concentration: 360 mg/L
Inland silverside 96-hour median lethal concentration: 15 mg/L
Green algae median effective concentration: 2600 mg/L

Environmental Fate: Benzyl alcohol
5-day Biological Oxygen Demand (acclimated microbial sludge): 70%
5-day Biological Oxygen Demand (sewage seed sludge): 48%
Bioconcentration factor: 4
Log Kow: 1.10

Environmental Summary: Benzyl alcohol—Slightly toxic to practically non-toxic in aquatic organisms. Material is expected to be mobile in soil. Material is expected to biodegrade rapidly and is not expected to bioconcentrate in aquatic organisms.

Section 13—Disposal Considerations

Waste Disposal: Dispose of any cleanup materials and waste residue according to all applicable laws and regulations.

Section 14—Transport Information

Regulatory Organizations:

DOT: Not Regulated

ICAO/IATA: Not Regulated

IMO: Not Regulated

Section 15—Regulatory Information

Below is selected regulatory information chosen primarily for possible Eli Lilly and Company usage. This section is not a complete analysis or reference to all applicable regulatory information. Please consider all applicable laws and regulations for your country/state.

U.S. Regulations
Heparin sodium and benzyl alcohol
TSCA—Yes
CERCLA—Not on this list
SARA 302—Not on this list
SARA 313—Not on this list
OSHA Substance Specific—No

Figure 11.1. *(Continued)*

EU Regulations
Not assigned an overall EC classification.

Section 16—Other Information

MSDS Sections Revised: MSDS Status.

Emergency Overview Sections Revised: Emergency Overview Status.

As of the date of issuance, we are providing available information relevant to the handling of this material in the workplace. All information contained herein is offered with the good faith belief that it is accurate. THIS MATERIAL SAFETY DATA SHEET SHALL NOT BE DEEMED TO CREATE ANY WARRANTY OF ANY KIND (INCLUDING WARRANTY OF MERCHANTABILITY OR FITNESS FOR A PARTICULAR PURPOSE). In the event of an adverse incident associated with this material, this safety data sheet is not intended to be a substitute for consultation with appropriately trained personnel. Nor is this safety data sheet intended to be a substitute for product literature which may accompany the finished product.

For additional information contact: For additional copies contact:
Eli Lilly and Company   Eli Lilly and Company
Hazard Communication   1-800-LILLY-Rx (1-800-545-5979)
317-433-7171

GLOSSARY:

ACGIH = American Conference of Governmental Industrial Hygienists
AIHA = American Industrial Hygiene Association
BEI = Biological Exposure Index
CAS Number = Chemical Abstract Service Registry Number
CERCLA = Comprehensive Environmental Response Compensation and Liability Act (of 1980)
CHAN = Chemical Hazard Alert Notice
CHEMTREC = Chemical Transportation Emergency Center
DOT = Department of Transportation
EC = European Community
EINECS = European Inventory of Existing Chemical Substances
ELINCS = European List of New Chemical Substances
EPA = Environmental Protection Agency
HEPA = High Efficiency Particulate Air (Filter)
IARC = International Agency for Research on Cancer
ICAO/IATA = International Civil Aviation Organization/International Air Transport Association
IEG = Lilly Interim Exposure Guideline
IMO = International Maritime Organization
Kow = Octanol/Water Partition Coefficient
LEG = Lilly Exposure Guideline
LEL = Lower Explosive Limit
MSDS = Material Safety Data Sheet

MSHA = Mine Safety and Health Administration
NA = Not Applicable, except in Section 14 where NA = North America
NADA = New Animal Drug Application
NAIF = No Applicable Information Found
NCI = National Cancer Institute
NIOSH = National Institute for Occupational Safety and Health
NOS = Not Otherwise Specified
NTP = National Toxicology Program
OSHA = Occupational Safety and Health Administration
PEL = Permissible Exposure Limit (OSHA)
RCRA = Resource Conservation and Recovery Act
RQ = Reportable Quantity
RTECS = Registry of Toxic Effects of Chemical Substances
SARA = Superfund Amendments and Reauthorization Act
STEG = Lilly Short Term Exposure Guideline
STEL = Short Term Exposure Limit
TLV = Threshold Limit Value (ACGIH)
TPQ = Threshold Planning Quantity
TSCA = Toxic Substances Control Act
TWA = Time Weighted Average/8 Hours Unless Otherwise Noted
UEL = Upper Explosive Limit
UN = United Nations
WEEL = Workplace Environmental Exposure Level (AIHA)

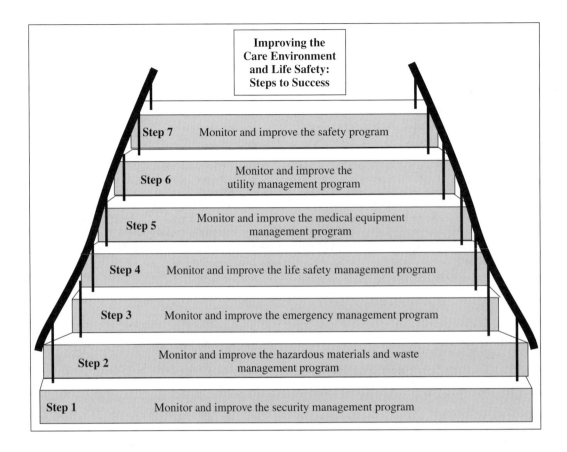

Step 3: Monitor and improve the emergency management program

Emergency management can help organizations better prepare for critical, major events such as a terrorist attack or earthquake. The emergency management program should provide a concise, preestablished plan to be implemented during an internal or external disaster or other emergency. The goal of the program is to ensure the continuity of patient care during emergencies. The possible types of disasters and emergencies encountered by healthcare organizations include bomb threats, civil disturbances, loss of communication networks, evacuation of industrial or other sites, internal fires, severe weather conditions, loss of utilities such as electricity and water, and major transportation systems accidents. The organization's emergency management plan should be integrated with that of community emergency response agencies. Cooperative planning with other organizations (including competitors) require they share or pool resources (communications, nursing, or medical staff) in the event of an influx of patients due to a major disaster.

Most healthcare organizations are responsible for conducting a hazard vulnerability analysis (HVA) which is intended to identify and prioritize potential emergencies that could affect the need for the organization's services or its ability to provide services. An example of a Hospital Hazard Vulnerability Assessment Tool is shown in figure 11.2. This tool assesses the organization in four areas of possible hazards: hazardous materials, human-related events, naturally occurring events, and technological events.

Figure 11.2. Hospital Hazard Vulnerability Assessment Tool

TECHNOLOGIC EVENTS

Event	PROBABILITY	SEVERITY = (MAGNITUDE − MITIGATION)						RISK
		HUMAN IMPACT	PROPERTY IMPACT	BUSINESS IMPACT	PREPAREDNESS	INTERNAL RESPONSE	EXTERNAL RESPONSE	
	Likelihood this will occur	Possibility of death or injury	Physical losses and damages	Interruption of services	Preplanning	Time, effectiveness, resources	Community/ Mutual Aid staff and supplies	Relative threat*
SCORE	0 = N/A 1 = Low 2 = Moderate 3 = High	0 = N/A 1 = Low 2 = Moderate 3 = High	0 = N/A 1 = Low 2 = Moderate 3 = High	0 = N/A 1 = Low 2 = Moderate 3 = High	0 = N/A 1 = High 2 = Moderate 3 = Low or none	0 = N/A 1 = High 2 = Moderate 3 = Low or none	0 = N/A 1 = High 2 = Moderate 3 = Low or none	0–100%
Electrical Failure	1	2	1	1	1	1	1	13%
Generator Failure	1	2	1	1	1	1	1	13%
Transportation Failure	1	1	1	1	2	1	2	15%
Fuel Shortage	1	1	1	1	1	1	1	11%
Natural Gas Failure	1	1	1	1	1	1	1	11%
Water Failure	1	1	1	1	1	2	1	13%
Sewer Failure	1	1	1	1	1	1	1	11%
Steam Failure	1	1	1	1	1	1	1	11%
Fire Alarm Failure	1	1	1	1	1	1	1	11%
Communications Failure	2	0	0	2	2	2	3	33%
Medical Gas Failure	1	1	1	1	1	1	0	8%
Medical Vacuum Failure	1	1	1	1	1	1	0	9%
HVAC Failure	1	1	1	1	1	1	0	9%
Information Systems Failure	2	1	1	3	1	1	1	30%
Fire, Internal	1	2	2	2	1	1	1	17%
Flood, Internal	1	0	1	1	1	1	1	9%
Hazmat Exposure, Internal	1	1	1	1	1	1	2	13%
Supply Shortage	1	1	1	2	1	1	1	13%
Structural Damage	2	2	2	2	2	2	2	44%
AVERAGE SCORE	1.16	1.11	1.05	1.32	1.16	1.16	1.11	15%

RISK = PROBABILITY × SEVERITY	
0.15	0.39 0.38

Threat increases with percentage.

(Continued on next page)

Figure 11.2. *(Continued)*

HAZARDOUS MATERIALS

Event	PROBABILITY	SEVERITY = (MAGNITUDE – MITIGATION)						RISK
		HUMAN IMPACT	PROPERTY IMPACT	BUSINESS IMPACT	PREPAREDNESS	INTERNAL RESPONSE	EXTERNAL RESPONSE	
	Likelihood this will occur	Possibility of death or injury	Physical losses and damages	Interruption of services	Preplanning	Time, effectiveness, resources	Community/ Mutual Aid staff and supplies	Relative threat*
SCORE	0 = N/A 1 = Low 2 = Moderate 3 = High	0 = N/A 1 = Low 2 = Moderate 3 = High	0 = N/A 1 = Low 2 = Moderate 3 = High	0 = N/A 1 = Low 2 = Moderate 3 = High	0 = N/A 1 = High 2 = Moderate 3 = Low or none	0 = N/A 1 = High 2 = Moderate 3 = Low or none	0 = N/A 1 = High 2 = Moderate 3 = Low or none	0–100%
Mass Casualty Hazmat Incident (From historic events at your MC with ≥ 5 victims)	1	1	0	1	2	2	2	15%
Small Casualty Hazmat Incident (From historic events at your MC with < 5 victims)	2	1	0	1	2	2	2	30%
Chemical Exposure, External	1	1	1	1	2	2	2	17%
Small–Medium-Sized Internal Spill	1	1	1	1	2	2	2	17%
Large Internal Spill	1	1	1	1	2	2	2	17%
Terrorism, Chemical	1	2	2	2	2	2	2	22%
Radiologic Exposure, Internal	1	1	1	1	2	2	2	17%
Radiologic Exposure, External	1	2	1	2	3	3	3	26%
Terrorism, Radiologic	1	2	1	2	3	3	3	26%
AVERAGE SCORE	1.11	1.33	0.89	1.33	2.22	2.22	2.22	21%

RISK = PROBABILITY × SEVERITY
0.21   0.37   0.57

Threat increases with percentage.

Figure 11.2. *(Continued)*

HUMAN RELATED EVENTS

Event	PROBABILITY	SEVERITY = (MAGNITUDE − MITIGATION)						RISK
	Likelihood this will occur	HUMAN IMPACT	PROPERTY IMPACT	BUSINESS IMPACT	PREPAREDNESS	INTERNAL RESPONSE	EXTERNAL RESPONSE	Relative threat*
		Possibility of death or injury	Physical losses and damages	Interruption of services	Preplanning	Time, effectiveness, resources	Community/ Mutual Aid staff and supplies	
SCORE	0 = N/A 1 = Low 2 = Moderate 3 = High	0 = N/A 1 = Low 2 = Moderate 3 = High	0 = N/A 1 = Low 2 = Moderate 3 = High	0 = N/A 1 = Low 2 = Moderate 3 = High	0 = N/A 1 = High 2 = Moderate 3 = Low or none	0 = N/A 1 = High 2 = Moderate 3 = Low or none	0 = N/A 1 = High 2 = Moderate 3 = Low or none	0–100%
Mass Casualty Incident (Trauma)	2	2	0	1	1	1	2	26%
Mass Casualty Incident (Medical/infectious)	1	2	0	1	1	2	2	15%
Terrorism, Biological	1	2	0	2	1	2	2	17%
VIP Situation	2	0	0	0	1	1	1	11%
Infant Abduction	1	1	1	2	2	2	0	15%
Hostage Situation	1	1	0	1	3	2	1	15%
Civil Disturbance	1	1	1	1	2	2	2	17%
Labor Action	1	0	0	3	2	2	0	13%
Forensic Admission	3	0	0	0	1	1	1	17%
Bomb Threat	1	1	1	2	2	2	2	19%
AVERAGE SCORE	1.40	1.00	0.30	1.30	1.60	1.70	1.30	21%

RISK = PROBABILITY × SEVERITY

0.21   0.47   0.44

Threat increases with percentage.

(Continued on next page)

Figure 11.2. (Continued)

NATURALLY OCCURRING EVENTS

Event / SCORE	PROBABILITY — Likelihood this will occur — 0 = N/A, 1 = Low, 2 = Moderate, 3 = High	SEVERITY = (MAGNITUDE – MITIGATION) HUMAN IMPACT — Possibility of death or injury — 0 = N/A, 1 = Low, 2 = Moderate, 3 = High	PROPERTY IMPACT — Physical losses and damages — 0 = N/A, 1 = Low, 2 = Moderate, 3 = High	BUSINESS IMPACT — Interruption of services — 0 = N/A, 1 = Low, 2 = Moderate, 3 = High	PREPAREDNESS — Preplanning — 0 = N/A, 1 = High, 2 = Moderate, 3 = Low or none	INTERNAL RESPONSE — Time, effectiveness, resources — 0 = N/A, 1 = High, 2 = Moderate, 3 = Low or none	EXTERNAL RESPONSE — Community/ Mutual Aid staff and supplies — 0 = N/A, 1 = High, 2 = Moderate, 3 = Low or none	RISK — Relative threat* — 0–100%
Hurricane	1	1	1	0	0	0	0	4%
Tornado	1	1	1	0	0	0	0	4%
Severe Thunderstorm	2	1	1	0	1	1	1	19%
Snow Fall	0	0	0	0	0	0	0	0%
Blizzard	0	0	0	0	0	0	0	0%
Ice Storm	0	0	0	0	0	0	0	0%
Earthquake	3	2	3	3	1	1	1	51%
Tidal Wave	1	1	1	0	0	0	3	9%
Temperature Extremes	1	1	0	0	3	0	3	13%
Drought	0	0	0	0	0	0	0	0%
Flood, External	0	0	0	0	0	0	0	0%
Wild Fire	0	0	0	0	0	0	0	0%
Landslide	0	0	0	0	0	0	0	0%
Dam Inundation	0	0	0	0	0	0	0	0%
Volcano	0	0	0	0	3	0	0	0%
Epidemic	1	2	0	2	2	2	2	19%
Average Score	0.63	0.58	0.44	0.31	0.63	0.25	0.63	3%

RISK = PROBABILITY × SEVERITY

0.03   0.21   0.16

*Threat increases with percentage.

Adapted from: http://www.hazmatforhealthcare.org/downloads.cfm. Reprinted with permission from Mitch Saruwatari, National Threat Assessment Manager, Kaiser Permanente, and Paul Penn, EnMagine.

The emergency management plan outlines the following information in four distinct phases (JCAHO 2003b):

Mitigation activities:

Activities an organization or facility puts in place to eliminate or lessen the effects of a potential emergency or hazard. "Mitigation begins with identifying hazards that may affect the facility and analyzing how vulnerable patients, personnel, facilities, telecommunications, and informational resources may be to those hazards" (JCAHO 2003b, p. 11). Mitigation activities include:

- Policies and procedures to be followed in response to a variety of internal and external emergencies
- Preparation, staffing, organization, activation, and operation of organizational response in the event of a communitywide emergency
- Means for minimizing suffering, loss of life, personal injury, and damage to property
- Triage system for handling incoming injured persons to identify the appropriate level of treatment

Preparedness activities:

Activities that "build individual and organization ability to manage the effects of hazards that actively affect a facility" (JCAHO 2003b, p. 11), including:

- Prearranged agreements with vendors, healthcare networks, and community response agencies
- Maintaining an ongoing planning process, holding staff orientation on basic response actions, and implementing facilitywide rehearsals

Response activities:

"Control[ling] the negative effects of emergency situations" (JCAHO 2003b, p. 11). Actions may include rehearsed responses by staff in the event of an emergency (such as a fire) or actions taken by management:

- Identification of a command structure that ties into the community structure
- Procedures for notifying the proper authorities within and outside the organization
- Procedures for notifying the organization's staff that the disaster plan is to be implemented and defining staff roles and responsibilities
- Internal and external emergency communications and information systems, ensuring that these tie into the community's ability and procedures for communication
- Procedures for the partial or total evacuation of the facility in conjunction with the community plan for evacuation defining the placement of patients by priorities established within the community emergency plan

- Management of patients, including schedule modification or discontinuation of services; release of information regarding patients; and admission, transfer, or discharge of patients

Recovery actions:

The process the facility utilizes to restructure or rebuild its normal activities after a disaster. "Recovery actions begin almost concurrently with response activities and are directed at restoring essential services and resuming normal operations" (JCAHO 2003b, p. 11).

Monitoring activities for emergency management primarily concern maintaining employees' knowledge of and ability to perform appropriate procedures to handle emergencies. Disaster drills must be carried out regularly so that employees have an opportunity to practice the processes they would need to put into effect in the event of a real emergency. They must also know how to coordinate efforts with other institutions and regional disaster management agencies and what communication paths are available in this kind of situation. Regular educational sessions should be held, and the abilities of each participant should be assessed after the training is complete.

Step 4: Monitor and improve the life safety management program

The life safety management program focuses on protecting patients, visitors, staff, and property from fire, smoke, and other products of combustion. According to the JCAHO's Environment of Care Standard EC.1.5, the program includes the following activities (JCAHO 2003a):

- Inspection, testing, and maintaining fire detection and protection equipment such as alarm systems, extinguishers, evacuation routes, and so forth

- Reporting and investigating fire protection deficiencies, failures, and user errors

- Evaluating proposed acquisition of bedding, window coverings, furnishings, room decorations, wastebaskets, and equipment in the context of fire safety

The other major function of the life safety program is fire safety training. All employees and clinical staff must be able to demonstrate their knowledge of and ability to perform fire alarm transmission, physical containment of smoke and fire, transfer of patients and those injured to areas of refuge, extinguishing of fire, preparation for evacuation of buildings, and location and use of equipment for evacuating and transporting patients.

Monitoring activities in the life safety preparedness area primarily concern maintaining employees' knowledge of and ability to perform appropriate procedures to handle fire emergencies. Fire drills must be carried out regularly so that employees have an opportunity to practice the processes they would need to put into effect in the event of a real fire.

Step 5: Monitor and improve the medical equipment management program

The medical equipment management program focuses on ensuring that all of the equipment used within the healthcare organization is functioning properly. It also ensures that employees know how to properly use the equipment necessary to their job duties as well as the equipment's capabilities, limitations, and special applications. Users must know the emergency procedures to be instituted in case of equipment failure.

According to JCAHO's Environment of Care Standard EC.1.6, planning includes identifying processes for (JCAHO 2003a):

- Establishing risk criteria for identifying and evaluating medical equipment and performing inventories of medical equipment

- Selecting and acquiring equipment

- Maintaining appropriate procedures for monitoring the equipment's functioning and providing maintenance strategies when necessary

- Tracking systems for all occurrences associated with each piece of equipment that resulted in increased exposure to risk of injury to patients or personnel, such as the clinical alarm systems used in the facility

In addition, **recall logs** must be maintained for each piece of medical equipment to document communications from manufacturers regarding problems.

The program must be carried out in conjunction with the biomedical engineering staff. The program can be staffed by regular, full-time employees or by contractors from an outside agency. It is important to recognize that most individual departments also perform quality control activities on a regular basis to ensure the proper equipment performance. For example, in the clinical laboratories, all of the instruments are calibrated at least daily to ensure that the equipment is performing well and providing accurate reports. Each piece of equipment is tested with calibrating reagents of known control values to ensure that the instrument can achieve results within acceptable limits. This type of testing is mandated by the federal government for all Food and Drug Administration–approved analyzers and is codified in the Clinical Laboratory Improvement Amendments (CLIA), regulations to which clinical laboratories must adhere. In addition, both the JCAHO and the College of American Pathologists publish additional standards regarding the calibration and maintenance of laboratory analyzers.

Other areas of healthcare organizations have the same kinds of issues. Each runs its own set of calibration routines. Two other important examples are the studies performed on the equipment in critical care areas and the studies performed on radiological imaging equipment. In both areas, equipment malfunctions can lead to severe patient injury or death. Performing appropriate inspection, testing, and maintenance of equipment is extremely important in direct patient care areas.

Step 6: Monitor and improve the utility management program

The utility management program focuses on ensuring that all utility resources used within the organization are functioning properly and that, when disruptions occur, employees are prepared to perform appropriate procedures to manage both the disruption and its effects on patient services. Utility resources include electricity, natural gas, water, sewer, and telecommunications services provided from outside the organization by community utilities companies.

Physical facilities within the organization where utilities are turned on or off and monitored are called **control areas.** Control areas for each of the utility resources must be adequately labeled so that employees know how to shut down the utility when necessary in an emergency. These systems must also be routinely monitored for proper functioning and necessary preventive maintenance. Failures and user errors must be identified through the occurrence reporting system.

An important component of the utility management program is the requirement that the organization be able to provide a source of emergency power should electrical service be interrupted in the community. Engineering services must routinely inspect, test, and maintain emergency power generators to provide electricity to critical areas any time the community's electrical service is interrupted. Common critical areas include alarm systems, exit illumination, elevators, emergency communication systems, blood/bone/tissue storage units, medical/surgical vacuum systems, air compressors and ventilators, postoperative recovery rooms, and special care units.

The organization must also meet state statutes and regulations from other entities, including OSHA, the National Fire Protection Association, and the Life Safety Code as well as the JCAHO.

Step 7: Monitor and improve the safety program

The safety program focuses on ensuring that all employees are aware of the safety risks encountered in the patient care environment and that they understand and can initiate the occurrence reporting system for risk management. Employees also must be able to take action to prevent, eliminate, or minimize safety risks. Each employee is provided department- and job-specific safety training that is documented in each employee's personnel file. All staff need to be aware of the National Patient Safety Goals as they apply to their area of employment. All facility safety programs should address these concerns in their routine performance improvement process and in the performance improvement plan for the facility.

Real-Life Example

One Saturday morning in April, a third-party construction company was laying cable under the street about a quarter mile north of Community Hospital of the West. The workers were using an auger to drill under the roadway so that they would not need to tear up the street. Although the gas lines and electric cables in the area had been marked with blue stakes, the construction crew inadvertently damaged the main telephone cable under the street. Communications in and out of the hospital and other businesses in the area were disconnected.

The hospital did not have a cellular telephone. Because they were not able to make or receive calls, the staff at the hospital faced major communication problems with physicians and in providing care support for emergency department patients. The staff had to find a quick solution. They decided to use employees' cellular telephones, and one pay phone was still inexplicably working. Crews from the telephone company worked all day to restore service, which was reestablished at 7:30 that evening.

Because the organization had not been prepared for this type of emergency, the hospital's safety council met later to initiate the following steps:

- Purchase cellular telephones to be kept by the telephone operator

- Establish and maintain an employee telephone directory including cellular telephone numbers

- Determine if and where cellular telephones interfere with hospital equipment and technology

- Develop a location map with safe zones for cellular telephone use throughout the hospital

QI Toolbox Technique

Postprogram assessments are the most common method used to evaluate employee knowledge of care environment and safety area issues. The assessments are given to employees after each training session. At minimum, every employee must attend training in each of the areas annually. Human resources systems track each employee's training attendance and record attendance in employee personnel files. Figure 11.3 shows an example of a posttraining assessment tool used at the Community Hospital of the West.

Some organizations also have safety fairs on an ad hoc basis. They set up carnival games or similar activities, then answer environment-of-care questions using these games. If questions are answered successfully, winners are entered into drawings or awarded small prizes. But many kinds of activities can be developed to keep these issues in the fore, such as participating in a communitywide disaster mock drill.

Case Study

Students should search the Internet for the material safety data sheet (MSDS) information for five common household products. List the item description information, effects of exposure, and safe handling and disposal information for each of the five products. Product information may be located at http://hazard.com/msds/index.php.

Project Application

Students should identify existing environmental and safety guidelines or regulations that have implications for their projects. If the student project recommends PI training, participants should consider how they would assess the training outcomes to determine whether individuals retained important aspects of the training.

Figure 11.3.   Example of a Posttraining Assessment Tool

Exposure to Blood-Borne Pathogens

1. Universal precautions means:
   A. All employees must be trained in the same safety procedures.
   B. Wearing protective personal equipment is not necessary when handling organic tissue, alive or dead.
   C. All blood and other potentially infectious body fluids are treated as though they were infected with blood-borne pathogens.

2. The following conditions must be present for you to catch an infectious disease:
   A. Port of entry
   B. Contact with contaminated blood or other body fluids
   C. None of the above
   D. A and B

3. The exposure control plan is:
   A. OSHA's standards regarding procedures for handling exposures to HIV and HBV
   B. Our organization's written plan to help employees eliminate or minimize exposure to blood-borne pathogens
   C. A policy on how to deal with infectious exposure to HIV and HBV

4. Hand washing is required:
   A. After removing gloves, before eating, but not before gloving
   B. Before eating, before gloving, but not before leaving the workplace
   C. Before leaving the workplace, after touching potentially contaminated surfaces, but not after every patient
   D. Before gloving, after each patient, before eating, and before leaving the workplace

5. Healthcare workers have developed AIDS following exposure to contaminated blood and body fluids through:
   A. Needlesticks
   B. Open cuts and wounds
   C. Mucous membranes
   D. All of the above

6. Following are some symptoms of hepatitis B infection:
   A. Extreme fatigue, excessive thirst, red eyes
   B. Nausea, learning difficulties, anorexia
   C. Dark urine, jaundice, abdominal pain
   D. Abdominal pain, rash, swelling limbs

7. When an exposure occurs:
   A. You should report the exposure to your supervisor immediately.
   B. When possible and with consent, the source individual is tested to determine whether he or she is infected with HIV or HBV.
   C. You will be provided medical evaluation, treatment, and counseling.
   D. All of the above.

Figure 11.3. *(Continued)*

Codes: Fires and Disasters

8. What does the PBX operator announce for fire?
   A. Code yellow
   B. Code blue
   C. Code green
   D. Code red

9. What does the PBX operator announce for internal and external disasters (check two)?
   A. Code red
   B. Attention, disaster #
   C. Attention, please, disaster and location
   D. Code yellow

10. In a hospital emergency, you dial the following number to get help:
    A. 911
    B. 5000
    C. O (Operator)
    D. 411

Security

11. The employee parking lot is located:
    A. In front of the hospital
    B. By the emergency department
    C. In front of the professional building
    D. West of the hospital

Chemical Hazcom

12. You need to know about the chemicals in your workplace because:
    A. This will test your knowledge.
    B. You will find it interesting.
    C. You may not be aware of the type of hazards for each chemical.

13. Your rights to know about hazardous chemicals are contained in:
    A. The Hazardous Communication manual
    B. Personnel memos
    C. Letter to supervisors only

14. Chemical hazard labels:
    A. Give you quick information
    B. Are in blue
    C. Are not required

15. Chemical material safety data sheets:
    A. Are available only for supervisors
    B. Replace the need for labels
    C. Give detailed information about hazardous materials

16. The hazard communications program includes training on:
    A. Stress control
    B. How chemicals can enter the body
    C. Lifting techniques

17. A safe work environment is:
    A. Not always possible
    B. Up to you
    C. A responsibility shared by employers and employees

(Continued on next page)

Figure 11.3.   *(Continued)*

Fire Extinguisher

18. Portable fire extinguishers are effective if you:
    A. Know how to use them and act fast
    B. Wait for instructions
    C. Use more than one

19. The "fire triangle" includes:
    A. Small, medium, and large
    B. You, supervisor, and firefighter
    C. Fuel, air, and heat

20. Multipurpose dry chemical extinguishers are effective on:
    A. Class A, B, and C fires
    B. Class A fires only
    C. Class B and C fires only

21. In case of fire, top priority is:
    A. PATE (patient, alarm, telephone, extinguish)
    B. Whatever you decide
    C. Going on a break

22. If a fire becomes larger:
    A. Get more extinguishers
    B. Get closer
    C. Get out and close doors

23. If the contents of a fire extinguisher are partially used:
    A. Put it back
    B. Report it; get it replaced
    C. Throw it out

Body Mechanics

24. Good body mechanics includes all of the following except:
    A. Twisting at the waist
    B. Tightening stomach muscles when you lift
    C. Lifting with your legs
    D. Holding the load close to the body

25. Basic prevention of back injury involves all of the following principles of lifting except:
    A. Using the squat position
    B. Bending at the waist
    C. Using the heavy muscles of the legs
    D. Maintaining the normal arch of the lower back

26. When a job requires staying in the same posture or position for long periods of time, you should:
    A. Twist at the waist occasionally to improve circulation
    B. Listen to the radio to take your mind off the pain
    C. Change positions as often as possible
    D. Just tough it out even when your feet get numb

27. Conditions that lead to low back pain include which of the following:
    A. Poor posture
    B. Lack of exercise
    C. Long exposure to sunlight
    D. Overeating

Summary

A significant challenge in PI activities for the environment of care and patient safety is maintaining employee knowledge and skill. Appropriate emergency response is crucial to minimizing risk to other employees, patients, and medical staff. Therefore, PI activity is organized around the assessment of potential hazards and training and retraining personnel so that emergency procedures are initiated without question or hesitation when necessary. The abilities of individual employees must also be consistently and frequently assessed to ensure that their knowledge base is adequate at all times. Equipment maintenance and security management are also critical areas for performance improvement.

References

Hinckley, C. M. 1997. Defining the best quality control system by design and inspection. *Clinical Chemistry* 43(5):873–79.

Joint Commission on Accreditation of Healthcare Organizations. www.jcaho.org.

Joint Commission on Accreditation of Healthcare Organizations. 2001. *Joint Commission Perspectives* (December).

Joint Commission on Accreditation of Healthcare Organizations. 2002. *Environment of Care News* (January/February).

Joint Commission on Accreditation of Healthcare Organizations. 2003a. Environment of Care. In *2003 Hospital Accreditation Standards.* Oakbrook Terrace, Ill.: JCAHO.

Joint Commission on Accreditation of Healthcare Organizations. 2003b. *Environment of Care: Essentials for Health Care,* 3rd ed. Oakbrook Terrace, Ill.: JCAHO.

Joint Commission on Accreditation of Healthcare Organizations. 2003c. 2003 National Patient Safety Goals. Accessed online at http://www.jcaho.org/accredited+organizations/patient+safety/npsg/npsg_03.htm.

Chapter 12
Developing Staff and Human Resources

Learning Objectives

- To recognize the need to integrate performance improvement and patient safety data into the management of the human resources function in healthcare

- To identify the tools commonly used to manage the recruitment and retention of human resources

- To outline the credentialing process for independent practitioners and employed clinical staff

Background and Significance

The environment in which healthcare organizations manage their human resources is dictated in large part by state and federal employment laws, regulatory agencies, and market trends. Variations in the way healthcare is delivered also affect the organization's recruitment, training, and retention practices. Potential legal action undertaken by employees against healthcare organizations must also be considered. This chapter introduces information on laws, regulations, shifting market trends, and organizational policy to demonstrate some of the processes and tools used to manage people in today's healthcare environment.

During the twentieth century, inequitable and unsafe employment practices were a driving force in the passage of legislation intended to improve the American work environment. Legislation that addressed issues such as safe working conditions (the Occupational Safety and Health Act of 1970), wages and hours (the Equal Employment Opportunity Act of 1972), disabilities (Americans with Disabilities Act of 1990), and medical leave (Family and Medical Leave Act of 1993) has played a role in shaping the workplace environment and standardizing employment practices. Other employment guidelines and standards of practice in the areas of state licensure and outside accrediting agencies have evolved to direct the employer–employee relationship.

For any healthcare organization, employees are both the most important resource and the biggest liability. Developing effective processes related to recruitment, retention, development, and continuing education of all staff members is a critical function of an organization's leadership, who must ensure that it employs a highly competent staff through appropriate human resources processes.

Effective leadership is defined by external agencies as how well an organization's leaders plan, direct, integrate, and coordinate services, and how well they create a culture that focuses on continual performance improvement (PI), while at the same time maintaining a balanced budget and fiscal viability in the market. These key leadership functions are demonstrated in human resources management in the following ways:

- Defining the qualifications and performance expectations for all staff positions
- Providing adequate staff
- Ensuring that the staff's qualifications are consistent with their job responsibilities
- Assessing, demonstrating, and improving staff competence

Today's cutting-edge work environment encourages and empowers employees to take responsibility for improving the environment in which they work, no matter what their position in the organizational structure. Performance improvement can be a powerful concept when woven into the culture of an organization. Organizations that embrace ongoing PI and patient safety do so because their leaders foster a culture of competency through staff self-development and lifelong learning. The competitive advantage for healthcare organizations today lies in their intellectual capital and organizational effectiveness. This style of motivating and developing employees is grounded in the ability of leaders to create, communicate, and model the organization's mission, vision, and values.

Steps to Success: Developing Staff and Human Resources

Historically, the determination of the actual number of FTEs (full-time equivalent staff) has been developed from licensing regulations that outline not only the number of staff, but also the type of license and degree required. In planning its budget, the organization would take into account the needed revenue generated and the numbers of FTEs required to run a facility that met licensing agreements and maintained fiscal viability.

In 2003, the Joint Commission on Accreditation of Healthcare Organizations (JCAHO) developed measurable standards related to staffing effectiveness in an attempt to make prescribed staffing levels a more realistic, dynamic process. It is recognized that effective staffing is complex, dynamic, and unique to each individual organization. At the core of staff effectiveness are competency and skill mix, as well as acuity ratios (how many and what kind of staff are needed to properly provide a certain level of care) for the population in question. Staffing has a direct impact on the quality and safety of patient care. An aging workforce and population, which points to potential shortages of key staff, will continue to challenge and impact staffing effectiveness.

The approach to effective hospital staffing includes the use of multiple clinical/service measures and human resources indicators as a screening tool to identify potential staffing issues. Each selected measure, for which there must be at least two clinical and two human resources in all patient populations, is defined by performance expectations. At a minimum, identified performance measures relate to processes appropriate to the care and services provided and to problem-prone areas. A clinical/service measure could be an adverse drug event, for example, and an example of a human resources measure could be staff vacancy rates. The rationale for indicator selection is based on relevancy and sensitivity to each patient area where staffing is planned. Figure 12.1 provides a list of JCAHO-approved screening indicators currently used in hospitals. The intent is that multiple indicators examined in combination with one another may provide information related to staff effectiveness. Because staff effectiveness is considered a critical process in healthcare, the standards require that a thorough analysis be initiated whenever an organization detects or suspects significant undesirable performance or variation.

Most organizations define their clinical staffing needs in two categories: direct and indirect caregivers. Direct caregivers include RNs, LPNs, physical therapists, chaplain, and social services. Indirect caregivers include clinical support professionals from the pharmacy, laboratory, housekeeping, dietary, and so on. Both direct and indirect caregivers should be included in the human resources screening indicators.

Step 1: Manage the recruitment process

A well-defined recruitment process provides both the organization and the applicant with an opportunity to evaluate and determine their potential compatibility. The recruitment process is usually initiated to fill a vacant position, to fill a newly created position, to staff a newly instituted service, or to support increases in patient acuity and/or census. Organizations sometimes retain the services of an outside recruitment agency when there are shortages in qualified applicants or because of an organization's remote geographic location.

Position requisitions, position (or job) descriptions, and employment applications are the primary tools used in the recruitment process. Position requisitions begin the formal communication between the human resources staff and the department or service area initiating the recruitment process. (See figure 12.2.) A detailed position requisition provides human resources staff with the information they need to prepare advertisements for open positions, including the position title, qualifications and experience required, work schedule, and salary. Position requisitions can also be used to monitor the appropriateness of personnel requests. For example, a nonbudgeted position or a new position request may require additional documentation and administrative approval before recruitment can begin.

Figure 12.1.   Screening Hospital Indicators in Measuring Staff Effectiveness

Human Resources Screens:

- Nursing care hours per patient day
- On-call or per-diem use
- Overtime
- Sick time
- Staff injuries
- Staff satisfaction surveys
- Staff turnover rate
- Staff vacancy rate
- Understaffing compared to staffing plan

Clinical Service Screens:

- Falls
- Injuries to patients
- Lengths of stay
- Medication errors
- Patient/family complaints
- Pneumonia
- Postoperative infections
- Pressure ulcers
- Shock/cardiac arrests
- Upper GI bleeding
- Urinary tract infections

Figure 12.2.	**Sample Position Requisition**

PERSONNEL REQUISITION

JOB TITLE	DATE OF REQUEST	DATE REQUIRED

DEPARTMENT

☐ Replacement for:

☐ Additional Personnel	Budgeted: ☐ Yes ☐ No			
☐ Temporary: End Date _____	☐ Full-Time	☐ Part-Time	☐ Exempt	☐ PRN	☐ Per-Diem

Shift: Hours per week: Suggested Pay Grade:

Qualifications needed: education, experience, training, duties, physical fitness		Actual:	Budgeted:	Prior Year:
	YTD FTEs			
	MH / STAT			
	SAL $/STAT			
	Reason for additional personnel or new position:			

APPROVALS (must be obtained prior to offering position)

DIRECTOR/MANAGER: DATE:	CEO: DATE:
CFO: DATE:	HUMAN RESOURCES: DATE:

NEW PERSONNEL

To be completed by human resources		Date Required
Employee's Name	Department	Employee Number
Address	Social Security Number	
Telephone Number	Birthdate	
Date of Hire	Temporary Date of Termination	

☐ New Hire	☐ Rehire	☐ Full-Time	☐ Part-Time	☐ PRN	☐ Per-Diem	☐ Exempt	☐ Temp.
RATE/HOUR	BIWEEKLY RATE			RACE	EEO CAT.	TAX EXEMP.	POS CODE

COMMENTS OR SPECIAL INSTRUCTIONS:

Position descriptions define the responsibilities of the job and the qualifications needed to fulfill those responsibilities. Qualifications may define education; licensure, certification, or registration status; and experience required. Qualifications may also include previous experience requirements, such as training or experience working with children and adolescents.

In healthcare organizations, **credentials** and **licenses** are qualifications of foremost importance. Both clinical professionals, such as physicians, psychologists, nurses, occupational therapists, physical therapists, respiratory therapists, as well as social workers, and most ancillary services professionals, such as health information managers, radiology technicians, laboratory technicians, and scientists, are required to hold a credential or license to practice their professions.

Credentials are usually conferred by national professional organizations dedicated to a specific area of healthcare practice. Licenses are conferred by state regulatory agencies. Credentialing and licensing processes usually require an applicant to pass an examination to obtain the credential or license initially and then to maintain the credential or license through continuing education activities thereafter.

Job responsibilities describe the major tasks of the position. Responsibilities should be stated in measurable terms, which is the standard in healthcare today. A clinical position, for example, may describe a position responsibility as "completing a nursing assessment on each admission within the time frame defined by policy."

In general, position descriptions provide guidance to human resources staff by defining the initial selection criteria for potential candidates. Position descriptions are also used to provide the applicant or new employee with information about position expectations. With the support of human resources specialists, managers generally maintain position descriptions that reflect local market and current position requirements.

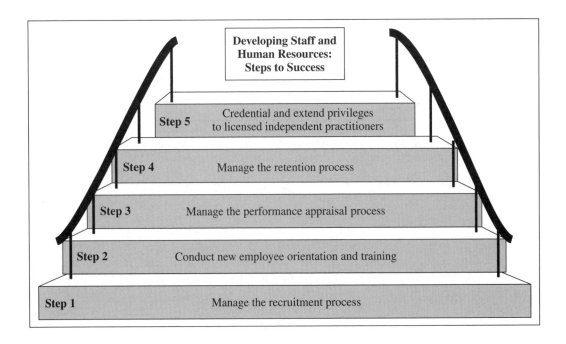

An employment application is provided to interested candidates who contact the organization in response to an advertised position. A review of the completed application and the applicant's resume provides human resources, and the recruiting department or service area manager, with enough information to determine whether the applicant meets the minimum position requirements. (See the example in figure 12.3.) Applicants who meet the initial position requirements are invited to interview for the position.

The interview gives the applicant and the organization an opportunity to exchange information not listed on the employment application or in the position advertisement. The organization may have additional questions specific to experience, skills, and qualifications that are not addressed in the application. The candidate, in turn, may have questions related to salary, benefits, or the organization's history. The interview provides an excellent opportunity for the interviewer to evaluate the applicant's communication skills, general personality, and demeanor. During the interview process, the interviewee may be asked to sign a release for information on a background check. Most states offer this service as a risk reduction and patient safety strategy for facilities that provide physical and emotional care for patients. This regulation was put into place to safeguard clients and facilities from individuals with documented histories of abuse. Potential employees must pass a background check before being offered a position.

Several types of interviews can be used to screen applicants. A structured interview is conducted using a set of standard questions that are asked of all job applicants, the purpose of which is to gather comparative data. The unstructured interview uses general questions from which other questions are developed over the course of the conversation. The purpose of this type of interview is to prompt the interviewee to speak openly.

The following examples of unstructured questions may be used during an interview:

- Tell us about your previous position. In general, what kinds of duties did you perform?

- Tell us about your personal understanding of the need for maintaining the patient's rights to confidentiality.

The stress interview uses specific questions to determine how the interviewee responds under pressure. Questions in this type of interview set up specific scenarios and ask the interviewee how he or she would handle the situation.

- If a newspaper reporter came to you and wanted you to reveal information from a prominent person's medical record, how would you respond?

- If an attorney came onto your nursing unit and wanted to review her client's medical record, what would you do?

Some types of questions should be avoided during an interview. Questions that might be interpreted as attempts to learn personal information about the applicant (such as his or her age, marital status, or family status) are prohibited by federal regulations. Figure 12.4 provides examples of questions that are not related to employment and would be considered discriminatory.

Some facilities will add a skills-testing segment to their interview process such as a coding test, medication exam, developmental age exam, or a computer literacy test. This testing adds additional information of a concrete nature to the screening for possible candidates.

Figure 12.3.   Sample Employment Application

INSTRUCTIONS TO THE APPLICANT:

Please complete the application in full. You can send a resume with your completed application.

Please apply for a specific position(s) and include that position's job title(s) on the application.

APPLICANT: The federal government requires employers to collect statistical information on job applicants. Providing this information is voluntary. The information is used for statistical purposes only and will not be included in your application file. Information will be maintained and utilized in accordance with applicable laws and regulations. Refusal to provide the information below will not subject you to any adverse treatment or affect the application process in any way.

1. SEX   __ Male   __ Female

2. RACE/ETHNIC BACKGROUND (check only one):

    ___ Asian or Pacific Islander (O)
    ___ American Indian/Native American (A)
    ___ Black (not of Hispanic Origin) (B)
    ___White (not of Hispanic Origin) (C)
    ___ Hispanic (a person whose cultural or linguistic origins are Spanish or Latin American, regardless
         of race, for example, Mexican, Puerto Rican, Cuban, Central or South American) (S)

3. DISABLED OR VETERAN STATUS (check all that apply):

    ___ Disabled, but not a veteran (I)
    ___ Veteran of Vietnam era (V)
    ___ Disabled veteran (D)
    ___ None of the above

    ___ I do not wish to provide this information.

PLEASE CONTINUE ON THE NEXT PAGE

(Continued on next page)

Figure 12.3. *(Continued)*

_____ App#
(For office use only)

APPLICATION FOR EMPLOYMENT

All applicants selected for employment must satisfactorily pass a preemployment drug screen and criminal background check to be eligible for employment.
CHW is an Equal Opportunity and Affirmative Action Employer.
CHW hires only individuals who are authorized to work in the United States.
This application is subject to the conditions set forth in the Certification and Agreement section on the last page.

PLEASE COMPLETE APPLICATION IN FULL
DATE OF APPLICATION _____
LAST NAME _____ FIRST NAME _____ MIDDLE INITIAL _____
CURRENT ADDRESS _____ CITY _____ COUNTY _____
STATE _____ ZIP CODE_____
SOCIAL SECURITY # _____ HOME PHONE (___) _____ WORK PHONE (___) _____

SOURCE OF REFERRAL
RECRUITING METHOD: (Please check one)
__ (60) Student Work Experience   __ (05) Walk-in   __ (10) Internal Referral _____
__ (30) Newspaper   __ (80) Past Experience   __ (06) Temp Employment Agency, Temp-to-hire
__ (13) Internet   __ (15) Job Line   __ (85) Community-Based Organization
__ (07) Job Service   __ (20) Professional Journal __ (52) Job Elimination
__ (90) School Organization Referral

Have you ever been convicted of a felony or a misdemeanor, or have you ever plead no contest to any criminal charges?   __ Yes   __ No
Provide date, city, state, and an explanation for any yes responses: _____
Criminal conviction is not an absolute bar to employment but will be considered in relation to specific job requirements.

Can you perform the functions of the job for which you are applying, either with or without a reasonable accommodation?   __ Yes   __ No
Do you have any relatives employed by Community Hospital of the West?   __ Yes   __ No
If yes, Where/Relationship_____
Have you ever been employed by or are you currently employed by Community Hospital of the West?   __ Yes   __ No
If yes, Where? _____ When? _____
Why did you leave? _____

POSITION(S) DESIRED (only two positions per application please)
TITLE: _____ Job Number:_____
TITLE: _____ Job Number:_____

WORK AVAILABILITY
MARK ALL THAT APPLY
Type of Employment   Work Schedule/Shift   Weekends __ Yes __ No   No Rotating Weekends __ Yes __ No
__ Full-Time   __ Days
__ Part-Time   __ Evenings
__ Temporary   __ Nights
__ On-Call   Hours Available _____
Current Salary: $ _____ Minimum Salary Requirement $ _____ Date Available to Work _____

JOB SKILLS
Check all that you have experience with: PC Graphics __ Yes __ No   Word Processor __ Yes __ No
Desktop Publishing __ Yes __ No   PC __ Yes __ No   Spreadsheet __ Yes __ No
LAN __ Yes __ No   Database __ Yes __ No   Windows __ Yes __ No
AS/400 __ Yes __ No   Tandem __ Yes __ No   Medical Terminology __ Yes __ No
List specific software programs used: _____
Typing Speed: _____ WPM   10-Key by Touch __ Yes __ No   _____ SPM

Figure 12.3.   (Continued)

EDUCATION

Have you graduated from high school or completed the GED equivalent? __ Yes __ No

List all degrees that you have received. List your HIGHEST DEGREE FIRST. Do NOT list degrees that you are currently working toward (see below)

MAJOR	DEGREE	SCHOOL	GRADUATION DATE
_____	_____	_____	_____
_____	_____	_____	_____
_____	_____	_____	_____

Are you currently enrolled? __ YES __ NO   Last year attended: _____   Major: _____

Check last level of school completed:

Years completed: Undergraduate: __ Freshman   __ Sophomore   __ Junior   __ Senior

Graduate: __ 1st year   __ 2nd year   __ 3rd year   __ 4th year

LICENSURE/REGISTRATION/CERTIFICATION

List all professional licenses, registrations, and certifications

Lic/Reg/Cert Type	License #	State	Expiration Date
_____	_____	_____	_____
_____	_____	_____	_____
_____	_____	_____	_____

Do you have any pending restrictions and/or suspensions on your current professional license/registration that would restrain you from performing in this position? __ YES __ NO

Have you ever been refused professional licensure, or had a license/registration suspended or revoked? __ YES __ NO

If yes, please explain: _____

List any trade or professional organizations of which you are a member, include offices held: _____
_____

List any special skills: _____

EMPLOYMENT HISTORY

Starting with your most recent employment, give a complete record of all employment and reasons for periods of unemployment.

How many years of experience do you have related to this position? _____

MAY WE CONTACT YOUR CURRENT EMPLOYER? __ YES __ NO   If no, why? _____
_____

NOTE: If your current or most recent employer is not contacted before an offer of employment is made, then any offer of employment that is made will be subject to CHW subsequently contacting such employer, and may be withdrawn based on the information received from such employer.

COMPANY NAME	ADDRESS	CITY	STATE	ZIP CODE	AREA CODE PHONE
_____	_____	_____	_____	_____	(____) _____

TYPE OF BUSINESS   SUPERVISOR'S NAME, TITLE & PHONE NUMBER:

_____   _____

DATE EMPLOYED: MO _____   YR _____   DATE LEFT: MO _____   YR_____

TITLE AND DUTIES: _____

REASON FOR LEAVING: _____

IF YOUR EMPLOYMENT RECORDS EXIST UNDER ANOTHER NAME, PLEASE SPECIFY: _____

FINAL SALARY: $_____

(Continued on next page)

Figure 12.3.   *(Continued)*

COMPANY NAME	ADDRESS	CITY	STATE	ZIP CODE	AREA CODE PHONE
_____	_____	_____	_____	_____	(_____) _____

TYPE OF BUSINESS   SUPERVISOR'S NAME, TITLE & PHONE NUMBER:

_____   _____

DATE EMPLOYED: MO _____   YR _____   DATE LEFT: MO _____   YR_____
TITLE AND DUTIES: _____
REASON FOR LEAVING: _____
IF YOUR EMPLOYMENT RECORDS EXIST UNDER ANOTHER NAME, PLEASE SPECIFY: _____
FINAL SALARY: $_____

COMPANY NAME	ADDRESS	CITY	STATE	ZIP CODE	AREA CODE PHONE
_____	_____	_____	_____	_____	(_____) _____

TYPE OF BUSINESS   SUPERVISOR'S NAME, TITLE & PHONE NUMBER:

_____   _____

DATE EMPLOYED: MO _____   YR _____   DATE LEFT: MO _____   YR_____
TITLE AND DUTIES: _____
REASON FOR LEAVING: _____
IF YOUR EMPLOYMENT RECORDS EXIST UNDER ANOTHER NAME, PLEASE SPECIFY: _____
FINAL SALARY: $_____

GIVE THREE ADDITIONAL WORK-RELATED REFERENCES

Name	Occupation or Title	Firm Name and Address (include city, state, and zip)	Phone
_____	_____	_____	_____
_____	_____	_____	_____
_____	_____	_____	_____

CERTIFICATION AND AGREEMENT

I certify that the information I provided in this application is complete and accurate to the best of my knowledge. I understand that any misrepresentation or omission of facts in this application disqualifies me from further consideration, or, if I am employed, is sufficient cause for dismissal. I understand that any alteration of this application in content or form may be considered cause for disqualification and/or termination.

I authorize investigation of all statements contained in this application and understand that I may be required to provide verification (diploma, license, transcripts, type tests, etc.) of information contained in this application.

I authorize any and all persons, companies or agencies to release to CHW any and all information they may have which is relevant to the application process. I also release all such parties from any liability that may result from furnishing information to CHW.

I understand that to be considered as a formal applicant, the position for which I am applying must be specifically identified as open, and recruitment for the position going on at the time this application is received by the Human Resources Department.

I understand that if I am employed with CHW, my employment will be at-will. As such, it can be terminated by me or by CHW with or without advance notice, at any time, and for any reason not prohibited by law. I agree that if I am employed by CHW, I will review the information contained in CHW's General Information Handbook.

I understand that any employment offer is contingent upon the following: (1) producing documents establishing my eligibility to work in the United States; (2) satisfactorily passing the preemployment drug screen, criminal background, and reference checks; and (3) complying with CHW's preemployment application procedures.

By either writing or typing my name and submitting this application to CHW, I acknowledge that I have read the certification and agreement and agree to abide by its terms.

NAME: _____   DATE :_____

Community Hospital of the West is an Equal Employment Opportunity/Affirmative Action Employer.

Whatever interview approach or combination of approaches is used, the information collected at this point in the recruitment process should be enough to determine who will advance to the next stage of the selection process. Reference checks; criminal background checks; and verifications of licensure, training, and educational requirements should be documented before any applicant is offered a position. In the past, checking past employment references sometimes yielded important new information about a candidate. Today, however, most organizations release only general information regarding an employee's past work performance, such as dates of employment and whether or not the applicant is eligible for rehire. While the information received from these contacts may be minimal in terms of a potential employee's skills, they do speak to the veracity of documentation in the resume in general and may be critical in a legal defense situation for a facility.

Some facilities utilize a point scoring process for interviews that awards points for education, knowledge at time of interview, and experience with the specific population of patients. This system helps to ensure that the top-qualified individuals will be in the running for a position.

Figure 12.4.   Appropriate and Inappropriate Interview Questions

Examples of Appropriate Job-Related Questions

- How does your education and work experience relate to the position we are discussing?
- What courses did you take in school?
- What was your grade point average?
- Are you willing to work on weekends?
- Are you willing to work overtime?
- What kind of work have you done in your previous positions?
- What kind of work do you enjoy the most? The least?
- What areas of your work skills would you like to improve?
- How do you perform under pressure?
- How many people were you responsible for supervising in your previous positions?
- Why did you leave your last position?
- How do you explain the gaps in your employment history?
- Would you require any additional training to perform this position?
- What are your career goals?
- What is your understanding of this area of the healthcare industry?
- What was your attendance record like in your previous positions?
- How long have you lived in this area?
- Are you able to work the hours required for this position?

Examples of Inappropriate Questions Unrelated to the Job

- Where were you born?
- What race are you?
- What does your husband (or father) do for a living?
- Are you married?
- Do you plan to get married?
- How old are you?
- Do you have any children? How many? What are their ages?
- Do you plan to have children?
- Is your husband likely to be transferred?
- Who cares for your children while you are at work?
- Are you involved in any church groups?
- Where do you live?
- Do you own or rent the place where you live?
- What organizations do you belong to?
- Have you ever been arrested?

Step 2: Conduct new employee orientation and training

The JCAHO reported in a Hospital Executive Briefings conference in 2002 that "inadequate orientation and training are the root causes of 63% of all sentinel events." Orientation should be provided to every new hire, including contracted staff, agency personnel, and volunteers. New staff members are required to receive an overview of the organization as well as details about their specific job responsibilities. The organization's leaders are expected to develop a process to ensure that all new staff are informed of performance expectations and specific responsibilities and that they are qualified to fulfill these expectations. When these issues are addressed before new staff members are allowed to perform independently, the orientation process promotes safe and effective job performance. In general, staff should be oriented to the organization's mission, vision, values, PI program, life safety procedures, infection control practices, specific job duties, and safety issues specific to their job assignments. Additional orientation and training may be specific to a particular type of healthcare organization, such as age-specific training for staff working with neonatal or pediatric age groups where skills at interpreting nonverbal communication is essential.

New staff should not work independently until orientation is complete and competency has been demonstrated. Testing to verify the competency of a clinical staff member may include having the employee insert a catheter, draw blood, or use a piece of medical equipment. This type of training and orientation may include having the new employee shadow a seasoned employee for a specified period of time, complete a self-directed study course, or attend classroom instruction and then demonstrate his or her competency by return demonstration a set number of times with a certified trainer.

As they are completed, orientation and demonstrated job competencies must be documented. Some organizations use a training checklist to document employee orientation. The checklist documents the names of the trainee and trainer(s), orientation dates, tasks, and demonstrated competencies. The orientation checklist should be maintained in the employee's personnel file so that it can be used in the performance appraisal process to monitor whether training and competency requirements have been met.

Aggregate data can also be compiled from training checklists to monitor organization-wide and care- or service-specific orientation and competency requirements. In hospitals, aggregate data on the levels of staff competence, relevant patterns and trends in training needs, and competence maintenance activities should be reported to the board of directors at least annually. Gathering data for the competence assessment process can be accomplished through observations, demonstrations, and PI reports such as occurrence report findings, peer review findings, and satisfaction surveys.

Policies and procedures facilitate the orientation and training of staff and form the basis for individual accountability. Policies and procedures should clearly represent and communicate the required functions and tasks the employee is expected to perform, as well as the procedures to initiate when an incident occurs that is not an accepted occurrence in the facility, such as a medication error. Employee training related to incident reporting is critical to all employees in the area of risk identification and reduction. Data collected from this process become the basis for much performance improvement in healthcare facilities. (Chapter 9 discusses the application of incident reports to performance improvement in greater detail.)

Step 3: Manage the performance appraisal process

Every healthcare organization should have a process in place to periodically assess each staff member's ability to meet performance expectations described in his or her job description. Healthcare organizations determine how often these assessments should occur. For example, competence can be assessed at hire; by the end of the orientation process; when new or updated technologies, products, procedures, or services are introduced; when job responsibilities change; or at regular intervals, such as once per year.

The employee performance appraisal process should be specific to the staff member's assigned responsibilities and assessed competencies. Ideally, the appraisal should include a one-on-one discussion between the employee and the supervisor or manager along with a written appraisal that provides room for employee response. The employee and supervisor or manager should work together to develop new performance goals and modify or enhance performance standards. The manager or supervisor should follow up with the employee to monitor his or her progress in meeting goals and performance standards through frequent, informal assessments as well as periodic formal appraisals. At a minimum, the periodic assessment should address each of the following areas of the employee's performance:

- Degree of compliance with written standards of performance as stated in the staff member's job description

- Participation in ongoing PI and patient safety activities

- Findings from competency assessment activities

Other areas such as a staff members' strengths, weaknesses, working relationships, morale, motivation, and attendance may also be addressed during the appraisal process.

The performance criteria used in the appraisal process should be directly linked to the employee's current position description and to the organization's mission. This process should also verify that the employee has maintained his or her credentials or licenses, as appropriate.

Organizations that have successfully implemented a PI culture have expanded the appraisal process to include team performance in tandem with individual performance. Teams may include a clinical treatment team, an administrative or leadership team, or a single department team. The appraisal process includes the team's definition of performance expectations and goals and a periodic assessment of its performance as a team rather than as individuals.

Ideally, organizationwide performance measures should be referenced and incorporated into the employee and team appraisal process. Provider- or employee-specific PI and patient safety monitoring activities may identify staff development needs. This same information reported in aggregate may identify trends in care that support the need for changes in staffing patterns, skill sets, or the care environment.

Other clinical service and human resources screening criteria that are used to monitor staffing effectiveness, as noted in figure 12.1, are also reviewed in preparation for the organization's annual strategic planning process.

Step 4: Manage the retention process

Commitment to the development and retention of human resources is of strategic importance to every healthcare organization. High retention rates communicate a strong message about the organization's values to existing and potential employees. Beyond traditional in-house orientation programs, many organizations offer tuition and professional education reimbursement as an incentive for ongoing staff development and retention. Maintaining a competitive salary and benefits package continues to be a primary leverage point in recruitment and retention programs. Organizations that go beyond traditional incentives look to employee input for the creation of reward structures in promoting employee retention. Incentives such as profit-sharing, job-sharing, sign-on bonuses, shared leadership, and cutting-edge technology resources are a few of the trends.

Certainly the most compelling link to employee retention is the creation of a work environment in which organizational and personal values mesh and a natural synergy develops. Teams working together with a common vision begin with a self-directed, motivated workforce. Many times, this team factor will directly influence employee retention even more than salary issues. Healthcare-related literature indicates that money is not the number one or two reason why people leave positions (Shaw, et al. 1998). Employees who have been prepared for their jobs through education and who feel that they are part of a vital team providing excellent care are willing to stay even if their salary is not tops in the field.

Step 5: Credential and extend privileges to licensed independent practitioners

Another category of staff who work in healthcare organizations is the medical staff. Generally referred to as independent practitioners, they include individuals permitted by law and the organization to provide patient care services without direction or supervision, within the scope of their license and individually granted clinical privileges. Physicians, dentists, and podiatrists are the most commonly credentialed and privileged independent practitioners working in healthcare organizations. However, physicians, certified nurse anesthetists, physician's assistants, nurse practitioners, registered nurse–midwives, speech pathologists, dieticians, clinical psychologists, and clinical social workers may also be privileged and credentialed by the healthcare organization, as either supervised employees or contract workers.

The credentialing and privileging process for the initial appointment and reappointment of independent practitioners should be defined in the healthcare organization's medical staff bylaws and uniformly applied. The credentialing process includes obtaining, verifying, and assessing qualifications of a licensed healthcare practitioner. State licenses, postgraduate degrees, residency and fellowship training, specialty board status, and other healthcare affiliations should all be primary source verified. Primary source verification involves contacting the original source of a specific credential to affirm qualifications reported by an individual healthcare practitioner. The primary source verification process is sometimes contracted out by the healthcare organization to a credentials verification organization (CVO), such as the American Medical Association Physician Master File, which provides verification of a physician's medical school graduation and residency completion, or the American Board of Medical Specialties for verification of a physician's board certification. Although using such agencies may relieve the organization from the process of gathering the information, it does not relieve the organization from the responsibility of having complete and accurate information on the applicant.

Each licensed independent practitioner who provides care under the auspices of a healthcare organization must do so in accordance with delineated clinical privileges. The delineation of clinical privileges should be based on the practitioner's training, experience, and proven clinical competence. The privilege lists sent out to applicants at initial appointment and at reappointment should be associated with the applicant's type of practice and limited to hospital-specific privileges.

The initial appointment process includes the independent practitioner requesting, completing, and submitting an application to the medical staff along with a request for delineated clinical privileges. The credentialing and privileging process is generally initiated upon completion and approval of an application for membership and request for privileges. The information submitted by the applicant should include the following:

- Education, both undergraduate and postgraduate with names of institutions

- Training, including residencies, fellowships, or any continuing medical education or training for new skills/privileges

- Previous and current healthcare affiliations (hospitals practiced in, private office locations)

- Specialty board certifications

- Current state licenses

- Drug Enforcement Administration (DEA) registration number with expiration dates

- Professional (peer) references who have personal knowledge of the applicant's recent professional performance and experience

- Information on current health status

- Professional liability insurance coverage

- Past and present professional litigation and liability history

- Clinical privileges being requested

Letters are generally sent to at least two peer references and one healthcare organization where the applicant currently holds privileges, requesting specific information about the applicant's qualifications and competencies in relation to the privileges requested, health status, and professional working relationships as observed by the reference source.

Healthcare organizations are required by law to query two additional sources of information on applicants requesting clinical privileges: the **National Practitioner Data Bank** (NPDB) and the **Healthcare Integrity and Protection Data Bank** (HIPDB). The NPDB maintains reports on medical malpractice settlements, clinical privilege actions, and professional society membership actions against licensed healthcare providers. The HIPDB maintains reports on civil judgments and criminal convictions of licensed healthcare providers.

Once applicant information has been source verified and references and data bank queries have been returned, the application and supporting documentation are reviewed by the organization's credentials and/or medical executive committee(s). Individual appointment and privilege delineation recommendations are then forwarded to the organization's board of directors for final determination. When the application is approved, the appointment period is generally for two years. A provisional period is generally required (the medical staff bylaws should specify a time limit) for all new staff members. The performance of new staff members should be observed and monitored by an assigned proctor during the provisional period.

In 2002, the JCAHO added a new standard that required hospital medical staffs to initiate a process to identify and manage issues of individual physician health. This must be done separately from the disciplinary function of the medical staff. The purpose of the standard is to help facilities address the physical health of practitioners who are at risk in their industry for such job-related illnesses as addiction, stress-related physical illness, and emotional illnesses. Regular staff education regarding these potential illnesses is another aspect of the new standard. Facilities will be required to facilitate the process for confidential diagnosis, treatment, and rehabilitation for any potentially impairing conditions such as the ones defined. These procedures must be done in compliance with any state or federal reporting regulations.

The reappraisal and reappointment process generally occurs every two years and includes a review of current licenses, DEA registration, professional liability coverage, national data bank queries, continuing medical education (as required by the state and/or for competency training) for new privileges, peer references, information related to claims and litigation, health, and changes in outside affiliations. Training, education, and certification, as well as the frequency with which a clinical privilege has been exercised, should show evidence of continuing proficiency for the privileges requested. An assessment of the provider's profile since the last staff appointment to verify peer review activities provides the final basis for supporting reappointment or reprivileging.

An ongoing provider profile includes both administrative aspects of staff membership (for example, meeting attendance statistics, medical record delinquency status, medical staff committee appointments, and practice volume statistics) and clinical performance data (for example, clinical outcome statistics, committee or department citations, peer review, and performance-monitoring reviews and actions). Again, the NPDB and HIPDB should be queried to determine whether any adverse information has been reported since the previous period of appointment.

Following review of the reappointment application information and provider profile information, the chief of staff or clinical department chairperson provides a written recommendation on reappointment and reprivileging to the credentials and/or executive committee of the medical staff. These bodies then make their own recommendation and forward it to the board of directors for a final decision. When a recommendation for continuing privileges is not made, or a restriction in privileges requested, the applicant must be offered due process. **Due process** provides for fair treatment through a hearing procedure that is generally outlined in the healthcare organization's medical staff bylaws. The procedure stipulates the means by which the applicant's application and supporting materials will be reviewed by an impartial panel to ensure objective assessment.

Real-Life Example

Western States University Medical Center recently completed a national search and recruitment effort to fill its chief clinical officer (CCO) position. The position had been vacant for six months and had experienced rapid turnover of two previous CCOs within the past two years. The members of the search team invited to participate in the recruitment effort included the chief executive medical director, the chief financial officer, the human resources director, the medical staff president, and the nurse managers from each clinical care area.

The search team followed internal policy requirements related to the recruitment process. The position description was reviewed and updated, and from it minimum position requirements were identified and included in the advertisement posted internally and published in the local newspaper and a national nursing journal. The search team also retained the services of a national recruitment firm in an effort to fast-track the recruitment process because the organization was less than a year away from its triennial survey with the Joint Commission on Accreditation of Healthcare Organizations.

The search generated about fifteen applicants. The team screened the applicants and narrowed the list down to the six that seemed to meet the minimum position requirements. These six applicants (five of whom were out-of-state candidates) were interviewed via conference call. The selection was then narrowed to two candidates, who were invited to fly in and meet with the search team.

Of primary importance to this search team's recruitment charge was finding a chief clinical officer who had a clinical background and experience that paralleled the services provided at Western States. The candidate's experience, strong knowledge of accreditation and licensure requirements, educational background with a master's degree, past employment patterns, personal demeanor, communication style, leadership philosophy, and availability were important factors in the selection. The final candidate selected by the search team met the important criteria and minimum position requirements. Reference checks and source documents verifying the applicant's credential confirmed the search team's decision to hire the final candidate as the CCO.

Five months into the new CCO's tenure, and six weeks before the triennial survey, Western States's director of risk and quality management resigned. Because the risk and quality position reported to the CCO, the new CCO was asked to facilitate survey preparations. Within days of this decision, the CCO requested that the leadership team recruit a local consulting group to assist in survey coordination efforts. Just two weeks before the anticipated survey, the CCO resigned.

Two main factors contributed to the CCO's resignation. First, the source documents confirming the CCO's graduate education were not in her human resources file. As personnel files were being reviewed in preparation for survey inspection, it was noted that the primary source verification of the CCO's graduate degree was incomplete. When the graduate program was contacted, it was revealed that the CCO had never completed the master's program. Second, although the CCO's curriculum vitae and references confirmed that she had twenty-plus years' experience in nursing administration and JCAHO survey work, she was unable to assume a leadership role in facilitating the nursing component of the JCAHO survey preparation process.

Western States University Medical Center has since defined what information it primary source verifies (verbal and written) when confirming education, experience, and licensure. It has also implemented a check-off list to be used by human resources to ensure that key action items are completed in the recruitment process. Western States University Medical Center has also developed sets of interview questions that better evaluate key competencies and skill sets of applicants.

QI Toolbox Technique

Summary profiles of physician performance provide the credentials committee with significant data about specific physicians scheduled for reappointment to the medical staff. With this information, the credentials committee has the information needed to make a sound decision on the performance of members of the hospital's medical staff.

The type of data captured on a physician profile summary should be unique to the healthcare organization's specific needs. The sample form in figure 12.5 is designed for a physician in the specialty area of obstetrics. It includes data about the physician's cesarean section rate, the number of vaginal births after a cesarean section, types of medications used, transfusion usage, record completion delinquency rates, risk management issues, and attendance at medical staff meetings. It also allows documentation of any disciplinary action that has taken place since the last credentialing process for the physician.

Case Study

The physician performance data for the Obstetric Service at Community Hospital of the West are provided in figure 12.5. In figure 12.6, an example of a physician profile for an individual physician, Dr. Jones, is listed. The physician index summary for Dr. Doe, an OB/GYN physician on the medical staff, is provided in figure 12.7. Using the physician index summary, students should complete as much of the information as possible on the blank physician profile (figure 12.8) for Dr. Doe. They should identify which information cannot be obtained from the physician index. Then they should determine what other sources of hospital data would have to be accessed in order to complete the physician profile for Dr. Doe. Finally, they should compare Dr. Doe's performance to the performance of the rest of the physicians on the Obstetric Service in the areas for which data are available. (See figure 12.5.)

Case Study Questions

1. How did Dr. Doe perform in comparison to the rest of the service?
2. Should Dr. Doe be reappointed to the medical staff? Why or why not?

Project Application

Students should identify any human resources issues that may have a bearing on their project. They should determine what types of issues these are and how they can obtain the data needed to evaluate them. Then they should develop recommendations for staffing.

Figure 12.5.   Physician Profile for OB/GYN Group

COMMUNITY HOSPITAL OF THE WEST
PHYSICIAN PERFORMANCE REVIEW SUMMARY FOR REAPPOINTMENT

		Profile Time Frame:	
SERVICE	*OB/GYN*	**From:**	1/1/
CATEGORY	*Active*	**To:**	12/31/

UTILIZATION:

Admissions	*1,400*	Procedures	*598*	
Patient Days	*5,143*	V-BACs	*154*	
Deliveries	*1,187*	Blood Given	*25*	
C-Sections	*137*			

OUTCOMES:

Category	#	%	Comments:
C-Section Rate	*137*	*11.5%*	
V-BAC Rate	*154*	*11%*	
Nosocomial Inf Rate	*21*	*1.5%*	
Surgical Wound Inf Rate	*5*	*0.36%*	
Mortality Rate	*1*	*0.07%*	

PERFORMANCE REVIEW:

Category	# Reviewed	# Appropriate/%	Comments:
Surgical/Inv/Noninvasive Procedures	*60*	*58/96.7%*	
Medication Use	*140*	*139/99.3%*	
Blood Use	*25*	*25/100%*	
Utilization Management	*140*	*135/95.7%*	
Other Peer Review	*140*	*130/92.8%*	*Clinical Pert.*

DATA QUALITY

Data Quality Monitoring			Comments:
Delinquency (>21 days)	*15*	*7.7%*	
Suspensions	*5*	*2.6%*	

RISK/SAFETY MANAGEMENT

Incidents reported by other professionals/ administration	*2*	*2/100%*	Comments:
Litigation	*1*	*0.07%*	

MEETING ATTENDANCE

Medical Staff Meetings	*26*	*93%*	Comments:
Committee Meetings	*40*	*85%*	

Figure 12.6.   Physician Profile for Dr. Jones

COMMUNITY HOSPITAL OF THE WEST
PHYSICIAN PERFORMANCE REVIEW SUMMARY FOR REAPPOINTMENT

PHYSICIAN	Bob Jones, MD		Profile Time Frame:	
SERVICE	OB/GYN		From:	1/1
CATEGORY	Active		To:	12/31

UTILIZATION:				
Admissions	175	Procedures	53	
Patient Days	540	V-BACs	28	
Deliveries	145	Blood Given	2	
C-Sections	22			

OUTCOMES:			
Category	#	%	Comments:
C-Section Rate	22	15.2%	
V-BAC Rate	28	80%	
Nosocomial Inf Rate	3	1.71%	
Surgical Wound Inf Rate	2	3.8%	
Mortality Rate	1	0.57%	

PERFORMANCE REVIEW:			
Category	# Reviewed	# Appropriate/%	Comments:
Surgical/Inv/Noninvasive Procedures	6	5/83%	
Medication Use	10	8/80%	
Blood Use	2	2/100%	
Utilization Management	18	18/100%	
Other Peer Review	N/A		

DATA QUALITY			
Data Quality Monitoring			Comments:
Delinquency (>21 days)	3	12.5%	
Suspensions	1	4.2%	

RISK/SAFETY MANAGEMENT			
Incidents reported by other professionals/ administration	0	0%	Comments:
Litigation	1	0.57%	

MEETING ATTENDANCE			
Medical Staff Meetings	4	100%	Comments:
Committee Meetings	10	80%	

Figure 12.6. *(Continued)*

FOR COMPLETION BY SERVICE CHAIRMAN OR CREDENTIALS COMMITTEE CHAIRMAN

CATEGORY	YES	NO	Comments:
Has the applicant been considered for or subject to disciplinary action since last reappointment?		√	
Have the applicant's privileges or staff appointment been suspended, revoked, or diminished in any way, either voluntary or involuntary, since last reappointment?	√		
Are there any currently pending challenges to any licensure or registration or the voluntary relinquishment of such?		√	
Are there any physical or behavioral conditions or limitations?		√	
Has the applicant exhibited satisfactory professional performance?	√		

APPROVALS: **APPROVED?**

REVIEWER	SIGNATURE	YES	NO	DATE
Service Chair				
Credentials Chair				

Figure 12.7.   Physician Index Summary for Dr. Doe

Patient Age	LOS	Discharge Status	Final Dx	Diagnosis Text	Final Proc	Procedure Text
52	3	Home	6262	EXCESSIVE/FREQUENT MENSTRUATION	684	TOTAL ABDOMINAL HYSTERECTOMY
			4254	PRIMARY CARDIOMYOPATHIES	6562	REMOVAL OF REMAINING OVARY AND TUBE
			4019	ESSENTIAL HYPERTENSION, UNSPECIFIED BENIGN		
			25000	DIABETES MELLITUS WITHOUT COMPLICATION,		
			2181	INTRAMURAL LEIOMYOMA OF UTERUS		
29	4	Home	65341	FETOPELVIC DISPROPORTION, DELIVERED	741	LOW CERVICAL CESAREAN SECTION
			66111	SECONDARY UTERINE INERTIA, DELIVERED	731	SURGICAL INDUCTION OF LABOR
			65841	INFECTION OF AMNIOTIC CAVITY, DELIVERED	7309	ARTIFICIAL RUPTURE OF MEMBRANES
			64501	PROLONGED PREGNANCY, DELIVERED		
			V270	MOTHER WITH SINGLE LIVEBORN		
27	2	Home	65421	PREVIOUS CESAREAN DELIVERY, DELIVERED	736	EPISIOTOMY (with subsequent repair)
			V270	MOTHER WITH SINGLE LIVEBORN	7359	MANUALLY ASSISTED DELIVERY
					7309	ARTIFICIAL RUPTURE OF MEMBRANES
79	4	Home	6185	PROLAPSE OF VAGINAL VAULT AFTER HYSTERECTOMY	7077	VAGINAL SUSPENSION & FIXATION
			9975	URINARY COMPLICATION, NOT ELSEWHERE CLASSIFIED	5459	OTHER LYSIS OF PERITONEAL ADHESIONS
			5990	URINARY TRACT INFECTION, SITE NOT SPECIFIED		
			9973	RESPIRATORY COMPLICATION, NOT ELSEWHERE		
			5180	PULMONARY COLLAPSE (ATELECTASIS)		
			78831	URGE INCONTINENCE		
			E8788	OTHER SURGICAL OPERATION, WITH ABNORMAL		
			4019	ESSENTIAL HYPERTENSION, UNSPECIFIED BENIGN		
			5680	PERITONEAL ADHESIONS (POSTOPERATIVE)(POS		
25	4	Home	65221	BREECH PRESENTATION WITHOUT VERSION, DELIVERED	741	LOW CERVICAL CESAREAN SECTION
			64661	INFECTIONS OF GENITOURINARY TRACT IN PRE		
			6169	UNSPECIFIED INFLAMMATORY DISEASE OF CERVIX		
			66311	CORD AROUND NECK, WITH COMPRESSION, COMP		
			V270	MOTHER WITH SINGLE LIVEBORN		
39	1	Home	66331	UNSPECIFIED CORD ENTANGLEMENT, WITHOUT C	7359	MANUALLY ASSISTED DELIVERY
			64201	BENIGN ESSENTIAL HYPERTENSION COMPLICATI		
			V270	MOTHER WITH SINGLE LIVEBORN		
			4019	ESSENTIAL HYPERTENSION, UNSPECIFIED BENIGN		

Figure 12.7. *(Continued)*

Patient Age	LOS	Discharge Status	Final Dx	Diagnosis Text	Final Proc	Procedure Text
34	2	Home	64881	ABNORMAL GLUCOSE TOLERANCE IN MOTHER COM	7359	MANUALLY ASSISTED DELIVERY
			V270	MOTHER WITH SINGLE LIVEBORN	7569	REPAIR OF CURRENT OBSTETRIC LACERATION
			66401	FIRST-DEGREE PERINEAL LACERATION, DELIVERED	7309	ARTIFICIAL RUPTURE OF MEMBRANES
20	1	Home	650	NORMAL DELIVERY	7359	MANUALLY ASSISTED DELIVERY
			V270	MOTHER WITH SINGLE LIVEBORN	736	EPISIOTOMY (with subsequent repair)
36	3	Home	6398	COMPLICATION FOLLOWING ABORTION/ECTOPIC/	684	TOTAL ABDOMINAL HYSTERECTOMY
			6259	UNSPECIFIED SYMPTOM ASSOCIATED WITH FEMA		
			311	DEPRESSIVE DISORDER, NOT ELSEWHERE CLASS		
			30503	ALCOHOL ABUSE IN REMISSION		
			30593	MIXED/UNSPECIFIED DRUG ABUSE IN REMISSION		
			6262	EXCESSIVE/FREQUENT MENSTRUATION		
			2182	SUBSEROUS LEIOMYOMA OF UTERUS		
			2181	INTRAMURAL LEIOMYOMA OF UTERUS		
			78701	NAUSEA WITH VOMITING		
48	3	Home	2181	INTRAMURAL LEIOMYOMA OF UTERUS	684	TOTAL ABDOMINAL HYSTERECTOMY
			57410	CALCULUS OF GALLBLADDER WITH CHOLECYSTITIS	6561	REMOVAL OF BOTH OVARIES AND TUBES AT SAM
			6170	ENDOMETRIOSIS OF UTERUS	595	RETROPUBIC URETHRAL SUSPENSION
			2189	LEIOMYOMA OF UTERUS, UNSPECIFIED	7052	REPAIR OF RECTOCELE
			6256	STRESS INCONTINENCE, FEMALE	7092	OPERATION ON CUL-DE-SAC
			6180	PROLAPSE OF VAGINAL WALLS WITHOUT MENTION	5123	LAPAROSCOPIC CHOLECYSTECTOMY
					8753	INTRAOPERATIVE CHOLANGIOGRAM
42	2	Home	65421	PREVIOUS CESAREAN DELIVERY, DELIVERED	734	MEDICAL INDUCTION OF LABOR
			V270	MOTHER WITH SINGLE LIVEBORN	7301	INDUCTION OF LABOR BY ARTIFICIAL RUPTURE
			66331	UNSPECIFIED CORD ENTANGLEMENT; WITHOUT	7359	MANUALLY ASSISTED DELIVERY
			64881	ABNORMAL GLUCOSE TOLERANCE IN MOTHER COM	7351	MANUAL ROTATION OF FETAL HEAD
			65291	UNSPECIFIED MALPOSITION/PRESENTATION OF		
22	2	Home	66421	THIRD-DEGREE PERINEAL LACERATION, DELIVERED	7562	REPAIR OF CURRENT OBSTETRIC LACERATION O
			65681	FETAL & PLACENTAL PROBLEM, AFFECTING MAN	7359	MANUALLY ASSISTED DELIVERY
			V270	MOTHER WITH SINGLE LIVEBORN		

(Continued on next page)

Figure 12.7. (Continued)

Patient Age	LOS	Discharge Status	Final Dx	Diagnosis Text	Final Proc	Procedure Text
37	14	Home	65971	ABNORMALITY IN FETAL HEART RATE/RHYTHM,	741	LOW CERVICAL CESAREAN SECTION
			67131	ANTEPARTUM DEEP PHLEBOTHROMBOSIS COMPLIC	6632	BILATERAL LIGATION AND DIVISION OF FALLO
			64821	ANEMIA IN MOTHER COMPLICATING PREGNANCY,		
			65821	DELAYED DELIVERY AFTER SPONTANEOUS/UNSPEC		
			65221	BREECH PRESENTATION WITHOUT VERSION, DEL		
			64421	EARLY ONSET OF DELIVERY, DELIVERED		
			65961	ELDERLY MULTIGRAVIDA, DELIVERED		
			64891	CURRENT CONDITION IN MOTHER COMPLICATING		
			2898	DISEASE OF BLOOD/BLOOD-FORMING ORGANS		
			7821	RASH AND NONSPECIFIC SKIN ERUPTION		
			2859	ANEMIA, UNSPECIFIED		
			V270	MOTHER WITH SINGLE LIVEBORN		
			V252	STERILIZATION		
28	1	Home	66331	UNSPECIFIED CORD ENTANGLEMENT, WITHOUT	736	EPISIOTOMY (with subsequent repair)
			65921	MATERNAL PYREXIA DURING LABOR, UNSPECIFIED	734	MEDICAL INDUCTION OF LABOR
			V270	MOTHER WITH SINGLE LIVEBORN	7301	INDUCTION OF LABOR BY ARTIFICIAL RUPTURE
22	2	Home	66331	UNSPECIFIED CORD ENTANGLEMENT, WITHOUT	7309	ARTIFICIAL RUPTURE OF MEMBRANES
			65921	MATERNAL PYREXIA DURING LABOR, UNSPECIFIED	7359	MANUALLY ASSISTED DELIVERY
			V270	MOTHER WITH SINGLE LIVEBORN	7569	REPAIR OF CURRENT OBSTETRIC LACERATION
			66411	SECOND-DEGREE PERINEAL LACERATION, DELIVERED		
42	2	Home	65961	ELDERLY MULTIGRAVIDA, DELIVERED	731	SURGICAL INDUCTION OF LABOR
			V270	MOTHER WITH SINGLE LIVEBORN	7359	MANUALLY ASSISTED DELIVERY
			66551	INJURY TO PELVIC ORGANS, DELIVERED	7569	REPAIR OF CURRENT OBSTETRIC LACERATION
67	3	Home	6271	POSTMENOPAUSAL BLEEDING	684	TOTAL ABDOMINAL HYSTERECTOMY
			4465	GIANT CELL ARTERITIS	6561	REMOVAL OF BOTH OVARIES AND TUBES AT SAM
			2765	VOLUME DEPLETION DISORDER (DEHYDRATION)	5459	OTHER LYSIS OF PERITONEAL ADHESIONS
			2181	INTRAMURAL LEIOMYOMA OF UTERUS		

Figure 12.7. *(Continued)*

Patient Age	LOS	Discharge Status	Final Dx	Diagnosis Text	Final Proc	Procedure Text
51	3	Home	6210 5680 V1251 49390 6270 2800	POLYP OF CORPUS UTERI PERITONEAL ADHESIONS (POSTOPERATIVE)(POS PERSONAL HISTORY OF VENOUS THROMBOSIS AN ASTHMA, UNSPECIFIED TYPE, WITHOUT STATUS PREMENOPAUSAL MENORRHAGIA IRON DEFICIENCY ANEMIA SECONDARY TO BLOOD LOSS	684 6561	TOTAL ABDOMINAL HYSTERECTOMY REMOVAL OF BOTH OVARIES AND TUBES AT SAM
33	1	Home	2181 6208 6200 6259 66702 V270 65281 64501	INTRAMURAL LEIOMYOMA OF UTERUS NONINFLAMMATORY DISORDER OF OVARY/FALLOP FOLLICULAR CYST OF OVARY UNSPECIFIED SYMPTOM ASSOCIATED WITH FEMALE RETAINED PLACENTA WITHOUT HEMORRHAGE, DEL MOTHER WITH SINGLE LIVEBORN MALPOSITION/MALPRESENTATION OF FETUS, DELIVERED PROLONGED PREGNANCY, DELIVERED	7309 7351 7359 736 754	ARTIFICIAL RUPTURE OF MEMBRANES MANUAL ROTATION OF FETAL HEAD MANUALLY ASSISTED DELIVERY EPISIOTOMY (with subsequent repair) MANUAL REMOVAL OF RETAINED PLACENTA
22	2	Home	66331 V270	UNSPECIFIED CORD ENTANGLEMENT, WITHOUT MOTHER WITH SINGLE LIVEBORN	7359 7309 736	MANUALLY ASSISTED DELIVERY ARTIFICIAL RUPTURE OF MEMBRANES EPISIOTOMY (with subsequent repair)
25	2	Home	65281 V270	MALPOSITION/MALPRESENTATION OF FETUS, DEL MOTHER WITH SINGLE LIVEBORN	7351 736 7359 734 7309	MANUAL ROTATION OF FETAL HEAD EPISIOTOMY (with subsequent repair) MANUALLY ASSISTED DELIVERY MEDICAL INDUCTION OF LABOR ARTIFICIAL RUPTURE OF MEMBRANES
31	2	Home	66411 V270	SECOND-DEGREE PERINEAL LACERATION, DELIV MOTHER WITH SINGLE LIVEBORN	7359 7569 7301	MANUALLY ASSISTED DELIVERY REPAIR OF CURRENT OBSTETRIC LACERATION INDUCTION OF LABOR BY ARTIFICIAL RUPTURE
40	3	Home	6172 2189 25000 6146	ENDOMETRIOSIS OF FALLOPIAN TUBE LEIOMYOMA OF UTERUS, UNSPECIFIED DIABETES MELLITUS WITHOUT COMPLICATION, PELVIC PERITONEAL ADHESIONS, FEMALE (POS	684 6561	TOTAL ABDOMINAL HYSTERECTOMY REMOVAL OF BOTH OVARIES AND TUBES AT SAM

(Continued on next page)

Figure 12.7. *(Continued)*

Patient Age	LOS	Discharge Status	Final Dx	Diagnosis Text	Final Proc	Procedure Text
32	3	Home	65221 66881 3490 64811 2449 V270	BREECH PRESENTATION WITHOUT VERSION, DEL COMPLICATION OF ANESTHESIA/SEDATION IN L REACTION TO SPINAL/LUMBAR PUNCTURE THYROID DYSFUNCTION IN MOTHER COMPLICATI UNSPECIFIED ACQUIRED HYPOTHYROIDISM MOTHER WITH SINGLE LIVEBORN	741 0395	LOW CERVICAL CESAREAN SECTION SPINAL BLOOD PATCH
81	4	Home	6185 5180 6256 4019 2449 71690 9110 78701 5640	PROLAPSE OF VAGINAL VAULT AFTER HYSTERECTOMY PULMONARY COLLAPSE (ATELECTASIS) STRESS INCONTINENCE, FEMALE ESSENTIAL HYPERTENSION, UNSPECIFIED BENIGN UNSPECIFIED ACQUIRED HYPOTHYROIDISM ARTHROPATHY, UNSPECIFIED, UNSPECIFIED SI TRUNK, ABRASION/FRICTION BURN, WITHOUT NAUSEA WITH VOMITING CONSTIPATION	7050 595	REPAIR OF CYSTOCELE AND RECTOCELE RETROPUBIC URETHRAL SUSPENSION
28	2	Home	65231 66622 66481	TRANSVERSE/OBLIQUE PRESENTATION OF FETUS DELAYED & SECONDARY POSTPARTUM HEMORRHAGE TRAUMA TO PERINEUM & VULVA, DELIVERED	7309 7351 7569	ARTIFICIAL RUPTURE OF MEMBRANES MANUAL ROTATION OF FETAL HEAD REPAIR OF CURRENT OBSTETRIC LACERATION
54	3	Home	V270 1820 4240 6182 6256 6271 6171 6170 27800 9095 E9470 E8490	MOTHER WITH SINGLE LIVEBORN MALIGNANT NEOPLASM OF CORPUS UTERI, EXCE MITRAL VALVE DISORDER UTEROVAGINAL PROLAPSE, INCOMPLETE STRESS INCONTINENCE, FEMALE POSTMENOPAUSAL BLEEDING ENDOMETRIOSIS OF OVARY ENDOMETRIOSIS OF UTERUS OBESITY, UNSPECIFIED LATE EFFECT OF ADVERSE EFFECT OF DRUG, M DIETETICS CAUSING ADVERSE EFFECTS IN THE INJURY OR POISONING OCCURRING AT/IN THE	7271 684 7052 595 6561	VACUUM EXTRACTION WITH EPISIOTOMY TOTAL ABDOMINAL HYSTERECTOMY REPAIR OF RECTOCELE RETROPUBIC URETHRAL SUSPENSION REMOVAL OF BOTH OVARIES AND TUBES AT SAM

Figure 12.7. *(Continued)*

Patient Age	LOS	Discharge Status	Final Dx	Diagnosis Text	Final Proc	Procedure Text
19	1	Home	64421	EARLY ONSET OF DELIVERY, DELIVERED	7309	ARTIFICIAL RUPTURE OF MEMBRANES
			V270	MOTHER WITH SINGLE LIVEBORN	7359	MANUALLY ASSISTED DELIVERY
			65841	INFECTION OF AMNIOTIC CAVITY, DELIVERED	736	EPISIOTOMY (with subsequent repair)
37	3	Home	65221	BREECH PRESENTATION WITHOUT VERSION, DEL	741	LOW CERVICAL CESAREAN SECTION
			V270	MOTHER WITH SINGLE LIVEBORN		
			65801	OLIGOHYDRAMNIOS, DELIVERED		
24	2	Home	66421	THIRD-DEGREE PERINEAL LACERATION, DELIVE	—	—
			66331	UNSPECIFIED CORD ENTANGLEMENT, WITHOUT C		
			V270	MOTHER WITH SINGLE LIVEBORN		
21	2	Home	650	NORMAL DELIVERY	736	EPISIOTOMY (with subsequent repair)
			V270	MOTHER WITH SINGLE LIVEBORN	7359	MANUALLY ASSISTED DELIVERY
					7309	ARTIFICIAL RUPTURE OF MEMBRANES
					734	MEDICAL INDUCTION OF LABOR
39	2	Home	65941	GRAND MULTIPARITY, DELIVERED	—	—
			65961	ELDERLY MULTIGRAVIDA, DELIVERED		
			66401	FIRST-DEGREE PERINEAL LACERATION, DELIVERED		
			V270	MOTHER WITH SINGLE LIVEBORN		
			V252	STERILIZATION		
33	1	Home	650	NORMAL DELIVERY	736	EPISIOTOMY (with subsequent repair)
			V270	MOTHER WITH SINGLE LIVEBORN	7359	MANUALLY ASSISTED DELIVERY
					7309	ARTIFICIAL RUPTURE OF MEMBRANES
					734	MEDICAL INDUCTION OF LABOR
22	1	Home	66331	UNSPECIFIED CORD ENTANGLEMENT, WITHOUT C	7359	MANUALLY ASSISTED DELIVERY
			V270	MOTHER WITH SINGLE LIVEBORN	7309	ARTIFICIAL RUPTURE OF MEMBRANES
			64891	CURRENT CONDITION IN MOTHER COMPLICATING	736	EPISIOTOMY (with subsequent repair)
			V0251	CARRIER OR SUSPECTED CARRIER OF GROUP B		
20	1	Home	65811	PREMATURE RUPTURE OF MEMBRANES, DELIVERED	736	EPISIOTOMY (with subsequent repair)
			65971	ABNORMALITY IN FETAL HEART RATE/RHYTHM,	7359	MANUALLY ASSISTED DELIVERY
			64421	EARLY ONSET OF DELIVERY, DELIVERED		
			66311	CORD AROUND NECK, WITH COMPRESSION, COMP		
			64891	CURRENT CONDITION IN MOTHER COMPLICATING		
			V270	MOTHER WITH SINGLE LIVEBORN		
			V0251	CARRIER OR SUSPECTED CARRIER OF GROUP B		

(Continued on next page)

Figure 12.7. *(Continued)*

Patient Age	LOS	Discharge Status	Final Dx	Diagnosis Text	Final Proc	Procedure Text
34	3	Home	65221	BREECH PRESENTATION WITHOUT VERSION, DEL	741	LOW CERVICAL CESAREAN SECTION
			65811	PREMATURE RUPTURE OF MEMBRANES, DELIVERE	6632	BILATERAL LIGATION AND DIVISION OF FALLO
			64841	MENTAL DISORDER IN MOTHER COMPLICATING P		
			3051	TOBACCO USE DISORDER		
			V270	MOTHER WITH SINGLE LIVEBORN		
			V252	STERILIZATION		
22	1	Home	65281	MALPOSITION/MALPRESENTATION OF FETUS, DE	7351	MANUAL ROTATION OF FETAL HEAD
			65971	ABNORMALITY IN FETAL HEART RATE/RHYTHM,	7569	REPAIR OF CURRENT OBSTETRIC LACERATION
			66331	UNSPECIFIED CORD ENTANGLEMENT, WITHOUT C	7309	ARTIFICIAL RUPTURE OF MEMBRANES
			66401	FIRST-DEGREE PERINEAL LACERATION, DELIVERED	757	MANUAL EXPLORATION OF UTERINE CAVITY, PO
			V270	MOTHER WITH SINGLE LIVEBORN		
41	2	Home	64881	ABNORMAL GLUCOSE TOLERANCE IN MOTHER COM	7309	ARTIFICIAL RUPTURE OF MEMBRANES
			V270	MOTHER WITH SINGLE LIVEBORN	7351	MANUAL ROTATION OF FETAL HEAD
			64421	EARLY ONSET OF DELIVERY, DELIVERED	721	LOW FORCEPS OPERATION WITH EPISIOTOMY
			65701	POLYHYDRAMNIOS, DELIVERED		
			65951	ELDERLY PRIMIGRAVIDA, DELIVERED		
			65981	INDICATION FOR CARE/INTERVENTION RELATED		
			78703	VOMITING ALONE		
			66111	SECONDARY UTERINE INERTIA, DELIVERED		
			65671	PLACENTAL CONDITION, AFFECTING MANAGEMENT		
59	3	Home	6182	UTEROVAGINAL PROLAPSE, INCOMPLETE	6859	VAGINAL HYSTERECTOMY
			2182	SUBSEROUS LEIOMYOMA OF UTERUS	7050	REPAIR OF CYSTOCELE AND RECTOCELE
			6210	POLYP OF CORPUS UTERI	5718	SUPRAPUBIC CYSTOSTOMY
			2409	GOITER, UNSPECIFIED		
33	1	Home	66031	DEEP TRANSVERSE ARREST & PERSISTENT OCCI	7309	ARTIFICIAL RUPTURE OF MEMBRANES
			66401	FIRST-DEGREE PERINEAL LACERATION, DELIVE	7569	REPAIR OF CURRENT OBSTETRIC LACERATION
			V270	MOTHER WITH SINGLE LIVEBORN	7351	MANUAL ROTATION OF FETAL HEAD

Figure 12.7. *(Continued)*

Patient Age	LOS	Discharge Status	Final Dx	Diagnosis Text	Final Proc	Procedure Text
30	2	Home	65281	MALPOSITION/MALPRESENTATION OF FETUS, DE	721	LOW FORCEPS OPERATION WITH EPISIOTOMY
			65921	MATERNAL PYREXIA DURING LABOR, UNSPECIFIED	7359	MANUALLY ASSISTED DELIVERY
			65421	PREVIOUS CESAREAN DELIVERY, DELIVERED	757	MANUAL EXPLORATION OF UTERINE CAVITY, PO
67	3	Home	V270	MOTHER WITH SINGLE LIVEBORN	7092	OPERATION ON CUL-DE-SAC
			6180	PROLAPSE OF VAGINAL WALLS WITHOUT MENTION	7050	REPAIR OF CYSTOCELE AND RECTOCELE
			6256	STRESS INCONTINENCE, FEMALE	595	RETROPUBIC URETHRAL SUSPENSION
			6186	VAGINAL ENTEROCELE, CONGENITAL OR ACQUIRED		
81	3	Home	6181	UTERINE PROLAPSE WITHOUT MENTION OF VAGI	684	TOTAL ABDOMINAL HYSTERECTOMY
			6227	MUCOUS POLYP OF CERVIX	6561	REMOVAL OF BOTH OVARIES AND TUBES AT SAM
			6210	POLYP OF CORPUS UTERI	7092	OPERATION ON CUL-DE-SAC
			2181	INTRAMURAL LEIOMYOMA OF UTERUS		
			25000	DIABETES MELLITUS WITHOUT COMPLICATION,		
			4019	ESSENTIAL HYPERTENSION, UNSPECIFIED BENIGN		
			2724	UNSPECIFIED HYPERLIPIDEMIA		
26	1	Home	650	NORMAL DELIVERY	7359	MANUALLY ASSISTED DELIVERY
			V270	MOTHER WITH SINGLE LIVEBORN	734	MEDICAL INDUCTION OF LABOR
					736	EPISIOTOMY (with subsequent repair)
18	1	Home	66411	SECOND-DEGREE PERINEAL LACERATION, DELIVERED	7569	REPAIR OF CURRENT OBSTETRIC LACERATION
			64841	MENTAL DISORDER IN MOTHER COMPLICATING P	7359	MANUALLY ASSISTED DELIVERY
			3051	TOBACCO USE DISORDER	7309	ARTIFICIAL RUPTURE OF MEMBRANES
			V270	MOTHER WITH SINGLE LIVEBORN		
72	3	Home	1820	MALIGNANT NEOPLASM OF CORPUS UTERI, EXCE	684	TOTAL ABDOMINAL HYSTERECTOMY
			2449	UNSPECIFIED ACQUIRED HYPOTHYROIDISM	6561	REMOVAL OF BOTH OVARIES AND TUBES AT SAM
			6202	OVARIAN CYST	403	REGIONAL LYMPH NODE EXCISION
			7912	HEMOGLOBINURIA		
38	3	Home	6185	PROLAPSE OF VAGINAL VAULT AFTER HYSTEREC	7077	VAGINAL SUSPENSION & FIXATION
50	3	Home	6170	ENDOMETRIOSIS OF UTERUS	684	TOTAL ABDOMINAL HYSTERECTOMY
			2800	IRON DEFICIENCY ANEMIA SECONDARY TO BLOOD	6561	REMOVAL OF BOTH OVARIES AND TUBES AT SAM

(Continued on next page)

Figure 12.7. *(Continued)*

Patient Age	LOS	Discharge Status	Final Dx	Diagnosis Text	Final Proc	Procedure Text
			6172	ENDOMETRIOSIS OF FALLOPIAN TUBE	595	RETROPUBIC URETHRAL SUSPENSION
			6171	ENDOMETRIOSIS OF OVARY		
			6256	STRESS INCONTINENCE, FEMALE		
			6262	EXCESSIVE/FREQUENT MENSTRUATION		
			2181	INTRAMURAL LEIOMYOMA OF UTERUS		
			6160	CERVICITIS & ENDOCERVICITIS		
			6173	ENDOMETRIOSIS OF PELVIC PERITONEUM		
			6259	UNSPECIFIED SYMPTOM ASSOCIATED WITH FEMALE		
			25000	DIABETES MELLITUS WITHOUT COMPLICATION,		
			5533	DIAPHRAGMATIC HERNIA		
28	1	Home	65801	OLIGOHYDRAMNIOS, DELIVERED	721	LOW FORCEPS OPERATION WITH EPISIOTOMY
			65921	MATERNAL PYREXIA DURING LABOR, UNSPECIFI	7351	MANUAL ROTATION OF FETAL HEAD
			65281	MALPOSITION/MALPRESENTATION OF FETUS, DE	7569	REPAIR OF CURRENT OBSTETRIC LACERATION
			66951	FORCEPS/VACUUM EXTRACTOR DELIVERY WITHOUT	754	MANUAL REMOVAL OF RETAINED PLACENTA
			66411	SECOND-DEGREE PERINEAL LACERATION, DELIV	734	MEDICAL INDUCTION OF LABOR
			66331	UNSPECIFIED CORD ENTANGLEMENT, WITHOUT		
			V270	MOTHER WITH SINGLE LIVEBORN		
24	2	Home	65651	POOR FETAL GROWTH, AFFECTING MANAGEMENT	734	MEDICAL INDUCTION OF LABOR
			65921	MATERNAL PYREXIA DURING LABOR, UNSPECIFIED	7301	INDUCTION OF LABOR BY ARTIFICIAL RUPTURE
			V272	MOTHER WITH TWINS, BOTH LIVEBORN	7351	MANUAL ROTATION OF FETAL HEAD
			65281	MALPOSITION/MALPRESENTATION OF FETUS, DEL	736	EPISIOTOMY (with subsequent repair)
			V043	NEED FOR PROPHYLACTIC VACCINATION AND IN	757	MANUAL EXPLORATION OF UTERINE CAVITY, PO
					7359	MANUALLY ASSISTED DELIVERY
39	2	Home	66331	UNSPECIFIED CORD ENTANGLEMENT, WITHOUT C	736	EPISIOTOMY (with subsequent repair)
			V270	MOTHER WITH SINGLE LIVEBORN	734	MEDICAL INDUCTION OF LABOR
					7301	INDUCTION OF LABOR BY ARTIFICIAL RUPTURE
30	1	Home	66331	UNSPECIFIED CORD ENTANGLEMENT, WITHOUT C	7301	INDUCTION OF LABOR BY ARTIFICIAL RUPTURE
			V270	MOTHER WITH SINGLE LIVEBORN	734	MEDICAL INDUCTION OF LABOR
					7351	MANUAL ROTATION OF FETAL HEAD
					736	EPISIOTOMY (with subsequent repair)
					7359	MANUALLY ASSISTED DELIVERY

Figure 12.8.   Blank Physician Profile for Case Study

COMMUNITY HOSPITAL OF THE WEST
PHYSICIAN PERFORMANCE REVIEW SUMMARY FOR REAPPOINTMENT

PHYSICIAN		Profile Time Frame:	
SERVICE		From:	
CATEGORY		To:	

UTILIZATION:				
Admissions		Procedures		
Patient Days		V-BACs		
Deliveries		Blood Given		
C-Sections				

OUTCOMES:			
Category	**#**	**%**	Comments:
C-Section Rate			
V-BAC Rate			
Nosocomial Inf Rate			
Surgical Wound Inf Rate			
Mortality Rate			

PERFORMANCE REVIEW:			
Category	**# Reviewed**	**# Appropriate/%**	Comments:
Surgical/Inv/Noninvasive Procedures			
Medication Use			
Blood Use			
Utilization Management			
Other Peer Review			

DATA QUALITY			
Data Quality Monitoring			Comments:
Delinquency (>21 days)			
Suspensions			

RISK/SAFETY MANAGEMENT			
Incidents reported by other professionals/ administration			Comments:
Litigation			

MEETING ATTENDANCE			
Medical Staff Meetings			Comments:
Committee Meetings			

(Continued on next page)

Figure 12.8.   *(Continued)*

FOR COMPLETION BY SERVICE CHAIRMAN OR CREDENTIALS COMMITTEE CHAIRMAN

CATEGORY	YES	NO	Comments:
Has the applicant been considered for or subject to disciplinary action since last reappointment?			
Have the applicant's privileges or staff appointment been suspended, revoked, or diminished in any way, either voluntary or involuntary, since last reappointment?			
Are there any currently pending challenges to any licensure or registration or the voluntary relinquishment of such?			
Are there any physical or behavioral conditions or limitations?			
Has the applicant exhibited satisfactory professional performance?			

APPROVALS: **APPROVED?**

REVIEWER	SIGNATURE	YES	NO	DATE
Service Chair				
Credentials Chair				

Summary

The effective management of human resources is critically important in healthcare organizations. Quality of care depends on the processes established and uniformly applied in its recruitment, appointment, and reappointment processes. Healthcare services are complex and require specialized knowledge and experience. The licenses and other credentials of the clinical professionals who provide services in healthcare organizations must be maintained and verified. The competence of licensed caregivers must be reevaluated on an annual or biennial schedule to ensure their continued ability to provide healthcare services. Healthcare organizations must also be able to demonstrate that they have contributed to the ongoing development of their employees and medical staff.

References and Suggested Readings

Berenson, R. 1995. Profiling and performance measures: what are the legal issues? *Medical Care* 33 (1, suppl.):JS53–59.

Delorese, Ambrose. 1995. *Leadership: The Journey Inward.* Dubuque, Iowa: Kendall-Hunt Publishing.

Field, R. I. 1995. Sharing clinical data for provider profiling: protection of privacy versus the public's need to know. *Behavioral Healthcare Tomorrow* 4(3):71–73.

Hendryx, M. S., et al. 1995. Using comparative clinical and economic outcome information to profile physician performance. *Health Services Management Research* 8(4):213–20.

Joint Commission on Accreditation of Healthcare Organizations. 2002 (September). Hospital Executive Briefings. Oakbrook Terrace, Ill.: JCAHO.

Joint Commission on Accreditation of Healthcare Organizations. 2003. *Joint Commission Survey Process Guide.* Oakbrook Terrace, Ill.: JCAHO.

Orsund-Gassiot, Cindy, and Sharon Lindsey. 1990. *Handbook of Medical Staff Management.* Rockville, Md.: Aspen Publishers.

Schachter, W. 1995. The role of provider credentialing in quality-of-care improvement and clinic risk reduction. *Behavioral Healthcare Tomorrow* 4(4):71–73.

Shaw, Jason D., John E. Delery, G. Douglas Jenkins Jr., and Nina Gupta. 1998. An organization-level analysis of voluntary and involuntary turnover. *Academy of Management Journal* 41(5):511–25.

Part III
Management of Performance Improvement Programs

Chapter 13
Organizing for Performance Improvement

Learning Objectives

- To identify both the role of an organization's leaders in performance improvement activities and the committee and reporting structures that integrate performance improvement within the organization

- To describe the various configurations of leadership responsible for performance improvement activities

- To explain types of continuing education that optimize the performance of an organization's board of directors, leaders, and employed staff

- To delineate the best ways to organize performance improvement data for effective review by a board of directors

Background and Significance

Performance improvement (PI) does not happen effortlessly. The term *continuous* is often attached to improvement efforts (as in *continuous quality improvement)* for a reason: it serves to remind healthcare workers that PI activities require organizational and individual commitment, and need to be practiced on a regular basis if the organization is to benefit from them.

Given that PI requires commitment and continuity, which types of employees make this happen? Many employees come from a traditional management environment, where managers direct their subordinates in exactly what to do and when to do it. The guiding principle of quality improvement in the Deming–Juran–Crosby–Donabedian model, however, is that all members of an organization must be empowered by its leadership to contribute to the PI program if the program is to be successful. (See the introduction to this textbook for a detailed historical discussion of the quality movement and its evolving models.)

Today, it is commonly recognized across the healthcare industry that unless the leaders are committed to maintaining PI activities in the organization and motivating employees to do the same, PI that reaps genuine benefits for the organization cannot take place.

Leadership endorsement of PI is a crucial and integral contribution to an organization's continuous development.

Leading Performance Improvement Activities

Healthcare organizations must put formal structures in place for meaningful PI. The leaders of a healthcare organization set expectations, develop plans, and implement procedures to assess and improve the quality of important functions. Leaders include the members of the board of directors (or trustees), the chief executive officer, the director of nursing, the medical director, and other senior directors or managers, as well as the leaders of the medical staff.

The board of directors bears the ultimate responsibility for the quality of patient care provided by its healthcare organization. The board implements appropriate systems to monitor and assess all services and to detect variations from acceptable standards of care or service.

As in any other corporation, the board designates an executive to be its agent or managing partner. The chief executive officer (CEO), sometimes known as the administrator or executive director, is responsible for implementing board directives and for acting as board representative in managing the operations of the organization. While proper functioning of the organization remains the responsibility of the board, the CEO is a partial or full voting member of the board and is responsible for ensuring that the board is offered regular educational opportunities. Educational topics for this may include risk reduction strategies, review of organizationwide functions, credentialing issues, clinical processes, infection control management and reporting, and other critical PI issues.

There are three main components of a healthcare organization: the board of directors, the managerial and employed staff, and the medical staff. The key to a healthcare organization's success is the coordination and cooperation of these three groups as they work together to identify community needs and pursue organizational goals. Although responsibilities vary from group to group, all share common interests and must work together on planning, budgeting, capital development and expenditures, PI, and patient satisfaction. Figure 13.1 shows an example of a hospital organization chart and its organizational components.

Because the board is responsible for quality, it must ensure the competence and integrity of these individuals. (Refer to chapter 10 for additional information regarding competency of staff.)

It is a widely accepted principle that the individuals most qualified to evaluate a provider are his or her professional peers. Medical staff structure serves as a means to this end. Individual physicians assembled as a medical staff are self-policing and can provide professional input to the healthcare organization on matters of clinical services through established channels.

By performing under specific bylaws, electing its own officers, and setting up an effective structure to accomplish its tasks, the medical staff is, in effect, an organization within an organization. (Refer again to figure 13.1 to see the organization of the medical staff and how it is integrated into the hospital.) However, because the board of directors is ultimately responsible for the activities of the healthcare organization, it has final say in the medical staff's organizational structure and bylaws. Planning for the provision of an organization's healthcare services, on the other hand, is a function of the organization's leaders.

Figure 13.1.   Example Hospital Organization Chart

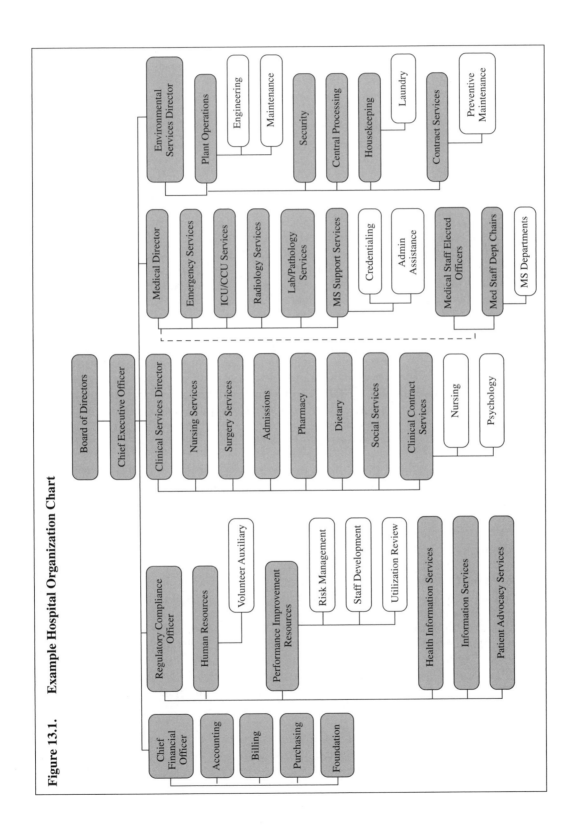

What does leadership mean in a healthcare organization? Ultimately, boards of directors are held accountable for the quality of the products and services provided. Through strategic planning, boards validate the mission, vision, and values for the organization as a whole and approve the direction of the organization as a business entity.

Leadership, including the organization's administrative officers, carefully considers the community that the organization serves, the technologies available to it, the expectations of customers, and the expertise of personnel and medical staff when crafting the mission, vision, and values. In turn, employees treat the mission, vision, and values as a product or a service. Operational personnel determine through their actions the quality of the product or service.

Both internal and external customers provide feedback about the products and services to the employed personnel and, ultimately, to the leadership through the performance improvement activities discussed in part II of this text. It is then the responsibility of the leadership to review the outcomes of the organization's mission and processes and to approve modifications that enhance the quality of the organization's products and services. Modifications are communicated to the organization via the organization's annual **strategic plan** and its administrators and managers. The cycle recommences in preparation for each succeeding year. This cycle follows the PI model discussed in chapter 1, and in this case, the PI team takes the leadership role.

Further reading on leadership is recommended for any student of PI in healthcare. A summary of the expectations of the leadership in healthcare organizations can be found in the leadership chapter of each setting-based set of the Joint Commission on Accreditation of Healthcare Organizations (JCAHO) accreditation standards. Leadership, however, is not an organizational body that meets and directs the actions of the organization every minute of every day. The administrators and managers of the organization are charged with day-to-day operations. How, then, is the strategic plan executed? A variety of approaches have been undertaken by different healthcare organizations.

Ideally, every individual in the organization understands the principles of PI and can contribute to PI activities when the need is recognized. In some organizations, most PI activities derive from initiatives of frontline employees, those responsible for providing products and services directly to healthcare consumers. Some refer to this as a "bottom-up improvement initiative." Figure 13.2 demonstrates how a healthcare organization has divided up specific accountabilities related to PI throughout the organization.

In other organizations, however, commitment to the PI philosophy and activities is not as widespread. In such organizations, leaders actively promote PI philosophies and encourage healthcare workers at all levels to participate in improvement efforts. The leaders are usually department managers or administrators.

In some organizations, an interdisciplinary PI council is developed. The council may include representation from the board of directors, the medical staff, administration, committee chairs from the organization's standing committees, and other interested and experienced individuals who lead PI efforts. In other organizations, a quality management department is developed to coordinate, oversee, and document PI activities. Some organizations retain the services of consultants to assist in developing an infrastructure that supports PI activities.

Finally, some approaches place the responsibility for performance improvement activities in the hands of top management. This is the most conservative and traditional PI management approach. Because improvements come by mandate from higher levels of management, this is often referred to as a "top-down improvement initiative."

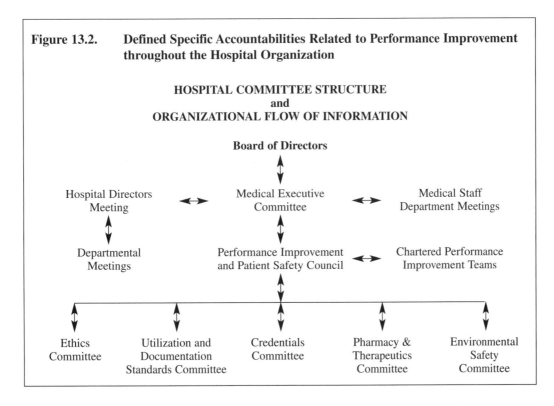

Figure 13.2.   Defined Specific Accountabilities Related to Performance Improvement throughout the Hospital Organization

HOSPITAL COMMITTEE STRUCTURE
and
ORGANIZATIONAL FLOW OF INFORMATION

Whatever approach an organization follows, the community and populations the organization serves expect that the organization will provide the highest-quality products and services possible. Accreditors and regulators want to see evidence of PI activities and the ways in which those activities have benefited the organization's customers. Ultimately, that expectation is what the organization's performance will be judged against.

Managing the Board of Directors's PI Activities

Members of healthcare boards of directors are appointed from the community at large. It is unlikely that the directors will have specific knowledge of healthcare operations or organizations when they are first appointed. Even after serving for several years, most directors will not have expertise in clinical processes. Therefore, it is important that the individuals leading PI activities in healthcare organizations know how to optimally manage the board of directors's PI oversight responsibility.

The board's oversight responsibility is a difficult task. Most directors feel awkward judging the work of physicians and other clinical staff. Coordinated systems of review are imperative to assist them in making decisions about the organization's quality of care and in taking appropriate action when action is necessary.

According to the American Hospital Association, four key elements affect the board's ability to carry out its PI responsibilities:

- The board's "understanding of the quality assessment and improvement system" followed in the organization.

- "Adequate reporting to the board by the staff on specific performance measures" (Figure 13.2 demonstrates one way that the results of PI monitoring can be reported.)

- The board's "oversight and approval of the process to ensure the continued competence of physicians and other clinical and technical staff"

- The board's "active questioning of the information" supplied on the quality of the care provided in the organization through performance monitoring and improvement activities (Umbdenstock 1992)

The board's ability to perform its responsibilities is thus of paramount importance. Carol Dye (1991) believes that education of the board should center on the following questions:

- What type of coordinated program does this facility have in place to integrate the review activities of all services for the purpose of both enhancing the quality of patient care [risk reduction] and identifying current and potential problems?

- How are the clinical and nonclinical activities of the institution monitored, and how are these two components integrated to ensure that the [PI] program is comprehensive?

- How is the [institution] organized to carry out the [PI] program? How are the services and departments organized for this function, and how are their activities coordinated?

- To whom do the various committees concerned with aspects of [PI] report, and how often?

- What are the institution's regular quality [monitoring] . . . activities and how often are they undertaken?

- What kinds of information [are] reported to the board, how often, and in what form? Because boards undertake much of their activities through committee structure, how [is the board] organized to receive and review information?

She goes on to say that "the board's ability to oversee [PI] functions depends on its access to timely and meaningful data. When boards are given relevant information in a form they can understand, they are in a better position to respond appropriately; that is, to review, question, and set policy in an effective manner. Providing data that are concise, appropriately displayed, and organized in a comparative format will maximize the use of the board's time and assist its members in accomplishing [its oversight] activities in an efficient and effective manner." Because many board directors come from the business community, they "have become adept at reviewing financial data presented in forms such as current ratios, liquidity ratios, cash flow to total debt ratios, and so forth. Thus, for the most part, [directors] will be comfortable reviewing quality-of-care information presented in similar statistical formats. Other [directors] may benefit from having the data put into formats such as bar graphs, pie charts, and so forth that translate the statistical information into a visual image of the institution's progress. . . . When sharing data with a board of directors, it is important to remember that few [board members] are health care professionals.

Thus, they may be unfamiliar with the data when first exposed to them. Careful attention should be paid to educating the trustees about certain concepts that govern data, such as validity, statistical confidence intervals, sample mean, and standard deviation."

Dye further suggests that the answers to several additional questions provide useful information for board members reviewing patient care quality data:

- How were the data collected and by whom?

- How often are data collected?

- How large is the sample?

- How do these data compare to data from previous assessments?

- What do the data suggest about the quality of care? Can specific patterns be identified? Are there individual sentinel events that require action?

- Are there other reviews that need to be undertaken in order to provide a more complete and/or accurate picture of the [organization's] quality of care?

- What should [the organization] strive for as a final goal in a particular area? What would be an appropriate objective for improvement?

- How does [the organization] compare to other similar organizations?

The board of directors should be provided information that answers the following questions when PI data identify adverse trends or occurrences:

- If a problem or a potential problem emerges, what actions will be taken to correct or eliminate the problem and prevent or reduce its recurrence?

- Is there an area (department, procedure) that requires focused review?

- What departments and which professionals will be notified of the findings, and what role will they have in follow-up?

- Will a report be provided to the board or an appropriate committee of the board and at what point in time?

- Have the problem, the review, and the plan of correction been documented? (Dye 1991)

Ordinarily, boards of directors meet at least quarterly, while the subcommittees of the board may meet more frequently. From the preceding comments, it should be obvious that performance monitoring and improvement activities should be a regular subject of inquiry and decision making at all board meetings.

Other Resources for Performance Improvement Programs

In addition to the leadership and the board of directors, three other important organizational resources should be considered by healthcare organizations in their planning of a PI program: standing committees of the medical staff, a new-hire and ongoing staff development program dedicated to understanding the approaches to and methods of PI, and formal quality management structures.

Standing Committees of the Medical Staff

Standing committees of the medical staff have made significant contributions to improving quality in healthcare for many decades. Standing committees are usually found in large healthcare organizations. Figure 13.2 provides an example of a hospital committee structure and demonstrates how important information is communicated throughout the organization.

Small organizations usually have small medical staffs that are not able to support a standing committee dedicated solely to the quality of patient care. Consequently, smaller organizations frequently use the executive committee as the "committee of the whole" to do the quality evaluation tasks traditionally executed by separate standing committees in larger organizations.

Commonly, standing committees are organized to review specific aspects of patient care services, such as ethical issues that may come up during the course of patient care; medication use—the selecting, prescribing, preparing, administering, monitoring; and the rate of nosocomial infections. Standing committees are usually chaired by a physician who specializes in an area related to the purview of the standing committee. Committee members are usually drawn from the medical staff, nursing units, and administrative areas related to that specialty.

Routine review of care is usually undertaken by standing committees according to a preestablished protocol stipulating the means by which cases are selected for review each month. Special review of particular cases is also undertaken when negative outcomes have been identified through administrative channels or through referral from other organizational care review processes. The intent of the review is to identify case-specific patterns of care that could have achieved better outcomes had the care processes involved been better designed or implemented. Review also seeks to identify the need for education among the clinicians or others involved in the particular area of care.

For example, one common area of care review currently focuses on cesarean section delivery of newborns. Contemporary obstetrical practice has clearly defined situations in which cesarean section delivery is appropriate because of the risk involved to infant and mother. When cesarean section is utilized in questionable circumstances, it is the responsibility of the obstetrical staff to determine appropriateness and make recommendations to the clinicians involved regarding better practice decisions. In healthcare settings, this type of review by colleagues is referred to as peer review. Findings of standing committee peer reviews are reported to the executive committee of the medical staff as well as to clinicians' professional files for consideration in the recredentialing process. (See the discussion of credentialing in chapter 12.) Findings are also considered by nursing and operations administrators when issues uncover opportunities for improvement of nursing or other functions supporting patient care. Because standing committee reviews and findings are clear evidence of the involvement of the medical staff in patient care improvement activities, it is important that the reviews and findings of standing committees of the medical staff be documented and reflected in the PI program documentation.

Performance Improvement Education

In addition to ensuring that the mission, vision, and values are communicated throughout the organization, the leaders should provide staff training in basic approaches to PI. Most healthcare organizations have a two-part training program that employees follow as part of the organization's new-hire orientation process and which then is repeated or supplemented annually for all staff. Generally, this training is a required competency. Frequently, organizations have linked the PI competency to the annual performance appraisal process where participation on PI teams and completion of training are rated.

The organization's leaders also need to meet established competencies for PI. Training for leaders is frequently conducted with the expectation that the leaders understand and can conduct training on the organization's established PI framework. Leaders are also expected to provide support and resources for staff involvement in PI activities.

Performance Improvement Resource Departments

Consideration of a formal quality management structure usually revolves around the question of whether an organization should develop a PI resource department. Answering this question is, in turn, dependent on a number of factors.

First, PI activities are data and information intensive. (See the discussion of information management in chapter 15.) A PI resource department can assist an organization in managing the information related to PI activities. It can become the centralized repository of information from PI activities, and its staff can coordinate the preparation of reports that turn data collection into information for committee and board analysis and decision making. The staff of this department can also be facilitators of the PI committee's responsibilities and the experts in the organization on facilitation of team activities and use of the QI toolbox techniques. They can take the lead in educating leaders and staff on the principles of PI, data collection, data analysis, and any regulatory changes as they occur. Finally, they can take the lead in the development of cross-functional and organizationwide reporting of PI activities to quality councils and boards of directors. When used effectively within an existing PI culture, these resource departments effectively support a healthcare organization's QI initiatives.

It is important to consider, however, whether an organization is considering developing a PI resource department to deflect and perhaps hide its unwillingness to commit to a more authentic PI culture. Some organizations develop PI departments to keep from having to develop organizationwide, multidisciplinary, integrated PI processes, and to keep from having to deal with issues inherent in developing a more empowered, nonmanagement staff.

Creating an authentic PI culture is a daunting task for many organizations because educating an entire organization in the application of PI techniques is a time-consuming and expensive proposition. Creating a PI resource department thus allows the organization to package its quality efforts in a less expensive box. Only a small number of people have to be trained in quality management activities, who then aid other departments in PI efforts. One possible problem with this approach, however, is that there is no accountability for improving patient care and services. Some individuals in the organization may want to point to the quality department and say to themselves that performance improvement is that department's responsibility. Such a mind-set is antithetical to the philosophical foundation and regulatory requirements of PI in healthcare.

Empowerment of frontline staff can provoke anxiety among managers as staff become more willing to express their opinions and ideas for improving work processes and functioning. In order for PI to be everyone's responsibility, managers in an organization have to be willing to let go of some authority for making decisions. They must be willing to listen to improvement suggestions from frontline staff and to support group decisions on how to improve processes. When a separate department is responsible for quality, managers can regard performance improvement as an external entity that they can either take or leave, cooperate with or resist. In an organization with a healthy and evolving PI culture, the PI department oversees PI efforts, ensures that individual departments complete their PI activities and meet reporting deadlines, and monitors the organizational culture to make sure that a healthy PI philosophy develops.

Case Study

When a large regional medical center became part of an integrated delivery system that had a central board of directors, the medical center's board began to struggle with its revised role. The new organizational environment included several outpatient clinics, multi-specialty physician practices, and an insurance entity. Many of the current board members had served the organization since the medical center was built, and board activities had always been performed in a certain way. Board meetings were rigidly controlled by the administration. No questions were asked, and the members routinely approved committee reports. The reports covered topics such as the organization's financial status and future financial plans, physician credentialing, care quality monitoring reports, new policies, and plans for a new hospital.

A new board member with a healthcare background was appointed after extensive screening and a personal interview with the executive committee. She was not part of the local business power structure, and the administration was concerned that her appointment might not be a wise move. During her first board meeting, two very interesting reports were given. One report detailed some reengineering projects. One of these involved redesigning nursing staffing patterns, decreasing the number of RNs, and replacing RNs with LPNs and CNAs. The current quality report documented a very high quality of care and positive patient satisfaction surveys. Data excerpted from this report can be seen in the left data column of table 13.1. Given that this was her first board meeting, the new board member remained silent and did not ask questions.

Within four months, the new nursing staffing pattern had been launched. Data excerpted from the quality indicators report presented to the board can be seen in the right data column of table 13.1. The new board member was very concerned and decided to ask the nurse administrator presenting the quality report whether the values, which were decreasing, were for the nursing units with the new nursing staffing patterns. The administrator reported that there was a direct correlation. This answer initiated discussion among other board members who were accustomed to using quality indicators in their businesses. This was the first substantive discussion at the board level that the new board member had seen. One board member wanted to know whether any data had been gathered from patient

Table 13.1.	Chapter 13 Case Study Data	
Quality Performance Measures	**1st Quarter**	**2nd Quarter**
Medication errors	3.20%	10.42%
Patient falls	4.21%	8.56%
Cesarean sections	14.21%	17.87%
Vaginal births after c-section rate	18.27%	15.72%
Nosocomial infections	1.78%	4.85%
X-ray discrepancies	0.15%	0.21%
Patient Satisfaction Measures	**1st Quarter**	**2nd Quarter**
Service overall	40.52%	20.74%
Clinical overall	86.72%	70.82%
Overall quality of service	45.40%	22.34%
Food	30.56%	32.54%
Overall cleanliness	85.89%	83.26%

focus groups. Another board member asked whether the average length of stay data had increased, and someone else asked about a cost–benefit analysis of the new staffing patterns. Following the usual process, the chair called for approval of the report and presentation of the next item on the agenda.

Case Study Questions

1. What changes or patterns do you see in the data? What remedies might be suggested for any problems?

2. Has the CEO carried out his or her responsibility for educating the board? Why or why not?

3. Depending on the answer to question 2, what strategies would you recommend at this point?

4. What quality data should be reported and utilized by this board of directors?

5. Given this administration's style and leadership approach, what would be your speculation on whether the minutes of the board meeting reflect actual board meeting discussions?

Summary

Several important aspects of performance improvement (PI) programs should be reflected in organizational structure. First and foremost, it is important for a healthcare organization to determine the roles of its leadership group and governing board within the PI program. A significant amount of board development is needed with respect to understanding healthcare processes and issues and to analyzing PI data and information. Some organizations use their standing medical staff committees as important adjuncts to team-based PI activities. Some develop formal PI resource departments in support of PI activities.

References

Arrington, B., K. Gautam, and W. M. McCabe. 1995. Continually improving governance. *Hospital and Health Services Administration* 40(1):95–110.

Dye, Carol. 1991. Quality assurance data management: the trustees' role. *Quantitative Methods in Quality Management.* Chicago: American Hospital Publishing.

Gardner, Karen, editor. 2003. *The Excellent Board: Practical Solutions for Healthcare Trustees and CEOs.* Chicago: American Hospital Publishing.

Holland, Thomas P., Roger A. Ritvo, and Anthony R. Kovner. 1997. *Improving Board Effectiveness: Practical Lessons for Nonprofit Health Care Organizations.* Chicago: American Hospital Publishing.

Joint Commission on Accreditation of Healthcare Organizations. 2003. *Comprehensive Accreditation Manual for Hospitals.* Oakbrook Terrace, Ill.: JCAHO.

Pointer, Dennis D., and James E. Orlikoff. 2002. *Getting to Great: Principles of Health Care Organization Governance.* San Francisco: Jossey-Bass.

Pointer, Dennis D., and James E. Orlikoff. 1999. *Board Work: Governing Health Care Organizations.* San Francisco: Jossey-Bass.

Umbdenstock, Richard J. 1992. *So You're on the Hospital Board!* Chicago: American Hospital Publishing.

Chapter 14
Navigating the Accreditation, Certification, or Licensure Process

Learning Objectives

- To differentiate between compulsory and voluntary reviews
- To explain the performance improvement perspectives of accreditation, certification, and licensure organizations
- To describe the various approaches of accreditation, certification, and licensure agencies to the site visit and survey
- To identify approaches that lead to success in the survey process

Background and Significance

The complexities of accreditation, certification, and licensure requirements can have a significant impact on healthcare organizations. Few healthcare professionals work in the industry for long without taking part in one of these three processes. This chapter provides a basic introduction to the concepts and processes involved.

Accreditation is the act of granting approval to a healthcare organization. The approval is based on whether the organization has met a set of voluntary standards that were developed by the accreditation agency. The Joint Commission on Accreditation of Healthcare Organizations (JCAHO) is an example of an accreditation agency. Accreditation confirms the quality of the services that healthcare organizations provide. For example, hospitals accredited by the JCAHO have a competitive advantage over nonaccredited hospitals in their geographical area.

Licensure is the act of granting a healthcare organization or an individual healthcare practitioner permission to provide services of a defined scope in a limited geographical area. State governments issue licenses on the basis of regulations specific to healthcare practices. For example, states issue licenses to individual hospitals, physicians, and nurses. It is illegal for unlicensed organizations and professionals to provide healthcare services in all fifty states.

Certification is grants approval for a healthcare organization to provide services to a specific group of beneficiaries. Healthcare organizations must meet the federal *Conditions* *of Participation* to receive funding through the Medicare and Medicaid programs. The Medicare and Medicaid programs are administered by the Centers for Medicare and Medicaid Services (CMS), which is an agency of the U.S. Department of Health and Human Services.

Healthcare Accreditation, Certification, and Licensure Standards

In healthcare today, many different agencies develop and monitor standards on the quality of healthcare services. These agencies accomplish their missions through a comprehensive review process. Some of the review processes are compulsory, and others are voluntary. **Compulsory reviews** are performed to fulfill legal or licensure requirements. **Voluntary reviews** are conducted at the request of the healthcare facility seeking accreditation.

Every accreditation, certification, and licensure agency develops written standards or regulations that serve as the basis of the review process. It is imperative that the healthcare facility monitor any changes and updates to the various standards and regulations and keep current sets of them on hand at all times to help maintain **compliance** status. *Compliance* is the process of meeting a prescribed set of standards or regulations in order to maintain active accreditation, licensure, or certification status. The materials may be provided in manuals (such as the JCAHO's *Comprehensive Accreditation Manual for Hospitals)* or state and federal regulations (for example, the Medicare *Conditions of Participation).*

Joint Commission on Accreditation of Healthcare Organizations

The JCAHO has been the organization responsible for accrediting healthcare organizations since the middle 1950s. The JCAHO (2003d) describes its mission as follows:

> The mission of the Joint Commission on Accreditation of Healthcare Organizations is to continuously improve the safety and quality of care provided to the public through the provision of health care accreditation and related services that support performance improvement in health care organizations.
>
> The Joint Commission evaluates and accredits more than 17,000 health care organizations in the United States, including hospitals, health care networks, managed care organizations, and health care organizations that provide home care, long term care, behavioral health care, laboratory, and ambulatory care services. The Joint Commission is an independent, not-for-profit organization, and the world's leading health care standards-setting and accrediting body.

The primary focus of the JCAHO at this time is to determine whether organizations seeking accreditation are continually monitoring and improving the quality of care they provide. The JCAHO requires that this continual improvement process is in place throughout the entire organization, from the board of directors down, as well as across all department lines.

Commission on Accreditation of Rehabilitation Facilities

In 1966, the Commission on Accreditation of Rehabilitation Facilities (CARF) was established. CARF is a private, not-for-profit organization committed to developing and maintaining practical, customer-focused standards to help organizations measure and improve the quality, value, and outcomes of behavioral health and medical rehabilitation programs. CARF accreditation is based upon the commitment that the organization has made to continually enhancing the quality of its services and programs and that its focus is on customer satisfaction.

American Osteopathic Association

The accreditation of healthcare organizations became a focus of the American Osteopathic Association (AOA) in 1945. Initially, the primary initiative of the AOA was to ensure that osteopathic students received their training through rotating internships and residencies in facilities that provided a high quality of patient care. The AOA has since developed accreditation standards for hospitals, ambulatory care facilities, ambulatory surgery centers, behavioral health facilities, substance abuse treatment facilities, and physical rehabilitation facilities.

National Committee for Quality Assurance

The National Committee for Quality Assurance (NCQA) began accrediting managed care organizations in 1991. Since then, the NCQA's activities have broadened to include accreditation of managed behavioral health organizations and credentials verification for physician organizations. As a private, not-for-profit organization, the NCQA is dedicated to improving the quality of healthcare by assessing and reporting on the nation's managed care plans. Its efforts are focused on the development of performance measurements in key areas such as member satisfaction, quality of care, access, and service. The NCQA uses the Health Plan Employer Data and Information Set (HEDIS) to accomplish these assessments of managed healthcare plans. The performance measures in HEDIS are related to significant public health issues and are a basis for purchasers and consumers to compare the performance of healthcare plans. HEDIS data along with NCQA accreditation provides organizations a method for selecting healthcare plans based on demonstrated value rather than simply on cost.

Accreditation Association for Ambulatory Health Care

The Accreditation Association for Ambulatory Health Care (AAAHC) surveys and accredits ambulatory medical facilities and managed care organizations. These include multi-specialty and single-specialty group practices, health maintenance organizations, dental groups, occupational health services, community and student health services, independent practice associations, diagnostic imaging centers, and radiation oncology centers. However, its primary focus for accreditation is ambulatory surgery centers.

Conditions of Participation

Every healthcare organization that provides services to Medicare and Medicaid beneficiaries must demonstrate its compliance with the CMS *Conditions of Participation*. The compliance process is known as certification, and it is usually carried out by state departments of health. The *Conditions of Participation* for healthcare facilities cover issues related to medical necessity, level of care, and quality of care. The CMS also contracts with non-governmental agencies across the country to monitor the care provided by independent healthcare practitioners. The agencies are called **quality improvement organizations (QIO).** QIOs retrospectively review patient records to ensure that the care provided by practitioners meets the federal standards for medical necessity, level of care, and quality of care.

State Licensure

Every healthcare facility must have a license to operate within the state in which it is located. This license grants the facility the legal authority to provide healthcare services within its scope of services. To maintain its licensed status, the organization must adhere to the state regulations that govern issues related to staffing, physical facilities, services provided, documentation requirements, and quality of care. The regulations are usually monitored and evaluated on an annual basis by the licensing agencies of state departments of health.

Development of Policies and Procedures to Meet Multiple Standards and Regulations

At a minimum, healthcare organizations must consider state licensure regulations when they develop policies and procedures that relate to the documentation and quality of their healthcare services. When a facility provides care to Medicare and Medicaid patients, it must also determine which of the *Conditions of Participation* apply. When a facility is accredited by the JCAHO, CARF, AOA, or some other accreditation organization, those standards must also be taken into consideration.

To ensure compliance, the healthcare organization must continuously review its operating policies and procedures to make sure they reflect ongoing changes to accreditation standards and to federal and state regulations that may affect scope of care and services provided. The most stringent standard or regulation on each aspect of care should be identified, and organizational policies and procedures should be based on that standard or regulation.

For example, when the healthcare facility is setting policy on charting by exception, the state licensure regulations must be reviewed for their requirements. Then the *Conditions of Participation* regulations must be reviewed for their requirements. Finally, all relevant accreditation standards that apply must have their requirements reviewed. The strictest of all standards should be written into the policy and procedure for chart documentation by exception.

Surviving the Survey Process

Ongoing foresight and planning are the keys to successful completion of accreditation and licensure reviews. The leaders of the healthcare organization must stay focused on the organization's accreditation, certification, and licensure status to ensure that these issues are not overlooked in the flurry of day-to-day operations. Preparation for accreditation and licensure processes cannot be accomplished a few weeks before the organization is due for review. A solid accreditation and licensure infrastructure must be built and maintained so that the organization is ready for an inspection at any time.

Some accreditation processes are scheduled and others are unannounced. JCAHO's progression over the years from scheduled triennial surveys to periodic self-assessments and unannounced surveys by 2006 represents a gradual paradigm shift in its attitude toward survey readiness, which has become increasingly stringent. State agencies representing state licensing or federal Medicare/Medicaid certification programs may also conduct unannounced surveys.

The CMS accepts accreditation by AAAHC, AOA, CARF, and the JCAHO in granting what it calls **deemed status.** When an organization is granted deemed status, CMS assumes that an organization meets the *Conditions of Participation* if the organization is currently accredited by these organizations. Regional offices of CMS, however, frequently review psychiatric facilities accredited by CARF for compliance with the *Conditions of Participation.* Additionally, with roughly ten percent of organizations that obtain deemed status through JCAHO accreditation, CMS will conduct a validation survey for *Conditions of Participation* compliance.

A review process with any of these agencies is more than just an issue of appearances. The reviewing agency becomes part of, and observes, the normal day-to-day operations of the healthcare organization. Surveyors from the JCAHO, for example, want to be assured that the facility's leadership and staff can successfully execute organizational policies and procedures. They want to be assured that the leadership and staff are continuously monitoring and improving performance in the organization and that those improvements are tied to the organization's strategic plan. Therefore, these aspects are the ones that surveyors will want to validate as operative in the organization when they arrive for a review.

Although state licensure surveys may be more focused on a facility's ability to meet department of health regulations, the emphasis is still on high-quality care. It is important to remember that the goal of reviewers is to help healthcare organizations consider their own performance. Members of the organization are often too close to everyday activities and events to recognize important trends or consequences of the organization's inaction.

There is no standard review process for accrediting and licensing agencies. Processes may change from year to year as philosophies change within the agencies. In general, voluntary accreditation processes, such as those of JCAHO and CARF, are more flexible and tailored to the type of organization that is being reviewed. Governmental processes tend to be more bureaucratic.

Following is an overview of the more prominent accreditation and licensure processes.

Accreditation of Acute Care and Other Facilities: Joint Commission on Accreditation of Healthcare Organizations

Currently, the JCAHO's accreditation survey process is scheduled in advance. The JCAHO is in the process of moving to completely unannounced surveys by 2006. This approach shifts the organization's focus from survey preparation to continuous survey readiness. Organizations interested in becoming accredited by the JCAHO must file an application that provides information on the type of organization it is, the services it provides, certain statistical characteristics, and the names of its executive officers.

Most organizations have undergone a site visit every three years; however, with the JCAHO's move to unannounced surveys, the new survey cycle will include a midpoint self-assessment. This self-assessment facilitates a more continuous, efficient accreditation process. The organization evaluates itself against applicable standards and submits a written progress report to the JCAHO. As part of this midpoint self-assessment, the organization must also submit a corrective action plan. The organization's chief executive officer must attest to the accuracy of this self-assessment and corrective action plan. At the time of the triennial survey, during the midpoint self-assessment, surveyor time will be devoted to validating the resolution of action plans submitted. In addition, there will be a review of compliance with other randomly selected standards.

Hospitals must notify the state licensure organization of the JCAHO site visit dates, as well as announce to the public through newspaper advertisement and postings throughout the facility of the impending site visit. This announcement allows the public the opportunity to schedule and meet with the survey team to discuss any issues of concern. Notification failure may impact the final accreditation decision.

The composition of the JCAHO survey team will vary depending on the size of the organization. The JCAHO survey team may include a physician, an administrator, a registered nurse, or other master's-level clinicians. The Joint Commission surveyors have many years of experience practicing in the healthcare industry. Every surveyor undergoes considerable training in JCAHO accreditation processes prior to being assigned to a survey team. In 2002, the JCAHO began requiring surveyors to successfully complete a surveyor certification examination in order to survey healthcare organizations for it.

Surveyors will tailor review activities to the characteristics and services of the organization under review. These activities are based on information, data, and the corrective action plan provided in the midpoint self-assessment, together with compliance assessment of other randomly selected standards. It is important to remember that JCAHO processes are not static from year to year because the JCAHO institutes regular improvements in its own survey processes. Any healthcare professional involved in preparations for a JCAHO accreditation survey must be aware of the current requirements, constraints, and expectations of surveyors.

The survey process lasts approximately three days, although the length of the process depends on the size and complexity of the organization. For a large and highly complex organization, such as a university medical center, accreditation survey activities may last for five days. Regardless, opening activities and document review take up the entire first day of the site survey.

The site visit begins with an **opening conference** and performance improvement (PI) overview with both the surveyors and the leaders of the organization. At the opening conference, the surveyors outline the schedule of activities and list any individuals whom they would like to interview.

At this time, the organization's leaders are expected to provide an overview of the organization's mission and vision, strategic goals and objectives, current experiences and outcomes, and performance monitoring and improvement activities. If the organization is experiencing significant challenges in any areas, it is expected that those areas will be identified for the surveyors. Information about the organization's recent achievements is very important as well because the surveyors want an accurate picture of the status of the organization.

Surveyors are interested in seeing how all members of the organization collaborate to provide healthcare services and products, how they measure the quality of those services and products, and how they identify and develop opportunities for improvement. Therefore, the opening conference is an important opportunity for the organization to set the tone for the survey process.

Surveyors will come with knowledge of the organization from its midpoint self-assessment action plan, any consumer complaints reported to the JCAHO, previous accreditation data, ORYX data, and other information related to the organization's performance. This will allow surveyors to tailor the onsite survey process to critical areas of focus, including processes, systems, and structures within the healthcare organization known to significantly impact patient safety and quality of care. The on-site survey utilizes a "tracer" method that permits assessment of operational systems and processes in relation to the actual experiences of selected patients who are currently under the care of the organization. As cases are examined in relation to the actual care processes, the surveyor may identify performance issues or trends in one or more steps of the process—or in the interfaces between processes.

The accreditation process of the future emphasizes a systems approach to evaluation of continuous improvement in key safety and quality areas. In the performance reports of the future, new information that can demonstrate an organization's commitment to quality and safety is likely to be included, such as achieving the National Patient Safety Goals, performing in ORYX core measures, or obtaining disease-specific care certification.

Surveyors then visit patient care settings and conduct interviews with the organization's leaders, program and department staff, and selected patients. During interviews, surveyors gain insight into the organization's recent successes and current challenges. They look at the management philosophy of the organization and the means by which the organization implements change as a result of performance improvement. As surveyors tour patient care settings, their mission is to verify the status of the organization as conveyed in presurvey documents and in the opening conference.

Trigger issues encountered during visits to care settings may be reviewed with staff. Some of the trigger issues on which JCAHO surveyors focus include patient rights, communication, medication use, information management, and general safety in the care environment. Whatever the trigger issues are, the organization's leaders must be aware of them and implement a corrective action plan.

Following completion of the interviews and visits to patient care areas, the survey team sequesters itself to consider its findings. It develops a preliminary report of the on-site survey at this time and identifies any deficiencies that it feels are evident in the organization.

Following development of the preliminary report, surveyors, members of the organization's leadership team, and a representative from the state's licensing agency reconvene for an **exit conference.** During the exit conference, JCAHO surveyors summarize findings and explain any deficiencies that have been identified during the site visit. Leaders have a short opportunity to discuss the surveyors' perspectives or to provide additional information related to any deficiencies. Finally, the surveyors report the probable accreditation decision of the JCAHO on the basis of the survey findings.

The JCAHO (2003a) uses seven categories to report its decisions on accreditation:

- **Accreditation with full standards compliance:** This category means that the organization has complied with all JCAHO performance standards.

- **Accreditation with requirements for improvement:** This category is assigned when an organization does not demonstrate sufficient compliance in one or more specific performance areas. The organization must meet the JCAHO's accreditation policy requirement within a specified time period in order to maintain its accreditation.

- **Provisional accreditation:** This accreditation decision category is assigned when the organization has demonstrated compliance with selected standards during an initial survey. A second survey conducted about six months later allows the organization enough time to demonstrate satisfactory performance and receive one of the other accreditation decisions.

- **Conditional accreditation:** An organization that receives this accreditation decision did not meet the JCAHO's standards at the time of the on-site survey but was given the opportunity to correct its performance or it experienced one or more adverse clinical events that reflected potential systems issues. Organizations that receive conditional accreditation must develop a plan for correcting the problem and demonstrate sufficient improvement in a follow-up survey conducted within a stipulated time period.

- **Preliminary denial of accreditation:** This decision category is assigned to organizations that were in significant noncompliance with JCAHO standards in multiple performance areas or with accreditation policy requirements. This accreditation decision is subject to subsequent review.

- **Accreditation denied:** This results when all available appeal procedures have been exhausted and the organization has been denied accreditation.

- **Accreditation watch:** When a sentinel event has occurred in an organization and a thorough and credible root-cause analysis of the sentinel event and an action plan have not been completed within a specified time frame, the JCAHO will place the organization on accreditation watch. This is not a separate accreditation decision but is a publicly disclosable issue in the organization's accreditation status. When the organization has conducted an acceptable root-cause analysis of the sentinel event, the accreditation watch is removed.

All other JCAHO-sponsored accreditations follow the same basic survey process as outlined for acute care. Site visits begin with an opening conference, then proceed to document review, care unit visits (which can include a client's home in home health accreditation), and an exit conference.

Certification and Licensure of Long-Term Care Facilities: State Departments of Health

Long-term care facilities are subject to government-directed certification and licensure programs. Licensure regulations are published by each state, and long-term care facilities are expected to comply with the regulations continuously. In addition to state regulations, the federal government developed its own set of regulations in 1974 as an attempt to improve the care provided in long-term care facilities that receive federal Medicare and Medicaid funds.

State departments of health usually conduct unscheduled reviews of long-term care facilities. All long-term care facilities must have a license to provide services in the states in which they operate, and usually licenses are renewed annually. Facility administrators understand that the state department of health will return for the next annual review within fifteen months of the previous on-site review.

In addition, for those long-term care facilities that receive Medicare/Medicaid funding, certification of compliance with the *Conditions of Participation* must be achieved as well. State departments of health also conduct Medicare/Medicaid certification reviews, which may or may not be performed concurrent with licensure review. Scheduling of certification reviews depends on the organizational structure of the state department of health. When there have been complaints from residents, families, or employees against a long-term care facility, department of health surveyors may visit to perform a special investigation at any time.

Department of health survey teams commonly consist of two surveyors who come from a nursing, pharmacy, dietetic, or clinical laboratory background. If the review is not conducted in response to a complaint, the survey process encompasses all aspects of facility operations. The surveyors determine at the time of the site visit which of those operations will be investigated.

The state review process is not standardized among states. The survey process in the state of California is presented here as an example. In California, the annual site visit begins with the posting of survey notification on facility doors and at every nursing station. The notification of survey requests that anyone—whether resident, staff member, or visitor—who has issues or perspectives that they wish to communicate to the surveyors make themselves known. An opening conference is held with the facility's administrators and the director of nurses to explain the purpose of the survey, and the sequence of the survey activities is outlined during the opening conference.

Long-term care site surveyors always look for evidence of three trigger issues in long-term care: excessive percentages of patients suffering from dehydration, decubitus ulcers in low-risk residents, and fecal impaction. Current federal regulations mandate the primary importance of these issues in the long-term care setting. Whether the facility has excessive percentages is determined on the basis of the Facility Quality Indicator Profile. The profile is compiled on the basis of information provided to state departments of health via the Minimum Data Set (MDS) for Long-Term Care. An MDS must be maintained for every resident of a long-term care facility. (See the earlier discussion of this subject in chapter 10.) Focused review of the health records of residents with these conditions is carried out when excessive numbers are identified.

Following the delineation of the survey objectives, the surveyors begin examination of facility operations. Members of the facility's residents' council are interviewed to determine whether any issues have been raised by residents in the facility since the last site review. Ancillary departments such as nutrition services are visited, and operations are reviewed for continuing adherence to public health standards. Nursing units are visited, and resident records are reviewed as deemed necessary by the surveyors. Records are reviewed for compliance with state regulations regarding such issues as annual care plan review by physicians, authentication of physicians' orders, proper administration of medication by nurses, and appropriate charting of care by nursing assistants.

When their review activities are complete, the surveyors reconvene with the facility's administrators and director of nurses to summarize their findings. Any deficiencies that require citations are also discussed at this time.

Accreditation of Psychiatric and Rehabilitative Care Facilities: Commission on Accreditation of Rehabilitation Facilities

Accreditation reviews by the Commission on Accreditation of Rehabilitation Facilities (CARF) are usually scheduled in advance. Organizations interested in CARF accreditation must file an application that outlines organization type, services provided, statistical and textual description of its characteristics, and names of individuals comprising its leadership. Most CARF-accredited organizations undergo a site visit every three years.

The CARF survey team commonly includes three members, although additional members may be added for special purposes unique to the applying organization. Typically, the team is made up of professionals from other CARF-accredited organizations. Their areas of expertise are similar to those in which the organization undergoing accreditation specializes. For example, when the organization under review is an inpatient psychiatric institution, the surveyors have psychiatric inpatient experience and have practiced as administrators or clinicians in that setting. Surveyors undergo considerable training in CARF accreditation processes before they are assigned to a survey team.

In contrast to the JCAHO process, the CARF accreditation process is much more flexible and is highly tailored to the patient care services and communities of interest of the organization. Although a template review schedule is followed, within each segment of the schedule the activities pursued depend on the characteristics of the organization applying for accreditation.

The CARF accreditation site visit begins with an opening conference. CARF requires that the opening conference be accessible to all the communities of interest in the organization. Interested participants may include payers, staff members, referring agencies, members of the community, and patients. It is the expectation of the surveyors that these constituencies will be allowed to voice concerns and issues during the opening conference. The survey team then outlines the activities that it wants to pursue over the ensuing two or three days of the site survey.

The second part of the CARF accreditation is the document review. The document review examines policies and procedures, administrative rules and regulations, administrative records, human resources records, and the case records of patients.

The third part of the survey involves interviews with program staff and patients. The surveyors seek to validate the information gathered from the document review and to determine whether staff or patients have any important issues regarding patient care services.

Finally, the CARF process ends with an exit interview with the organization's leaders. Surveyors identify any deficiencies that have been uncovered and present an overall summary of their findings.

Certification: Compliance with the Medicare/Medicaid *Conditions of Participation*

Some healthcare organizations in the United States undergo no accreditation process. Others have undergone accreditation with an accrediting agency but have been identified by federal Medicare officials as requiring specific review for compliance with the Medicare/Medicaid *Conditions of Participation.*

Surveys to determine a facility's compliance with the *Conditions of Participation* are carried out by state healthcare certification and/or licensure agencies. As with the state certification/licensure processes discussed earlier, state department of health reviews are typically unannounced. The survey team drops in to the healthcare facility as necessary either on an annual basis or in response to complaints from patients or employees. In addition to the surveyors commonly used by the department of health in a given state, regional Medicare agencies may provide one or two Medicare officials from the regional office.

After the opening conference during which the Medicare officials make it known that the review will be for the purpose of determining compliance with the *Conditions of Participation,* Medicare officials generally leave and do not participate in the on-site survey activities. Judgments about compliance are left to the state certification surveyors.

Real-Life Example

Table 14.1 provides an example of how applicable standards can be reviewed as a basis for developing an organization's policies and procedures to meet multiple standards. The hospital for which this analysis was developed treats patients at many different levels of care (inpatient hospitalization, partial hospitalization, outpatient group home environment, and so on). Hospital administrators wanted to develop a policy for charting by exception that would meet all applicable regulations and standards. Each regulatory and accrediting agency's standards were reviewed, and the standards were organized in a tabular format. This format allowed for easy viewing of all standards to determine which standard set the strictest requirements. None of the standards consulted prohibited the use of charting-by-exception methodologies.

Case Study

Henry McConnell has been an administrator surveyor with the JCAHO for five years. He currently serves on a survey team reviewing a large midwestern tertiary care facility. The survey is going well, and he and the nurse member of the team are making visits to the patient care areas of the facility. The chief operating officer (COO) and director of nursing (DON) accompany the two surveyors to various nursing units. They decide to visit the inpatient psychiatric unit.

This particular inpatient psychiatric unit cares for persons with psychotic and other severe emotional disturbances. Many of the patients on the unit frequently suffer hallucinations. Others have prehospital episodes of violence toward others. The unit is known in psychiatric medicine as a locked facility, meaning that special keys are necessary to enter or exit the unit.

As Mr. McConnell and his fellow nurse surveyor approach the unit with the COO and DON, the COO comments on the level of acute psychiatric patients that the institution commonly housed on the unit. She points out that the double doors are made of metal with wired glass windows and that the doors are locked from both sides. She makes a production of getting out her set of keys to the unit so that they could enter, making sure that the two surveyors saw that the doorknobs would not open the doors and that one could only enter with a key. After they all went through the door, she turned around to show them that the door had closed securely behind them.

Table 14.1.	Regulations Pertaining to Charting by Exception	
Regulatory Body	**Regulation**	**Comments**
Medicare *Conditions of Participation*	"All records must document the following as appropriate. . . . All practitioners' orders, nursing notes, reports of treatment, medication records, radiology and laboratory reports, and vital signs and other information necessary to monitor the patient's condition."	Regulations do not require specific documentation for progress notes and other information. The documentation must, however, be sufficient to follow the care process.
	Special medical record requirements for psychiatric hospitals: "The special medical record requirement applicable to psychiatric hospitals was designed so that 'active psychiatric treatment' could be identified. The clinical records, therefore, should provide evidence of individualized treatment or a diagnostic plan that could reasonably be expected to improve the patient's condition."	These standards are more specific to your treatment setting but still do not appear to prohibit charting by exception.
	"The treatment received by the patient must be documented in such a way to assure that all active therapeutic efforts are included." Surveyors are to verify that all treatment profiled is recorded by the team member(s) providing services. The treatment provided should be clearly documented as well as the patient's response to the treatment.	The standard does not prohibit charting by exception. It does define what charting must be able to accomplish; that is, it must describe the treatment (what was done) and how the patient reacted. This can be done in charting by exception if carefully defined and consistently formatted.
	Surveyors are instructed to verify that progress notes indicate how the patient is responding to the treatment being carried out. Specifically, the progress notes recorded by the professional staff responsible for the patient's treatment must give a chronology.	See comments above.
CARF	"The records of the persons served should communicate appropriate information in a form that is clear, complete, and current. The record of each person served should include: . . . Reports of initial and ongoing assessments, . . . signed and dated reports from each care giver."	The standard does not prohibit charting by exception. It does define what charting must be able to accomplish; that is, charting must communicate appropriate information in a form that is clear, complete, and current. This can be done in charting by exception if carefully defined and monitored.

Regulatory Body	Regulation	Comments
JCAHO	"(IM.7.2) The clinical record contains enough information to identify the individual, support the diagnosis, justify the treatment, document the course and results, and facilitate continuity of care among health care providers."	
	"(IM.7.2.14) . . . Progress notes made by the clinical staff and other authorized individuals and used as the basis for treatment and habilitation plan development and review. . . . (IM.7.2.15) All reassessments, when necessary. . . . (IM.7.2.16) Clinical observations. . . ."	
State regulations	"Information contained in the medical record shall be complete and sufficiently detailed relative to the patient's history, examination, laboratory and other diagnostic tests, diagnosis and treatment to facilitate continuity of care."	
	Medical records service or department: "Progress notes: Shall give a chronological picture of patient's progress and shall delineate the course and results of treatment. Patient's condition shall determine frequency."	
	Special requirements for inpatient psychiatric services	Standards address assessments, written individualized treatment plans, and written aftercare plans when appropriate. They do not address progress notes and other documentation specifically.

Table 14.1.　*(Continued)*

SUMMARY: None of the regulations reviewed above have specifically defined time frames for documentation of care or how progress notes need to be completed. Charting by exception, if well planned and implemented, can be used while maintaining compliance.

Then, they turn to go onto the unit to do the review. The COO and DON walk carefully out in front of the surveyors toward the nursing station, the surveyors following a little way behind. One of the first things Mr. McConnell observes, about five feet inside the doorway and up near the ceiling, is a three-foot-long, red-handled fireman's axe.

Case Study Questions

1. How did the axe get there?

2. What common characteristic of healthcare organizations discussed in the background and significance section of this chapter is exhibited in this case?

3. How could these kinds of situations be avoided?

Project Application

Community Hospital of the West is evaluating its medical staff rules and regulations in the area of physician documentation, specifically, dictated reports. This hospital has an accredited rehabilitation unit, so CARF regulations apply. Students should review the state licensure rules, the Medicare *Conditions of Participation,* the JCAHO standards, and the CARF standards for these documentation requirements. Then they should prepare a comparative report of the standards and make a recommendation as to what the new policy should be.

Summary

Accreditation, licensure, and certification activities are a significant component of contemporary quality management programs in healthcare organizations. A variety of accreditation and licensing agencies exist, including the Joint Commission on Accreditation of Healthcare Organizations, the Commission on Accreditation of Rehabilitation Facilities, and state departments of health. All of these agencies publish standards that organizations must meet in order to be awarded or maintain accreditation or licensure. Compliance with these accreditation standards and licensure regulations should be folded into the healthcare organization's operating policies and procedures, and performance improvement activities.

References

Accreditation Association for Ambulatory Healthcare. 2002. *2003 Accreditation Handbook for Ambulatory Care.* Wilmette, Ill.: AAAHC.

Accreditation Association for Ambulatory Healthcare. 2002. *2003 Self-Assessment Manual.* Wilmette, Ill.: AAAHC.

Accreditation Association for Ambulatory Healthcare. http://www.aaahc.org.

American Osteopathic Association. http://www.am-osteo-assn.org.

Brennan, T. A. 1998. The role of regulation in quality improvement. *Milbank Quarterly* 76(4):709–31.

Centers for Medicare and Medicaid Services. www.cms.hhs.gov.

Commission on Accreditation and Rehabilitation Facilities. 2002. *CARF Accreditation Sourcebook,* 2003 ed. Tucson, Ariz.: CARF.

Commission on Accreditation and Rehabilitation Facilities. http://www.carf.org.

Grant, Peter N., and W. Reece Hirsch. 2002. *Medicare Provider-Sponsored Organizations: A Practical Guide to Development and Certification.* San Francisco: Jossey-Bass.

Jencks, S. F. 1994. The governments' role in hospital accountability for quality of care. *Joint Commission Journal of Quality Improvement* 20(7):364–69.

Joint Commission on Accreditation of Healthcare Organizations. 2003a. Accreditation Decisions. Accessed online at http://www.jcaho.org/ accredited+organizations/publicizing+your+accreditation/accreditation+ decisions.htm.

Joint Commission on Accreditation of Healthcare Organizations. 2003b. *Comprehensive Accreditation Manual for Hospitals.* Oakbrook Terrace, Ill.: JCAHO.

Joint Commission on Accreditation of Healthcare Organizations. 2003c. *Continuous Standards Compliance: Survey Planner for Hospitals.* Oakbrook Terrace, Ill.: JCAHO.

Joint Commission on Accreditation of Healthcare Organizations. 2003d. Facts about the Joint Commission on Accreditation of Healthcare Organizations. Accessed online at http://www.jcaho.org/about+us/ index.htm.

Joint Commission on Accreditation of Healthcare Organizations. 2002. *The Complete Guide to the 2003 Hospital Survey Process.* Oakbrook Terrace, Ill.: JCAHO.

Joint Commission on Accreditation of Healthcare Organizations. 2002. *Joint Commission Perspectives* October: 1–15.

Joint Commission on Accreditation of Healthcare Organizations. http://www.jcaho.org.

Kelly, M. A. 1993. Thorough preparation key to successful surveys. *Health Facility Management* 6(2):38–44.

National Committee for Quality Assurance. www.ncqa.org.

Chapter 15
Implementing Effective Information Management Tools for Performance Improvement

Learning Objectives

- To identify the reasons that contemporary information technologies are important to quality improvement in healthcare

- To describe the information management tools commonly used in the performance improvement process

- To describe current developments in healthcare information technologies that will enhance performance improvement activities in the future

- To enumerate how information resources management professionals can help performance improvement teams pursue their improvement activities

Background and Significance

Performance improvement (PI) in healthcare is an information-intensive activity. Because PI models are based on the continuous monitoring and assessment of performance measures, the effective management of the data and information collected is crucial to the success of the PI program. To develop effective data and information management systems, one must have a clear picture of the ramifications of data and information management for PI activities.

Healthcare organizations collect all kinds of data in routine, day-to-day patient care, operations, and administrative activities. According to Elliott (1999, p. 210) and Johns (1997, p. 53):

> The basic unit of recording in healthcare is an event. An event is the observation of an occurrence, subjective characteristic, or objective measurement relevant to an individual's health status that can be described by numeric values, words, character strings, images, or sounds. The observation may be made by a healthcare worker, healthcare professional, diagnostic/therapeutic instrument, or a patient/client and family/associates. Commonly, events are further described by type, date and time, observer, and individual observed. It is the web of these recorded events regarding patient after patient that forms the basic data of the healthcare information system.

> Data, however, are facts, simple facts. They are singular units of *knowledge* that never provide a reliable and valid knowledge-picture of an entity—in this case patients and their various health

aspects—in its entirety. They rarely provide a competent knowledge-picture of even one aspect of an entity. In addition, data must often be couched in the context of their observation before their real meaning is understood. This understanding of the meaning of data in context transforms the data into information. Johns notes that this transformation comes by means of formatting, filtering, and manipulation: changing the configuration in context, selecting pertinent aspects or recombining aspects of the data to more clearly delineate its meaning. Johns also notes that after the data-to-information transformation the result is "useful to a particular task"; that is, it helps individuals to make decisions. What then, is knowledge? In information sciences, knowledge is commonly held to consist of the collection of information about an entity the abstract concepts of which have been validated by the consensus of multiple interpreters. Here, Johns refers to "a combination of rules, relationships, ideas, and experiences" that, again, facilitate decision-making.

Management of information resources for PI purposes must facilitate the transformation from data to information and from information to knowledge. Data collection in healthcare organizations falls into one of three categories:

1. **Patient specific,** pertaining to the care services provided to each patient

2. **Aggregated,** summarizing the experiences of many patients regarding a set of aspects of their care

3. **Comparative,** using aggregate data to describe the experiences of unique types of patients with one or more aspects of their care

But it must be emphasized that data are only meaningful in context, and they must be formatted, filtered, and manipulated to be transformed into information and knowledge that can be used in PI programs.

Transformation of Data into Knowledge

How does the formatting, filtering, and manipulation occur? Fortunately, today's information management technologies enhance that transformation when they are deployed appropriately for that purpose by a healthcare organization. Careful consideration must be given to the support that information technologies can provide.

First, the QI toolbox techniques discussed in the chapters in part II of this text can help with the task of manipulating and interpreting data. The QI toolbox techniques showcase excellent tools that make it possible for performance improvement teams to see what the data are really showing. Looking at a mass of numbers and picking out the salient points and trends is usually difficult. The QI toolbox techniques make it easier to organize data, work through them to uncover meaningful information, and present them in a way that other people can understand.

Aggregating and performing basic statistical analysis can be facilitated with computer-based spreadsheet applications. Performance improvement teams can download data from computer-based healthcare information systems and perform ratio and correlation analysis or other statistical analyses on the data or compare the outcomes of different groups.

Standardized reporting formats can be developed to track important measures that the organization has selected for periodic review. Figure 15.1 is an example of a PI report that a hospital uses to track its important measures on a quarterly basis. Using spreadsheet applications, potential scenarios can also be developed to examine possible outcomes of changes in healthcare processes. Presentation development and word-processing applications make it easy for PI teams to communicate effectively with others in the organization and to document progress on a project as it occurs.

Figure 15.1.   Example of a Routine Performance Improvement Report for a Community Hospital

General Statistics

Statistics	Jan–March 2002	April–June 2002	July–Sept 2002	Oct–Dec 2002	2002 Average	2001 Average
Admissions	625	711	802	775	728.25	747.25
Discharges	789	690	766	759	751.00	789.25
Patient days	1,657	1,671	1,623	1,611	1,640.50	1,910.00
Lost work hours	0*	33*	22*	17*	18*	17*
Observation patients	146	125	137	144	138.00	153.25
Inpatient mortality rate	1.32%	1.35%	1.03%	0.81%	1.13%	1.11%
Deliveries	234	221	232	245	233.00	268.25
Emergency department visits	5,523	5,683	5,890	5,789	5,721.25	6,232.50
Inpatient/outpatient operative encounters	687	664	665	676	673.00	728.75
Outpatient operative encounters	546	524	530	568	542.00	585.25

Patient Care

Measure	JCAHO Standard	Benchmark	Jan–March 2002	April–June 2002	July–Sept 2002	Oct–Dec 2002	2002 Average	2001 Average
		Operative, Other Invasive, and Noninvasive Procedures That Place Patients at Risk						
Discrepancies: preop/postop/pathological (op report indicates specimen removed)	PI.3.1.1	100% (I)	37%	47%	86%	85%	69%	40%
Procedure appropriateness: criteria met	PI.3.1.1	100% (I)	100%	100%	100%	100%	100%	100%
Patient preparation for procedure: adequate	PI.3.1.1	100% (I)	86%	79%	87%	89%	85%	83%
Procedure performance and patient monitoring: intraoperative incidents	PI.3.1.1	0% (I)	0.13%	0.0%	0.38%	0.29%	0.40%	0.21%
Procedure performance and patient monitoring: unplanned returns to OR (M)	PI.3.1.1	0% (M)	0.13%	0.13%	0.38%	1.1%	0.43%	0.27%

(Continued on next page)

Figure 15.1. *(Continued)*

Measure	JCAHO Standard	Benchmark	Jan–March 2002	April–June 2002	July–Sept 2002	Oct–Dec 2002	2002 Average	2001 Average
Postprocedure care: complications of postprocedure care	PI.3.1.1	0% (I)	0.26%	0.40%	0.25%	0.43%	0.33%	0.03%
Postprocedure patient education completed	PI.3.1.1	100% (I)	51%	67%	75%	74%	67%	67%
Medication Use								
Prescribing or ordering: orders changed as result of MD clarification	TX.3.5.2	NI	NI	89%	95.4%	93.8%	65.7%	89%
Preparing and dispensing: dispensing errors	TX.3.4, TX.3.5	0% (I)	0%	≤1%	0%	0%	≤1%	≤1%
Preparing and dispensing: medication delivery time—preop antibiotics within 2 hours of surgery (hips, knees, appendectomies, hysterectomies)	TX.3.5	≤2 hours (N)	89%	93%	93%	91%	91%	87%
Administering: DUE—appropriateness of dosage	TX.3.3, TX.3.5, TX.3.6	90% (I)	73%	67%	73.8%	81.0%	73.1%	89.0%
Monitoring the effects on patients: adverse reactions	TX.3.9	0.1% (I)	0.04%	0.09%	1.0%	0.22%	0.12%	0.36%
Monitoring the effects on patients: drug–drug interactions	TX.3.9	NI	23	2	4	1	7.5	16.0
Monitoring the effect on patients: drug–food interactions	TX.4.5	NI	7	23	44	62	26.5	20.0
Adverse effects during anesthesia	PI.4.5.2	NI	NI	NI	NI	NI	NI	NI

Figure 15.1. *(Continued)*

Measure	JCAHO Standard	Benchmark	Jan–March 2002	April–June 2002	July–Sept 2002	Oct–Dec 2002	2002 Average	2001 Average
Use of Restraints								
Documented evidence of less-restrictive measures used	TX.7.1.3.2.3	NI	3	3	11	6	6	NI
Use of Blood and Blood Components								
Ordering: blood usage appropriate	TX.5.1.5	100% (I)	100%	100%	100%	100%	100%	100%
Distributing, handling, and dispensing	TX.5.1.5	NI	NI	NI	NI	NI	NI	NI
Administering: blood slips completed and on chart	TX.5.1.5	100% (I)	88.2%	81%	82.6%	86.4%	84.9%	85.7%
Administration: cross-match: transfusion ratio	TX.5.1.5	≤2:1 (I)	1.8:1	1.6:1	2.2:1	1.8:1	1.8:1	2.0:1
Monitoring blood and blood component effects on patients: potential transfusion reactions	TX.5.1.4 TX.5.1.2	NI	0.0%	0.0%	0.0%	0.0%	0.0%	0.0%
Monitoring blood and blood component effects on patients: confirmed transfusion reactions	TX.5.1.4	0% (I)	0.0%	0.0%	0.0%	0.0%	0.0%	0.0%
Continuum of Care								
Utilization management: patients admitted as inpatients not meeting appropriateness criteria on initial review	CC.2	100% (I)	NI	NI	NI	NI	NI	NI
Utilization management: continuing-stay criteria met	CC.4	100% (I)	NI	NI	NI	NI	NI	NI
Utilization management: patients remaining inpatients after discharge criteria met	CC.2.1	100% (I)	NI	NI	NI	NI	NI	NI

(Continued on next page)

Figure 15.1. *(Continued)*

Measure	JCAHO Standard	Benchmark	Jan–March 2002	April–June 2002	July–Sept 2002	Oct–Dec 2002	2002 Average	2001 Average
Autopsy results: number performed/number met criteria	PI.3.1	100% (I)	NI	NI	NI	NI	NI	NI
Critical occurrences	PI.4.3	—	0	1	0	0	0.80	0
C-section rate (O)	PI.3.1	12% (M); 17% (N)	15.3%	12.8%	11.8%	12.4%	13.1%	12.1%
VBAC rate (O)	PI.3.1	50% (M); 36% (N)	37.5%	38%	38%	34%	37%	50%
Primary C-section rate (O)	PI.3.1	6.5% (O)	9.7%	7.8%	5.7%	6.1%	7.3%	7.9%
Percentage of total C-section (O)	PI.3.1	50% (O)	56.9%	55.8%	42.5%	43.6%	34.7%	59.2%
Repeat C-section rate (O)	PI.3.1	65% (O)	62.5%	62.1%	62.2%	65.9%	63.3%	50.0%
X-ray discrepancies resulting in change of care (O)	PI.3.1	1.0% (O)	0.7%	0.42%	0.36%	0.4%	0.5%	NI
Unplanned return to emergency department within 72 hours (M)	PI.3.1	0.6% (O)	0.53%	0.5%	0.7%	0.7%	0.6%	0.6%
Patients in emergency department more than 6 hours (M)	PI.3.1	12% (I)	14.7%	10%	13.1%	8.8%	11.6%	12.0%
Unplanned return to special care unit (M)	PI.3.1	0.0% (I)	1.9%	2.3%	1.8%	2.5%	2.1%	NI
Unplanned admits from outpatient surgery (M)	PI.3.1	2.0% (I)	2.3%	1.9%	2.4%	6.9%	3.6%	1.8%
Cancelled surgeries (M)	PI.4.3	1.4% (I)	1.0%	1.0%	1.9%	1.6%	1.4%	1.3%
Cancelled endoscopies (M)	PI.3.1	1.1% (I)	1.2%	1.9%	1%	1.4%	1.4%	1.1%

Figure 15.1. *(Continued)*

Measure	JCAHO Standard	Benchmark	Jan–March 2002	April–June 2002	July–Sept 2002	Oct–Dec 2002	2002 Average	2001 Average
				Quality Control Activities				
Clinical lab: number of QC functions completed/number required	PI.3.3.3	100% (I)	0.0%	1.54%	0.78%	2.0%	1.1%	1.2%
Diagnostic radiology: number of QC functions completed/number required	PI.3.3.3	100% (I)	NI	NI	NI	NI	NI	NI
Dietary: number of QC functions completed/ number required	PI.3.3.3	100% (I)	NI	NI	NI	NI	NI	NI
Equipment used to administer medication: number of QC functions completed/number required	PI.3.3.3	100% (I)	NI	NI	NI	NI	NI	NI
Pharmacy equipment used to prepare medication: number of QC functions completed/ number required	PI.3.3.3	100% (I)	NI	NI	NI	NI	NI	NI
Equipment malfunctions	EC.1.9	0% (I)	NI	1	0	6	7	3

(Continued on next page)

Figure 15.1. (Continued)

Measure	JCAHO Standard	Benchmark	Jan–March 2002	April–June 2002	July–Sept 2002	Oct–Dec 2002	2002 Average	2001 Average
Patient Rights								
Overall patient satisfaction	PI.3.1	68% (I)	67.2%	63.8%	68.3%	73.8%	68.3%	63.0%
Advance directives: patients asked whether they have an advance directive	RI.1.2.4	100% (I)	67.7%	68.6%	66.7%	82.6%	70%	NI
Advance directives: patients provided information about advance directives	RI.1.2.5	100% (I)	100%	100%	100%	100%	100%	NI
Human Resources								
Employee satisfaction: overall annual employee satisfaction rate	HR.4.3	NI	NI	NI	NI	NI	NI	NI
Annual turnover rate	HR.4.3	NI	8.6%	9.7%	5.6%	8.6%	32.5%	41.0%
Complete new-hire orientation	HR.4.0	100% (I)	100%	100%	100%	100%	100%	100%
Management of Information								
Data quality monitoring: documentation appropriateness	IM.3.2.1	90% (I)	95.3%	97.1%	95.4%	97.8%	96.4%	NI
Medical record delinquency: overall	IM.3.2.1	50% (J)	8.7%	8.4%	8.8%	11.1%	9.1%	11.1%
Medical record delinquency: history and physicals	IM.3.2.1	≤2% (I)	1.0%	0.3%	0.3%	0.12%	0.6%	0.6%

Figure 15.1. *(Continued)*

Measure	JCAHO Standard	Benchmark	Jan–March 2002	April–June 2002	July–Sept 2002	Oct–Dec 2002	2002 Average	2001 Average
Medical record delinquency: operative reports	IM.3.2.1	≤2% (I)	3.0%	4.1%	3.6%	4.8%	3.9%	2.9%
Verbal orders countersigned	IM.7.7	100% (I)	68.0%	NI	81.3%	95.6%	81.3%	NI
Medical records dated (all entries)	IM.7.8	90% (I)	48%	NI	53%	NI	50%	NI
Surveillance, Prevention, and Control of Infection								
Nosocomial surgical site infection rate	IC.2	2.5% (I)	0.8%	1.2%	0.6%	2.2%	1.2%	1.2%
Postop nosocomial pneumonia rate	IC.2	1.0% (I)	0.0%	0.3%	0.3%	0.6%	0.3%	0.13%
New Programs								
Measures of new program effectiveness	PI.3	NI	NI	NI	NI	NI	NI	NI

Benchmarking Key: N = national; J = JCAHO; S = state; L = local; I = internal; C = corporate; O = ORYX; M = Maryland Quality Indicator Project; NI = no information

*Number of incidents per 200,000 hours worked

Internet access and various indexing search engines provide information to PI teams on the resources contained in the journal and professional literatures in healthcare. With Internet access, benchmarking against other organizations' PI accomplishments is a simple process.

Ideally, the healthcare information resources supporting PI activities are based in an environment that has enhanced communications and information management technologies already implemented. Most healthcare performance measures are used to assess everyday healthcare service activities, and so the easiest and most effective places to find data about those measures is in the already-deployed information systems that support those service activities. For example, if a PI team wanted to assess the effectiveness of blood transfusion services, probably the best place to find data about those services would be in the blood bank component of clinical laboratory systems. If a PI team wanted to assess the effectiveness of wound care, the most likely place to find data about that service would be in the nursing component of clinical information systems.

However, healthcare organizations have found over the past few years that stand-alone information systems are more difficult to use for PI activities. The examination and improvement of a process often involves the analysis of data and information from a variety of organizational information resources. Thus, the objective has become to provide integrated configuration and access to information resources from a variety of systems across the organization. Important developments that can assist an organization in this objective are discussed in the following sections of this chapter.

Data Repositories

Late in the 1990s, some healthcare organizations began to develop data repositories to facilitate PI activities and long-range strategic planning. Organizations that have deployed this technology are copying every instantiation of every datum collected in the course of providing healthcare services to customers. In addition, they are collecting the secondary data acquired in the course of using the technologies to support care, such as those collected to provide audit trails and to support other administrative aspects of running information systems. Which users entered data, which access terminal was used, and which date and time of entry are key examples of this type of administrative data. As these repositories become more common in healthcare information systems implementations, they will provide healthcare professionals involved in PI activities with timely data and information that can be used continuously to monitor the quality of many different aspects of the care they provide.

Many healthcare organizations, however, have not yet implemented such information technology resources. In their absence, organizations must design information collection and transmission systems that can effectively support PI initiatives. Many routine reports produced in healthcare organizations should be made available across departments and across PI teams. The medical records of specific patients contain immense stores of data about all aspects of patient care. These paper-based records are repositories as well. But they cannot be accessed as easily as a computer-based repository. Systems must be developed to make these data available in spite of the fact that the organization does not have computer-based patient records.

Intranet-Based Communication Technologies

Everyone in the organization concerned with quality issues must be kept apprised of the current status of PI activities. A PI team working on one issue in a specific work unit of the organization may discover important information that could be used by another PI team working on a different issue in a different work unit. Without good communication of PI activities throughout the organization, the second team might need to recapture information or reanalyze collected data, thereby wasting time and money.

Deploying intranet-based communication technologies can help keep everyone in the organization apprised of the current status of PI projects. Intranets are wide-area or local-area network-based resources that allow members of a healthcare organization access to information resources from a variety of contexts within the organization. Only valid users from within the organization should have access to intranet-based materials. Commonly, the presentation of materials uses the standard presentation of the Internet and its Web browsers, but other presentations may be supported by the intranet.

Using the Web site configuration, however, can facilitate communication about PI activities. Each PI project can be accorded a Web on the intranet, where PI team members can post project documentation or presentations as the various milestones of the project are achieved. Team members can also present data that have been collected for their project so that other teams can use the data for their projects. At any time, the organization's members can look up the status of PI projects within the organization and review what projects have been completed.

Standardization and Support of Information Management Tools

Another important aspect of utilizing information systems in PI activities is standardization. A common response to the need for information management technologies is for the leadership of each PI team to want to use his or her personal favorites. This is understandable, given that each leader may have developed expertise in using particular products. However, statistical analysis applications are available from several vendors.

This situation points to the need for organizations to standardize data collection and analysis technologies across all PI activities. Standardization facilitates the use of data by multiple individuals and multiple teams. Sharing can decrease the time and cost to the organization for PI activities. To accomplish sharing ability, organizations should carefully consider the kind of information technology support that is most appropriate.

Today, of course, most organizations use office application packages that include both word-processing and spreadsheet applications. Software should be standardized across all departments. Statistical analysis and graphing applications should also be acquired to facilitate analysis of data sets collected during PI team activities. All staff involved in PI activities should be provided with some Internet access to periodical literature and scientific journal literature search capabilities. The National Library of Medicine now provides nationwide access to its clinical and scientific journal index through its PubMed search engine online. Other indexes for a variety of social science, biological science, physical science, and healthcare professions publications are available through the home pages of any university or public library.

User support must be provided for all information technology resources. Expert users should be identified and made available to PI teams to optimize the use of these technologies. Training sessions should be held periodically to acquaint staff across the organization with techniques for data analysis and statistical packages.

Information Warehouses

A recent development related to the issues of standardization and duplication in health information resources management is the deployment of information warehouses. Information warehouses allow organizations to store reports, presentations, profiles, and graphics interpreted and developed from stores of data for reuse in subsequent organizational activities. For example, a report developed by a PI team on the occurrence of methicillin-resistant Staphylococcus aureus infection in a neonatal intensive care unit could subsequently be used by the perinatal morbidity and mortality committee in a monthly review of infant morbidity. A marketing report developed on the need for services pertinent to women and children in an organization's locale could be used by a PI team that wants to delineate the important aspects of customer satisfaction with women's and children's services.

Providing access to warehouses using intranets and Web browser user interfaces facilitates information resource distribution. Browser search engines allow users within the organization to search the warehouse for previously compiled and interpreted information on any subject contained in the warehouse. Any materials available in the warehouse can be printed and redistributed to PI team members. Materials in electronic formats can be downloaded and reused in appropriate information technologies: word processing, spreadsheets, graphics, presentations, or statistical applications.

Comparative Performance Data

As discussed in part II of this textbook, **benchmarking** is an important tool for effective PI in healthcare organizations. Benchmarking is so important that the Joint Commission on Accreditation of Healthcare Organizations (JCAHO) has developed a standard (IM.10) that requires healthcare organizations to contribute data on significant performance measures to national data collections. In turn, the organization has access to data from the collection. The organization then can compare its performance to the performance of similar organizations. The comparison can assure the organization that it is performing up to industry standards or help the organization to identify opportunities for improvement.

Currently, the JCAHO has identified several sets of core measures, patient care characteristics that the JCAHO considers reflective of the quality of care an organization can provide for particular diagnoses. Pneumonia, congestive heart failure, and myocardial infarction are examples of current core measure sets about which accredited organizations are currently required to collect data. The data can be transmitted by the organization to the JCAHO directly or through the use of a vendor recognized by the JCAHO. Some vendors have developed data-entry applications on the Internet. In this case, the organization identifies in its medical records the values of each given measure as defined by the JCAHO and enters the values for each measure on the Web page for that diagnosis on the vendor's Web site. After completion of entry of the necessary number of cases, the vendor analyzes the data, develops summary reporting, and forwards the data and analysis to the JCAHO. The summary reporting details how the organization performed with respect to the measure in terms of percentage of compliance with the measure and shows how the organization performed with respect to other similar organizations. The JCAHO then utilizes the organization's performance on the core measures to design its areas of inquiry at the time of accreditation survey. It also aggregates the findings across multiple organizations to identify best practices and shares this information with the industry as a whole.

Information Resources Management Professionals

Regardless of the configuration of a healthcare organization's technical infrastructure, the staff involved in PI activities should recognize important resources already developed within the organization—health information services managers, information systems managers, and knowledge-based librarians. These information management professionals are usually already working in the facility and possess a wealth of professional expertise that can be extremely useful in PI.

Information resources management professionals can assist quality improvement programs in a variety of ways. First, they can assist in training PI teams to utilize appropriate sources for data regarding an improvement opportunity. They can assist the team in evaluating the quality of the data harvested from internal data sources and the reliability of data harvested from external sources. When software applications are acquired to support specific PI activities, they can assist in the development of system requirements and requests for proposal. When new information technology applications are developed to track PI measures, information resources management professionals can be called upon to oversee development with an organization-appropriate cost–benefit analysis.

JCAHO Information Management Standards

Effective information management for PI entails an understanding of the information management standards of the JCAHO. The information management chapter of the accreditation standards was developed during the mid-1990s to focus healthcare organizations on the importance of information systems issues in the provision of high-quality patient care. Any healthcare organization that uses accreditation as a component of its PI program must ensure that it meets these standards. Even healthcare organizations that do not use JCAHO accreditation as a component of their PI programs would be wise to consider the standards in developing PI systems and procedures.

The JCAHO information management standards focus on information systems issues, not on information systems. The systems implemented may be computer based or paper based. Either way, solutions to these issues must be developed and consciously implemented in healthcare organizations so that information systems can contribute to high-quality patient care in the ways that they should.

Information management standards address ten areas regarding the contribution of information resources management to high-quality patient care and the improvement of patient care. Each first-level standard is cited and followed by relevant elaboration of its intent as developed by the JCAHO (2003b).

IM.1   The healthcare organization plans and designs information-management processes to meet internal and external information needs.

Intent: [Healthcare organizations] vary in size, complexity, governance, structure, decision-making processes, and resources. Information management systems and processes vary accordingly. The [organization] bases its information management processes on a thorough analysis of internal and external information needs. The analysis considers what data and information are needed within and among departments, the medical staff, the administration, and the governing body, as well as information needed to support relationships with outside services, companies, and agencies. Leaders seek input from staff in a variety of areas and services. Appropriate individuals ensure that required data and information are provided efficiently for patient care, research, education, and management at every level.

IM.2   Confidentiality, security, and integrity of data and information are maintained.

Intent: The [healthcare organization] maintains the security and confidentiality of data and information and is especially careful about preserving the confidentiality of sensitive data and information. The balance between data sharing and data confidentiality is addressed. The [healthcare organization] determines the level of security and confidentiality maintained for different categories of information. Access to each category of information is based on need and defined by job title and function.

IM.3   Uniform data definitions and data capture methods are used whenever possible.

Intent: Standardizing terminology, definitions, vocabulary, and nomenclature facilitates comparison of data and information within and among organizations. Abbreviations and symbols are also standardized. Uniformly applied and accepted definitions, codes, classifications, and terminology support data aggregation and analysis and provide criteria for decision analysis. Quality control systems are used to monitor data content and collection activities, and to ensure timely and economical data collection. Standardization is consistent with recognized state and federal standards. The [healthcare organization] minimizes bias in data and regularly assesses the data's reliability, validity, and accuracy. [. . .] Medical records address the presence, timeliness, legibility, and authentication of . . . data and information as appropriate to the organization's needs.

IM.4   The necessary expertise and tools are available for the analysis and transformation of data into information.

Intent: Education and training enables individuals to understand security and confidentiality of data and information; use measurement instruments, statistical tools, and data analysis methods for transforming data into relevant information; collect unbiased data, gathered with a control for confounding or corrected on the basis of acceptable methodologies;

assist in interpreting data; use data and information to help in decision making; educate and support the participation of patients and family in care processes; and use indicators to assess and improve systems and processes over time.

IM.5   Transmission of data and information is timely and accurate.

Intent: Internally and externally generated data and information are accurately transmitted to users. The integrity of data and information is maintained, and there is adequate communication between data users and suppliers. The timing of transmission is appropriate to the data's intended use.

IM.6   Adequate integration and interpretation capabilities are provided.

Intent: The information management process makes it possible to combine information from various sources and generate reports to support decision making. Specifically, the information management process coordinates collection of information; makes information from one system available to another; organizes, analyzes, and clarifies data; and generates and provides access to longitudinal data. In addition, the information management process provides the capability to link [data and information from a variety of sources internal and external to the organization and from the clinical and management literatures].

IM.7   The healthcare organization defines, captures, analyzes, transforms, transmits, and reports patient-specific data and information related to care processes and outcomes.

Intent: Information management processes provide for the use of patient-specific data and information to facilitate patient care, serve as a financial and legal record, aid in clinical research, support decision analysis, and guide professional and organizational performance improvement. . . . Administrative and direct patient care providers produce and use this information for professional and organization improvement. [. . .]

IM.8   The [healthcare organization] collects and analyzes aggregate data to support patient care and operations.

Intent: The [healthcare organization] aggregates and analyzes clinical and administrative data to support patient care, decision making, management and operations, analysis of trends over time, performance comparisons over time and with other organizations, and performance improvement. . . . Aggregate performance improvement information includes information from risk management, utilization review, infection control, and [hazard and] safety management.

IM.9   Knowledge-based information systems, resources, and services meet the hospital's needs.

Intent: Knowledge-based information, often referred to as "literature," includes journal literature, reference information, and research data. . . .

Knowledge-based information is authoritative and up to date. It supports clinical and management decision making, performance improvement activities, patient and family education, continuing education of staff, and research. . . . Appropriate knowledge-based information is acquired, assembled, and transmitted to users . . . [is accessible and in appropriate formats].

IM.10   Comparative performance data and information are defined, collected, analyzed, transmitted, reported, and used consistent with national and state guidelines for data set parity and connectivity.

Intent: The [healthcare organization uses and contributes] to collections of performance data from multiple institutions. As part of its information management activities, the [healthcare organization] exchanges clinical and knowledge-based data and information with other healthcare organizations. These activities help the [organization] develop its future capabilities and goals. The [organization] uses external data and information to identify areas in which its own performance deviates from expected patterns. The [organization] also contributes its own information to external reference databases. To ensure that the data is comparable across institutions, the [organization] follows national and state guidelines on form and content.

The JCAHO is in the process of rewriting the information management standards. The projected new standards cover the 2003 standards IM.1 through IM.6 in revised standards IM.1 through IM.5. Four new requirements have been added to IM.2 to be consistent with the Health Insurance Portability and Accountability Act of 1996 (HIPAA), stressing its security and privacy provisions. IM.7, which describes the requirements of medical record systems in the 2003 standards, will be moved to IM.8 in the revised standards. Requirements for the availability of aggregate information about care processes in IM.8 of the 2003 standards can be found in the revised standards in IM.6. IM.9 of the 2003 standards has been moved and incorporated into revised IM.7. IM.10, which requires comparative data, has been deleted entirely. IM.10 has been seen as redundant to standards in the improving organization performance chapter and other requirements necessary to participation in the accreditation process (JCAHO 2003c).

Each of the principal standards has multiple subdivisions that discuss and provide examples of the issues evident around the standard. The student of PI in healthcare should not forget these standards and should refer to current and future JCAHO accreditation manuals for a more detailed discussion.

Case Study

The following excerpt is from a consultant's report on the status of information technologies at Community Hospital of the West:

Infrastructure: As is true with many organizations trying to keep up with the rapid developments in information technology, Community Hospital of the West has a variety of hardware and software that is used in its departments. Computer workstations are widely used across the organization, but there are many departments that have computer workstations too old to provide an adequate platform for the later versions of software that would most effectively support departmental reporting responsibilities. There is no organizationwide local-area network in place. Software applications and versions are not standardized across the organization, and so members of different departments cannot share data and information in electronic formats. This forces members of the organization to duplicate effort in report generation when reports contain the same or similar data. An office suite application served to users on a local-area network could help to solve this problem.

The absence of a local-area network and an administrative database accessible to department managers means that reports must be prepared in the generating department, output on paper, and then input again in administrative departments to be utilized in administrative applications. An administrative database served by a network to all departments would mean data would be gathered only once and then be available for subsequent users and purposes without redefinition or reprocessing. All of the logs that the organization currently generates, many of which are on an hourly or daily basis, could be more effectively administered and accessed if they were in electronic formats.

Organizational Knowledge of Computing Applications: Although there is broad distribution of hardware and software across the organization, organizational knowledge regarding the use of applications software available to it has not been optimized. Many of the respondents in the interviews stated that they had access to a computer workstation and applications but had not had access to adequate training on the application. Consequently, most reporting is done without the use of software applications (some even on a typewriter) where the use of software could accomplish the process more effectively and efficiently. This situation is intensified because there are so many different versions in existence across the organization. There appears to be no organizational expert with reference to software application use to assist personnel in solving their information *processing* problems. Maintenance of all the different versions of all the different applications must be essentially impossible for Information Systems personnel.

Interfacing and Use of Existing Databases: Several respondents felt that access to existing mainframe databases would improve the performance of their administrative reporting activities. This issue is also one that many organizations face as they try to make database information available and accessible, yet maintain data integrity and security. In particular, four of the respondents noted that they felt their productivity reporting could be more effectively accomplished if it were pulled from the payroll database. Some felt that information should be made available from financial and patient care systems and shared directly to an administrative database.

Archival of Administrative Reporting: The organization does not appear to have an archiving policy. Departmental staff decide for themselves how long they should keep reports that they generate or receive. The archival period varies widely from no archival at all to decades. Many departments receive and archive reports for which they have no use, the author of which they do not know, and the purpose of which is unknown. Most administrative reporting is archived on paper. If the organization had an administrative database, archival of many, if not all, reports could be accomplished electronically, increasing the availability, reliability, and security of administrative information for long-term use. There is no formal distinction made at this time between information that is valuable in the long term and that for short-term monitoring purposes only.

Case Study Questions

1. What issues does the consultant's report raise that may have an impact on PI activities in this facility?

2. Compare the case study situation to the JCAHO information management standards. What issues does this analysis raise?

Summary

Because PI activities are information intensive, healthcare organizations must pay special attention to the management of information resources to support improvements. Ideally, the organization would make available common business-oriented applications such as spreadsheets and word-processing software as well as statistical analysis and presentation packages. Many organizations make information available across the organization via intranets and archive clinical information system data permanently in data repositories. In this way, clinical data are made available for PI activities. Information resources must also include access to national comparative data collections for organizations accredited by the JCAHO, and organizations must ensure that they meet the other JCAHO information management standards as well. Information resources management personnel such as directors of health information services and information systems and institutional librarians with expertise in healthcare literatures should also work to support performance improvement activities.

References

Armoni, Adi, editor. 1999. *Healthcare Information Systems: Challenges of the New Millennium.* Hershey, Penn.: Idea Group.

Dearmin, J., J. Brenner, and R. Migliori. 1995. Reporting on QI efforts for internal and external customers. *Joint Commission Journal on Quality Improvement* 21(6):277–88.

Elliott, Chris. 1999. Introducing the electronic health record user community. In *Electronic Health Records: Changing the Vision,* Gretchen F. Murphy, et al., editors, pp. 209–30. New York City: W. B. Saunders.

Johns, Merida. 1997. *Information Management for Health Professions.* Albany, N.Y.: Delmar.

Joint Commission on Accreditation of Healthcare Organizations. 2003a. *2003 Comprehensive Accreditation Manual for Hospitals.* Oakbrook Terrace, Ill.: JCAHO.

Joint Commission on Accreditation of Healthcare Organizations. 2003b. *2003 Comprehensive Accreditation Manual for Hospitals (Automated), Update 2.* Oakbrook Terrace, Ill.: JCAHO.

Joint Commission on Accreditation of Healthcare Organizations. 2003c. 2004 Hospital Accreditation Standards. Accessed online at http://www.jcaho.org/accredited+organizations/hospitals/standards/ new+standards/2004+standards.htm.

National Library of Medicine. www.ncbi.nlm.nih.gov/entrez/query.fcgi.

Rosen, L. S., et al. 1996. Adapting a statewide patient database for comparative analysis and quality improvement. *Joint Commission Journal of Quality Improvement* 22(7):468–81.

Shapiro, Joe. 1998. *Guide to Effective Healthcare Information and Management Systems and the Role of the Chief Information Officer.* Chicago: Healthcare Information Management Systems Society.

Stegwee, Robert, and Ton Spil. 2001. *Strategies for Healthcare Information Systems.* Hershey, Penn.: Idea Group.

Chapter 16
Developing Effective Performance Improvement Teams

Learning Objectives

- To identify key aspects of organizational culture that promote effective performance improvement teams
- To describe the contributions that team charters, team roles, ground rules, listening, and questioning can make to improve the effectiveness of performance improvement teams

Background and Significance

The team approach to performance improvement (PI) activities in healthcare is helpful because it uncovers and reflects a variety of perspectives and a more complete knowledge base than do improvement approaches dominated by one or two individuals.

Effective team function often becomes an issue crucial to the success of PI programs in healthcare organizations. Program leaders can take a variety of initiatives to help teams function more effectively. There appears to be one major issue, however, that predicts more than any other the likelihood of team effectiveness in the PI realm: the organization's expectations.

Some consultants believe that every team needs development, and consequently, many use team-building exercises as a matter of course. The single most important way for an organization to achieve effective teams is to make team problem solving and team PI part of the organizational culture. From the moment an individual is hired to work within a healthcare organization, it should be overtly communicated by the organization that the individual is expected to participate in team projects and that participation in teams is part of everyone's job description. By making team participation part of the organizational culture, no individual employees will feel exempt or that they do not need to cooperate in team approaches to organizational issues.

That is not to say, however, that occasionally there will not be individuals on a team who are uncooperative or have their own agendas. Individuals who are not team players can be directed into positive team production, however, with knowledgeable and experienced facilitation.

Team Charters

In many healthcare organizations, the process begins with a data review and identification of a problem in the quality council or leadership meeting. The team makes a recommendation for the implementation of a PI team, and this is then formalized with a team charter.

Team charters explain what issues the team was implemented to address, describe the goal or vision, and list the initial members of the team and their respective departments. Team charters are helpful because they keep the team's objective in focus. Usually, team charters will also identify any mitigating factors that may limit the PI process, such as financial limitations, full-time employee restrictions, or time constraints. They keep the organization focused on the opportunity for improvement and the team focused on its mission. See figure 16.1 for an example of a team charter.

Team Roles

The role of team leader was discussed in chapter 3. In this chapter, other possible roles in PI teams that may enhance performance are included.

First, PI teams may want to identify a team facilitator. The facilitator should be someone who knows the PI process well and has facilitated such a team in the past. The facilitator may also be required to train the team in the PI process and quality improvement (QI) tools. The facilitator is primarily responsible for ensuring that an effective PI process occurs. The responsibilities of the team facilitator include the following:

- Serving as advisor and consultant to the team
- Acting as a neutral, nonvoting member
- Suggesting alternative PI methods and procedures to keep the team on target and moving forward
- Managing group dynamics, resolving conflict, modeling compromise
- Acting as coach and motivator for the team
- Assisting in consensus building when necessary
- Recognizing team and individual achievements

The role of team recorder (or scribe) is vital to the team's success. This team member keeps minutes of the team's work during the meetings, including any documentation required by the organization. The recorder performs the following functions:

- Recording information on a flip chart for the group
- Creating appropriate charts and diagrams
- Assisting with notices and supplies for meetings
- Distributing notices and other documentation to team members
- Developing meeting minutes within the facility policy time line and utilizing a reporting format that assigns duties with time frames

Figure 16.1.   Example of a Team Charter

PERFORMANCE IMPROVEMENT TEAM CHARTER
(Page 1 of 2)

Team Name	Date Submitted to Performance Improvement Council
Clinical Laboratory Services	*February 15th*

Statement of the Problem, Issue, or Concern to Be Addressed by the PIT

Safety issues or other problems concerning the hospital labs increased 207% over a one-year time frame.

Statement of the Goals, Objective, and Desired End State

Identify specific problem areas with laboratory services, conduct a baseline study to assess each area, analyze results, develop an action plan, implement improvements, and evaluate results.

Proposed Team Members

Name	Title	Department
Roger Jones	*Chief Clinical Officer*	*Administration*
Jill Andrews	*Lab Manager*	*Laboratory*
Ben Carlson, M.D.	*Emergency Physician*	*E.R.*
Sandy Johnson	*Director of Clinical Ser.*	*Administration*
Kathy Smith, R.N.	*Director of Nursing*	*Nursing*
John Rasmussen	*Lab tech*	*Laboratory*
Sue Holt	*Lab tech*	*Laboratory*
Pam Richards	*Coordinator*	*Quality Management*

Project Resources

Planned Start Date	Planned Completion Date	Planned Frequency of Meetings
February 20th	*June 1st*	*weekly*

Administrative/PIC Support Needed (if any)	Estimated Cost of Team's Work
	$1000.00

(Continued on next page)

Figure 16.1. *(Continued)*

PERFORMANCE IMPROVEMENT TEAM CHARTER
(Page 2 of 2)

What Important JCAHO Functions Will the Project Measure or Improve (check all that apply)?

☐ Rights, Responsibilities, and Ethics   ☐ Leadership

☐ Continuum   ☑ Management of the Environment of Care

☐ Assessment   ☐ Management of Information

☐ Care   ☑ Management of Human Resources

☑ Education   ☐ Surveillance, Prevention, and Control of Infection

☑ Improving Organizational Performance

What Dimensions of Performance Will the Project Improve (check all that apply)?

☐ Efficacy   ☐ Continuity

☐ Appropriateness   ☑ Safety

☐ Availability   ☑ Efficiency

☑ Timeliness   ☐ Respect and Caring

☐ Effectiveness

Project Benefits
How Will the Project Support the Mission/Values and/or Achieve the Organization's Strategic Goals?

☐ Improved Patient Outcomes   ☐ Time Savings

☑ Cost Savings   ☐ Other _____

☑ Improved Service   _____

How?   *reduce safety violations in the laboratory and rework*

_____

_____

_____

_____*Jill Andrews*_____   _____*Feb. 15th*_____
Signature of Applicant   Date

To Be Completed by Performance Improvement Council

Comments

_____

_____

Performance Improvement Council Recommendations _____

_____

_____

_____   _____
Signature   Date

Finally, teams may assign someone to be a timekeeper. The timekeeper helps the team manage its time and notifies the team during meetings of time remaining on each agenda item in an effort to keep the team moving forward on its PI project.

Ground Rules for Meetings

Establishing ground rules for meetings helps a team maintain a level of discipline. Ground rules include a discussion of attendance, time management, participation, communication, decision making, documentation, room arrangements, and cleanup. Ground rules will not be the same for every team, as each team should decide how it wants to proceed. But the ground rules should be well known to everyone on the team, and everyone should have participated in their development. Most teams that use ground rules allow for periodic review and revision, particularly when team membership changes. New members must be brought up to speed on the ground rules when they begin coming to team meetings. (See figure 16.2.)

The attendance discussion should establish who will schedule meetings, arrange for a meeting room, and notify members. The ground rules should also cover the team's expectations regarding absences, including whether team members can be removed from the team and replaced for absenteeism and whether or not substitutes can attend meetings.

Cancellation of meetings should be discussed in the ground rules, as well as how the team addresses issues of tardiness. This discussion should also include how the time allotted to agenda items will be monitored.

Figure 16.2.   Sample Meeting Ground Rules Worksheet

Ground Rules Worksheet

1. Every individual has a viewpoint that is valuable, every individual can make a unique contribution, and every individual can speak freely.

2. All team members must listen attentively and respectfully without interrupting. Only one person should speak at a time.

3. All team members must be willing to accept responsibility for assignments and complete any assigned tasks between meetings.

4. The organizational positions/levels of team members will not be recognized during team meetings. Every member of the team is an equal participant.

5. Solutions must be created with resources that are currently available. Money and additional staff are not considered issues.

6. _____

7. _____

8. _____

9. _____

10. _____

Discussion of team member participation should include the team's expectations regarding advance preparation. The team should plan the means by which equal contribution from all members can be ensured, how activities will be monitored to ensure productive meetings, how assignments and expectations for their completion are made, and how ad hoc members will be invited and prepared for their input.

Communication ground rules are imperative for team effectiveness, particularly regarding how candid members may be and whether information discussed in the team process must remain confidential. Performance improvement data are considered confidential in most facilities and may be legally protected from reproduction or use outside the facility or agency.

Another aspect of the facilitator's role may be to clearly define for the team how information will be managed and protected during the project. The team should also decide what will happen when discussions get off track, how interruptions or side conversations will be handled, what listening skills are expected, how differences of opinion and conflict among members will be expressed and resolved, and how creativity will be encouraged and negative thinking discouraged. The team must also decide whether consensus or majority decisions will be taken on issues that require a vote.

Other questions that may require discussion include the following:

- How will breaks be handled, if there are to be any?

- Who is responsible for setup and cleanup of the meeting room?

- Does the team require information technology support? If so, who will coordinate it?

- How, when, and why should administration be involved?

- How will department managers be notified of the need for department employees to participate on a team?

- Is overtime necessary for this team to complete its assignment?

Problem-Solving Techniques, Listening, and Questioning

Encouraging team productivity can be a major issue in many organizations. This is an outgrowth of the common management styles that most organizations exhibit. Organizations tend to gravitate to one of two management styles. One style is inclusive: all viewpoints are considered, each with respect to its potential contribution to solving the PI issue at hand. The other style is exclusive: its goal is to get to a result as quickly as possible. Each style has positive and negative aspects. People who operate by the inclusive style can get mired in detail and discussion and achieve results only after extensive processing. People who operate by the exclusive style can fail to perceive important details in their rush to implement a solution. A combination of the two styles is more effective than either style alone. Each style can be employed at appropriate points in the development of a team process.

The concept of facilitation was developed to help move teams along. The cyclical PI methodology was developed to give teams a structure by which to proceed in problem solving. QI toolbox techniques were developed to give teams an easy way to organize and analyze data.

Another area that is extremely important in the development of good team interaction and functioning is the ability to listen and question. Performance improvement team members need to be able to do both, and this may take some development on the part of team leaders. Commonly, in human communication in organizations, individuals tend to be either active communicators or passive listeners. Active communicators can quickly dominate a team meeting. They are accustomed to expressing themselves and being heard. Sometimes their listening skills have been eclipsed by their own volubility. They may have to be reoriented to practice listening more often, allowing the quieter individuals on the team an opportunity to express their perspectives. Similarly, the quieter members may have become accustomed to listening to other people and not voicing their opinion. They may have to be reoriented to contribute, sharing their knowledge and expertise with the group so that important details are not ignored.

Often, too, in human communication in organizations, individuals become invested in their own perspectives and ways of seeing and interpreting situations. This can happen among active communicators as well as passive listeners. The active communicators often react by trying to persuade everyone else on the team of the justness of their perspectives. The passive listeners may say nothing but internally retain their commitment to their own perspectives. Neither of these tactics moves the team to resolution of the problem it was convened to solve. Team members may have to be reoriented to listen carefully to others' perspectives and seek a common understanding of those perspectives through questioning techniques.

The power of the question lies in the fact that it compels an answer. When one asks the right questions, one acquires important answers in terms of information, experience, reactions, perspectives, and attitudes. When one fails to ask questions, one is left with only one's own perspectives, which may or may not reflect the reality of various situations. An individual can never know as much about an opportunity for improvement as the collective members of the team. When the team tries to make decisions concerning the opportunity without sufficient information, the likelihood that the new solution solves the problem decreases. Using questioning techniques effectively becomes an important tool for PI team members.

In using questioning methods, it is important that team members have a positive attitude on the importance of asking rather than telling and a conviction that people, because of their unique experience, background, and training, can potentially contribute unique information. It is also important to recognize that there is more than one type of questioning. Different styles of questioning can be used to accomplish different types of information gathering. Note the types of questions shown in figure 16.3 and the purposes for which each can be used.

People Issues

Finally, in discussing the effectiveness of PI teams, one must specifically recognize the effects that individuals have on PI processes. In reality, all of the team development techniques are intended to help teams function *through* the people issues and become effective teams. Effective teams typically succeed in the following objectives:

- To establish goals cooperatively with all members participating who have perspectives on the issues.

Figure 16.3.	**Types of Questions and Their Purposes**	
Type	**Purpose**	**Examples**
Factual	To get information To open discussion	How and all of the *W* questions: what, where, why, when, and who
Explanatory	To find reasons and explanations To broaden discussion To develop additional information	In what way would this help solve the problem? What aspects of this issue should be considered? Just how would this action be done?
Justifying	To challenge old ideas To develop new ideas To find reasons and proof	Why do you think so? How do you know? What evidence do you have?
Leading	To introduce a new idea To advance a suggestion	Should we consider this idea as a possible solution? Would this idea be a feasible alternative?
Hypothetical	To develop new ideas To suggest another, possibly unpopular opinion To change the course of discussion	What would happen if we did it this way? Would it be feasible for us to do this the way company X does it?
Alternative	To choose an alternative To obtain agreement	Which of these solutions is better? Does this solution represent our choice in preference to other alternative solutions?
Coordinating	To develop consensus To obtain agreement To take action	Can we conclude that this is the next step we should take? Is there general agreement on this plan?
Direction of Questions		
Overhead: directed to the group	To open discussion To introduce a new phase To give everyone a chance to comment	How shall we begin? What shall we consider next? What else might be important?
Direct: addressed to a specific individual	To call on an individual for specific information To get an inactive individual involved in the discussion	George, what are your suggestions? Gracie, have you had any experience in this area?
Relay: referred back to another individual or to the group	To help the leader avoid giving his or her own opinion To get others involved in the discussion To call on someone who knows the answer	Would someone like to comment on Peter's question? Mary, how would you answer Paul's question?
Reverse: referred back to the individual who asked the question	To help the leader avoid giving his or her own opinion To encourage the questioner to think for himself or herself To bring out opinions	Well, Bing, how about giving us your opinion first? Heddie, tell us first what your own experience has been in this area?

Source: Burns, Robert K. 1960. *The Questioning Techniques*. Chicago: Industrial Relations Center, University of Chicago. Reprinted with permission.

- To communicate in a two-way mode. All members participate, and members who do not spontaneously communicate are encouraged and held responsible for doing so.

- To value open expression of both ideas and feelings as important perspectives on organizational issues.

- To distribute leadership and responsibility among all team members, with each member responsible for tasks that make important contributions to team accomplishments.

- To distribute power among all team members. Power is distributed on the basis of information, ability, and contribution to team activities, not on the basis of place in the formal organizational structure.

- To match decision-making techniques to the types of decision-making situations. Important decisions are usually made through consensus, meaning that the group as a whole comes to agreement about the appropriate course of action.

- To view periodic controversy and conflict among team members as a positive aspect of team growth and team understanding regarding the processes the team has been initiated to improve.

- To focus on the issues for which they have been organized to address and to keep at the heart of their work the mission, vision and values of the overall organization and PI team.

- To be cost-conscious in their PI efforts and management of project design or tools.

Even those new to the team concept generally can see the importance of these factors. However, in real situations there is conflict between the roles of individuals in formal organizational structure and the roles necessary to an effective team structure. Formal organizational roles often require authority for various functions and responsibilities. Effective teams share authority for the team performance. Therefore, many individuals coming to team approaches to problem solving have to get reoriented, and for some, this reorientation is difficult.

Most individuals coming to a PI team for the first time are unfamiliar with the data collection and analysis aspects of team functioning. Some may not want to be involved with such detailed activities and may not have the mathematical skill necessary to perform these activities with ease. The team may have to spend some time helping that individual to accomplish his or her team tasks in this area.

Many individuals in healthcare, particularly clinicians, managers, and administrators, may be comfortable with decision making. However, effective teams make decisions as groups, often by consensus, acknowledging perspectives of all participants. For many, giving up the right to make individual decisions is difficult. The team or the leadership of the PI initiative may have to assist that person in learning this new, team-oriented, decision-making style.

Conversely, team members sometimes come to the team with little or no management experience. These individuals often have made few decisions in their work outside of day-to-day job procedures. Dealing with PI issues without carrying out someone else's orders may be difficult for these people. Encouragement to participate and mentoring through the process until this kind of individual develops some new skills are very helpful.

Case Study

Dave Richards was the chief financial officer of Community Hospital of the West. The organization had recently embarked on a PI initiative in response to increased competition and new industry regulations. The administrator had hired a consultant to assist in the implementation of the initiative, and the consultant had been training the administrative group in PI methodologies for two months. All administrators were being pressured by the board of directors to find something to improve in their divisions, and so Dave had convened a PI team of his managers and their assistants to discuss possible improvement opportunities.

"So what do you think?" asked Dave. "What is there in our areas that needs improvement?"

The managers sat looking at the pictures on the walls of the conference room or doodling on their note pads, silent. They had been sitting there for 20 minutes before Dave had arrived to chair the meeting.

"There has to be something we can improve, doesn't there?" asked Dave. "Nobody's perfect." Again, the managers were silent.

Finally, Marilyn, the director of patient accounting, spoke up. "Well, I don't see why we should spend all this time and money on these phony meetings, when we could just as well be back at our desks getting some real work accomplished. The suspense account report is longer than I've ever seen it since I've been working here. We're waiting on all kinds of accounts to be coded by medical records, and we're sitting here wasting time on this administrative boondoggle. Doesn't the board know we have better things to do?"

Bristling at Marilyn's condemnation of her department, Peggy, the director of health information services, retorted, "You know we don't have enough coders, Dave. I've begged you a multitude of times in the last few months to let me hire some more coders. You just don't seem to understand that when the census decreases, that means more work for us with all those discharged cases, not less. And are Marilyn's billers all working as hard as they can anyway? Half the time I see them sitting in the cafeteria drinking coffee!"

"We're not here to discuss the suspense account report. I know you've got lots of work to do, but I need us to come up with something to improve that the board can see. I've got to be seen as cooperating in this initiative, or we'll all have hell to pay. Now put on your thinking caps. I think we could spruce up the cashier's area a little. It's kind of dull over there. Probably doesn't make much of an impression on the patients when they come to make payments on their bills. That old, yellowed paint is not very appealing, and those fluorescent lights are hideous! I say we come up with a new look for the cashier's window. How's that?"

Several of the managers still just sat there doodling. A couple murmured, "Uh-huh," or "Yeah, okay."

"So let's have another meeting next week to discuss this some more," continued Dave. "I want each of you to go by there before the next meeting and take a good look at that area. Come up with some ideas about how we could make it look better. Then I'll have something to show for quality improvement. Next week same time, okay?" He looked around the table. No one said anything. "Good!" He rose to his feet and ambled out the door of the conference room, leaving the rest to quietly get up and leave.

Case Study Questions

1. Summarize the organizational dynamics at work in this scenario. What is the nature of Dave Richards's leadership style? Does his style work well for performance improvement activities?

2. Is the group performing effectively as a performance improvement team? What recommendations could be made with regard to team performance?

Summary

Developing effective teams is never an easy task for the leadership of a healthcare organization's performance improvement initiatives. It can be done, however, when undertaken with a positive philosophy and firm commitment at the outset and the recognition that much development may have to be done as the program evolves. Important team roles include the facilitator, the recorder/scribe, and the timekeeper. Team members may also need to learn effective listening and questioning techniques and to accept responsibility for participating in collaborative problem solving and other team activities.

References

Barczak, N. L. 1996. How to lead effective teams. *Critical Care Nursing Quarterly* 19(1):73–82.

Burns, Robert K. 1960. *The Questioning Techniques.* Chicago: Industrial Relations Center, University of Chicago.

Byham, William C., with Jeff Cox. 1998. *Zapp! The Lightning of Empowerment: How to Improve Quality, Productivity, and Employee Satisfaction.* New York City: Random House.

Byham, William C., with Jeff Cox and Greg Nelson. 1996. *Zapp! Empowerment in Health Care.* New York City: Random House.

Leeds, Dorothy. 2000. *Smart Questions: The Essential Strategy for Successful Managers.* New York City: Berkley Publishing Group.

Lynch, Robert F., and Thomas J. Werner. 1992. *Continuous Improvement: Teams and Tools.* Atlanta: Qual Team.

Purser, Ronald, and Steven Cabana. 1998. *The Self-Managing Organization: How Leading Companies Are Transforming the Work of Teams for Real Impact.* New York City: The Free Press.

Renneker, J. A. 1996. Team building for continuous quality improvement. *Seminar in Perioperative Nursing* 5(1):40–46.

Schwarz, Roger. 2002. *The Skilled Facilitator.* San Francisco: Jossey-Bass.

Chapter 17
Managing Healthcare Performance Improvement Projects

Learning Objectives

- To describe the function of project management in performance improvement programs
- To identify specific knowledge and skills required for team leadership
- To describe project life cycles and the group dynamics of team life cycles
- To identify the steps a team leader should follow to successfully implement and complete a project
- To describe the importance of closure with regard to reporting back to organizational leadership

Background and Significance

When a performance improvement (PI) program is initiated, the project requires a project team, who will be responsible for formulation and implementation of the program. Thus, to perform effectively, PI team members need to develop project management skills.

Project management is defined in a number of ways, from a narrowly focused approach with a small, task-oriented team, to the much broader, organizationwide philosophy that is reflected in organizational culture behavior, and structure. Project management as a discipline is rooted in engineering and is oriented toward quantitative application methods. Over time, project management has embraced organizational behavior as a critical element of knowledge and skill necessary for successful implementation of a project.

Performance improvement projects in modern healthcare organizations range from small efforts involving only a few departments to larger ones that impact the organization in very significant ways. Healthcare professionals are likely to be assigned to project teams and, in some cases, may be responsible for leading them.

When a PI team is formed (as discussed in chapters 1 and 3), the life cycle of the team and the project begins. Generally, the organization leadership first determines the composition of the team. Then team roles are established. The mission of the team should be developed in alignment with the organization's overall mission and vision.

Project Management and Organizational Structure

Organizational culture and structure are critical to the success of project management. Bureaucratic organizations with highly structured hierarchies are less accepting of the project management concept than a more dynamic and flexible organization is. Organizations where employees interact across organizational boundaries on a regular basis are going to be more open to project management, and their employees will perform better on a project team.

Project Life Cycle

The length of time a project will take over its entire life cycle varies depending on the scope and size of the project. Large building projects may take months or years, but most projects will last from a few weeks to a few months. The life cycle of a project is composed of several phases; the number of phases and their definitions vary depending on who is outlining the phases and the industry involved. Most projects range from four to six phases.

A Guide to the Project Management Body of Knowledge (PMBOK® Guide), updated in 2000 by the Project Management Institute (PMI) Standards Committee, compares several life-cycle phases representative of different industries. Some experts in the field of project management have chosen to focus on a series of processes from the *PMBOK® Guide* that groups project management processes into four life-cycle phases: initiation, planning, execution (implementation), and closure (Globerson 2002; Keeling 2002). These four phases are appropriate for projects in service industries such as healthcare delivery. For the sake of simplicity, these four phases are used in this chapter.

Initiation

The initiation phase begins with the determination that there is a gap between organization performance and expected outcomes. The leadership then identifies an opportunity for improvement and assesses the feasibility of the project.

Sponsorship
One or more individuals in an organization normally sponsor a project. The personal commitment a sponsor brings to a project coincides with the degree of empowerment a project manager will have. Sponsorship by top leadership, therefore, must be characterized by commitment and clear articulation of expectations.

Team Member Selection
Leadership will select members for the project team and identify any other resources needed to complete the project. Selection of team members should be based on the identification of individuals who possess a variety of skills and expertise. If all project team members are selected from the same department with similar experience and skills, the team runs the risk of overlooking viable alternative solutions that might be raised in a team with more diverse experience and skills. Organizational leadership usually completes much of the initiation phase. Preliminary definitions of the project objectives, activities, and expectations are prepared. Once formed, the team refines these processes.

Mission Statement

If a mission statement has not been articulated by leadership, the project team needs to establish one as a first priority at the beginning of the team's life cycle. A clear mission and vision statement will serve as a guide in the development of objectives and goals.

Project Teams

Once a project team has been formed, project management steps similar to the cycle of team PI processes are followed. (See figure 1.2 in chapter 1.) There are seven steps in the cycle of PI team processes:

1. Identify an improvement opportunity.

2. Research and define performance expectations.

3. Design and redesign process/education.

4. Implement process/education.

5. Measure performance.

6. Document and communicate findings.

7. Analyze and compare internal/external data.

A project progresses through a series of steps from one phase to the next. Figure 17.1 lists the processes that each of the four phases includes. These steps parallel the process steps found in the phases of a project.

Figure 17.1.	**The Performance Improvement Process Cycle**
Phases	**Processes (Steps)**
Initiation	1. Identify a performance improvement opportunity. 2. Determine feasibility of the project. 3. Define project objectives and scope. 4. Select team members. 5. Create vision and mission statement.
Planning	6. Identify the activities (tasks) the team will perform and estimate the duration of activities. 7. Develop final system requirement and criteria for standards of success. 8. Develop a schedule and cost estimates. 9. Perform tasks and track progress.
Execution (Implementation)	10. Present recommendations to leadership. 11. Execute implementation plan. 12. Begin training. 13. Track and monitor progress. 14. Revise project as needed.
Closure	15. Communicate results (final report). 16. Celebrate successes. 17. Continue evaluation and control and identify new opportunities for improvement.

Team Group Dynamics

Although a project life cycle is often defined by phases, the project team will go through a series of stages and adjust at various times through the phases of the project. The project team leader and members will be better prepared to complete the project if they understand the group dynamics of team development. A newly formed team will normally go through all stages of team development, regardless of how well the members have known one another.

Models of team development uniformly define four stages of progression (Montebello 1997 and Buzzotta 1993):

1. **Cautious affiliation:** This is the forming stage, in which team members tend to be very polite as they get to know one another. This is also a time in which the team members assess one another's strengths and weaknesses.

2. **Competitiveness:** This is the storming stage, in which conflicts emerge. Without effective leadership and the ability to resolve conflicts, it is difficult for teams to get past this stage. They will either stay in conflict or revert back to a phony politeness. Regardless of how they react, the productivity of the team is limited during this stage.

3. **Harmonious cohesiveness:** This is the norming stage, in which team members learn to communicate and collaborate. They become more focused on the task at hand. Members begin to feel that they are a contributing part of the team. They also begin to establish rules of engagement with one another.

4. **Collaborative teamwork:** This is the performing stage where a group of individuals begins to collaborate as a team. Team members come to understand group norms, and communication becomes more efficient and effective. In highly effective teams there is less conversation and more action. Individuals take pride in the results produced by the team.

The team leader needs to be prepared for the natural shift in dynamics of a group as it matures and be able to facilitate team development. Team leaders should allocate time for forming, storming, norming, and performing every time the team meets (Glacel 1997). This progression of stages is not necessarily linear. Even though a team has matured to norming or performing, events may occur that cause the team to revert to storming.

Allotting a few minutes at the beginning of each meeting to "check in" can help to move the team along toward greater maturity. "Checking in" can be as simple as asking each team member to tell the team what he or she is prepared to bring to the meeting for the day. If there is a change of even one team member, the team returns to the forming stage and the progression through the four stages begins again.

Some authors add a fifth stage to this process, called adjourning. Adjourning marks the dissolution of a group. If handled appropriately, it provides positive closure for the team members. This is a time to celebrate successes and recognize team member contributions and accomplishments.

Leadership

Situational leadership is a useful model for understanding and leading project teams (Hersey and Blanchard 1988). When a team leader understands which level of maturity his or her team has reached, he or she can select an appropriate, effective leadership style. The team leader should be more task oriented and directive with newly formed groups, and more relationship oriented and supportive of team members as they mature.

The project team leader is usually an employee from a functional area of the organization who is assigned responsibility for leading the team to completion of a project. This may put the leader in a position that divides attention and loyalty between the project team and the parent organization if the vision, goals, and objectives of the two are not aligned. A key role for the team leader is to bring these three elements into harmony.

Planning

Organization leadership should make clear to the project team members the importance of the project and the expected impact on the organization. However, once objectives for the project are established, the team should feel free to proceed without interference from leadership. Periodic feedback through reports and briefings can be scheduled to keep leadership informed about the progress of the project. (Chapter 5 discusses meeting minutes and reports that can be effective in keeping leadership informed regarding the team's progress.)

A critical element of the planning phase is the identification of final system requirements or criteria that set standards for measuring success. Without these standards in place, determining whether a project has succeeded becomes difficult.

Design

The most important contribution team members can make during the design phase is the development of alternative solutions. If the organizational culture truly embraces PI and problem solving, a team will be able to develop alternative solutions and work through a step-by-step process of deciding which alternative is the optimal solution.

As alternatives are developed and discussed, the cost of implementing a recommended solution should be considered. Costs should be divided into two categories: fixed and operating. Fixed costs are one-time expenses associated with buying new equipment and getting started. Operating costs are incurred to sustain the project on an ongoing basis.

Once the team decides to recommend an optimal solution, it needs to develop a schedule for implementation. This is a critical element of the planning function in project management. Planning must identify tasks, their duration, and who will be responsible for them.

Gannt Charts

An effective tool for planning and tracking the implementation of a project is the Gannt chart. The Gannt chart (discussed in chapter 7, page 95) is often used by PI teams to plan and track the progress of a project.

PERT Charts

If a more quantitative approach is required, the Program Evaluation and Review Technique (PERT) may be used. This is also called Critical Path Method (CPM). The PERT technique provides a structure that requires the project team to identify the order and projected duration of activities needed to complete a project. The most helpful element of PERT is that it identifies those critical activities that must be completed on time in order for the entire project to meet its final deadline.

PERT charts depict a network of activities represented by arrows, as shown in figure 17.2. The numbers above the arrows represent time required to complete the activities. Duration time is usually measured in days or hours. In order to construct a PERT chart, the PI team must identify all activities required and determine which activities precede one another. The letters below the arrow represent the activity. The circles, called events, represent the beginning of an activity.

Some activities may be concurrent. These are called parallel activities. By following any path of arrows through the network from start to finish and adding the duration times of each activity, one can determine the total amount of time that series of activities will require. The path with the greatest total duration time is called the critical path and represents the shortest amount of time required to complete the total project. The critical path in figure 17.2 is the sequence *a-d-g-h-i,* which will require 23 days. PERT and Gannt charts both require the planner to identify critical tasks, the duration of each task, and the expected completion dates. This scheduling process usually includes costs associated with tasks and the resources needed to complete them.

Execution

Once a plan is completed, execution (or implementation) begins. This is where installation of equipment or construction begins, and any policy or procedure manuals should be prepared for distribution. Specifications that may have been developed in the design phase should be finalized. Any new systems or processes should be tested for performance.

Individuals involved in implementation and continued operation of a new or reorganized system need training. Thus, the implementation plan should include training and identification of who will be trained. The training portion of the implementation plan should identify the content of the training, training objectives, and expected outcomes. If any QI toolbox techniques are going to be implemented, they should be part of the training plan.

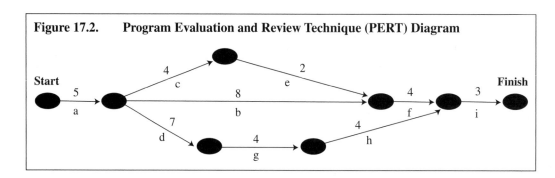

Figure 17.2.   Program Evaluation and Review Technique (PERT) Diagram

Measurement Techniques

A number of the QI toolbox techniques discussed throughout part II of this text can be used to continuously to measure performance and to evaluate the success or failure of newly implemented processes or systems. The charts and graphs discussed in chapter 4 can be useful for organizing data collected into a meaningful presentation when providing feedback to organizational leadership.

Closure

In closure, stage four of project management, the new system or process is used by the customer. This is the phase where the project shifts to become an integrated part of organizational operations. During the operational phase, management must continually monitor performance and determine whether or not it is meeting established performance criteria.

Evaluation and Control

As a project shifts from planning to execution to closure, testing performance results against finalized standards must not be ignored. Too often, this is where the organization can become distracted with other issues or new events and not follow up on the success or failure of the project. This is one reason why a project can fail.

The project becomes the standard way of doing things in the closure phase. Team members return to their functional roles or move on to newly assigned roles if the project is one that changes their old functional roles. This is the phase where lessons learned are cataloged and documented. As the established project continues, it must be continually evaluated to determine if performance is meeting established criteria and standards. During the process of evaluating results and outcomes, the organization looks for new improvement and innovation opportunities. When a new opportunity for PI is identified, a new project is initiated.

Real-Life Example

In the aftermath of September 11, 2001, the Joint Commission on Accreditation of Healthcare Organizations (JCAHO) has broadened accreditation requirements for specific plans that respond to bioterrorism. The Centers for Disease Control, Health Resources and Services Administration, and the Department of Homeland Security have begun to work together to ensure that state and local health departments, hospitals, and other health agencies are able to mount a collective response to bioterrorism.

A 150-bed community hospital, located approximately 35 miles from a large medical center in another county, has been approached by the county health officer to develop a bioterrorism response plan. Although the hospital has conducted annual disaster drills in compliance with previous JCAHO standards, the focus of the drills were mass casualty exercises based on the scenario of a major fire in an industrial setting, natural disasters, or a transportation accident involving large numbers of injured patients. Previous exercises have not addressed exposure of large numbers of patients to biological or chemical agents as a result of a terrorist act.

The hospital administrator and the county health officer agree that there is a need to enhance and broaden the scope of disaster planning to include issues that would require significant decontamination of patients or strict quarantine of patients exposed to biological or chemical agents. The hospital administrator has agreed to appoint a project team that will work with the county health officer to expand the existing disaster plan to include bioterrorism threats.

The county health officer has authorization to spend funds from the Department of Homeland Security to purchase supplies and equipment the hospital would need in the event of an attack. This includes items such as decontamination equipment and protective suits for hospital personnel assigned to bioterrorist response teams. In order to qualify for the funds, the hospital must coordinate with the county health officer and develop a plan for responding to a bioterrorist attack.

Because federal funding is involved, the plan must be complete and in place within twelve weeks or the funds will be withdrawn and reallocated to other regions of the country. The hospital is scheduled for a JCAHO survey in about nine months, an additional incentive to complete the project in a timely manner.

During the initiation phase of the project, the hospital administrator and the county health officer agreed upon the mission to develop an implementation plan for policies and procedures that prepare the hospital to effectively respond to a bioterrorist attack involving biological or chemical agents. The composition of individuals selected for the project team was based on ensuring that all potentially affected departments would be represented. The hospital administrator has decided to appoint the director of emergency services as the project team leader, who will be expected to work closely with the county health officer in directing the team's activities. The county health officer will be part of the project team as a consultant.

The director of health information services will play a critical role on the team. In past disaster drills, the administrative functions of admitting emergency patients and creating medical records were accomplished within the emergency room. The need to establish a decontamination area and triage outside the hospital building following implementation of the Health Insurance Portability and Accountability Act (HIPAA) of 1996 has created new requirements regarding the manner in which patients will be processed for admission.

The infection control nurse's activities become even more important in a bioterrorist attack, in regard to collecting data and for training staff on procedures that will control further spread of disease and illness from a patient infected by a biological agent.

The public affairs officer will have the challenging task of controlling press inquiries if a bioterrorist attack occurs. Reporters from the media can be overly zealous in trying to get the story and meet deadlines, and their movement in and around the facility must be controlled and coordinated. Additionally, with the implementation of stricter HIPAA guidelines on release of medical information, the organization must be more vigilant than ever before.

The director of nursing plays an important role in providing trained staff that can respond effectively to large numbers of casualties from biological or chemical agents. The director must also plan the use and location of hospital beds in case the hospital must expand to respond to caring for an unexpectedly large number of patients in need of specialized care. The director must also collaborate with the directors of emergency services and health information to establish procedures for evacuating existing patients and rerouting incoming patients to avoid overwhelming the hospital's capacity.

The director of physical facilities is responsible for ensuring that the heating, ventilation, and air-conditioning systems are functional and capable of handling a bioterrorist threat to the hospital environment. The director is also responsible for handling hazardous material (HAZMAT) decontamination, including storage of and access to equipment needed to respond to a threat.

As a part the initiation phase, the hospital administrator will bring the selected members of the project team together for orientation. At this point, the planning phase will begin. The team will then meet to brainstorm and identify problems. Figure 17.3 is a Gannt chart that depicts the four phases of the project and the tasks to be accomplished in each phase. In the planning phase, the subtasks are assigned to individual team members based on their areas of expertise and responsibility.

As team members bring information and expertise back to the team, the plan begins to take shape and the team prepares to move into the execution phase. The predominate tasks of the execution phase are writing the final draft of the plan and training staff in preparation for future exercises and the possibility of an actual terrorist attack.

During the last phase, closure, the team and hospital leadership can evaluate the results of the project. The rehearsal exercise will be documented so that lessons learned can be applied to opportunities for continued improvement and to demonstrate to the JCAHO surveyors that an annual exercise meeting the new standards has been conducted.

Why Projects Fail

The Center for Project Management publishes seven deadly sins of project failure on its Web site:

1. Mistaking half-baked ideas for viable projects

2. Dictating unrealistic project deadlines

3. Assigning underskilled project managers to highly complex projects

4. Not ensuring solid business sponsorship

5. Not monitoring project vital signs

6. Failing to develop a robust project process architecture

7. Not establishing a comprehensive project portfolio

Some of the sins are self-explanatory. Failure regarding solid business sponsorship places responsibility squarely on senior leadership of an organization. As mentioned earlier, executive-level commitment and support of a project is critical to its success. Without the backing of leadership, implementing a new project will be difficult, if not impossible, because of the natural resistance to change that occurs in most organizations.

Developing a viable project process shows that the organization has developed guidelines for standardized and repeatable processes of beginning and completing projects. Project team leaders should not have to start from scratch with each new project. A comprehensive project portfolio consists of a set of files organized into at least six areas, which leadership and project team leaders are able to reference. The portfolio should

Figure 17.3.	Gannt Project Chart													
Phases							Week							
Tasks	**Responsible Parties**	1	2	3	4	5	6	7	8	9	10	11	12
Initiation		◆▬▬◆											
First meeting of team leaders	Hospital administrator and county health officer	▼											
Mission statement	Hospital administrator	▬											
Team member selection and appoint team leader	Hospital administrator	▬											
Team orientation meeting	County health officer	▼											
Planning				◆▬▬▬▬▬◆									
Planning meeting to brainstorm and identify problem(s)	Emergency services	▼											
Assign individual tasks	Emergency services	▼											
FEMA coordination planning	County health officer	▬▬▬▬▬▬											
Admission and medical information management	Health information	▬▬▬▬											
Equipment storage and access	Physical facilities	▬▬▬											
Infection management	Infection control	▬▬▬											
Identify locations suitable for patients	Physical facilities and nursing	▬▬▬											
HVAC issues	Physical facilities	▬▬▬											
Media management	Public affairs	▬▬▬											

Figure 17.3.   *(Continued)*

| Phases | | Week | | | | | | | | | | | |
Tasks	Responsible Parties	1	2	3	4	5	6	7	8	9	10	11	12
Develop training plan	Emergency services					▬	▬	▬					
Execution									◆	▬	◆		
Collecting team input and writing the final plan	All team members								▬	▬			
Training	All team members								▬	▬	▬		
Closure										◆	▬	◆	
Submit plan for federal funding											▼		
Rehearsal exercise	All team members and hospital staff										▼		
Prepare and submit final report	Emergency services										▬	◆	
Begin ongoing evaluation												▼	

include a file of ideas, charters (or proposed projects), projects in execution, completed projects, suspended projects, and cancelled projects. With this collection of information, a project manager can compare his or her project to past successes and failures and other activity going on throughout the organization.

It is also just as important to plan and organize a team for success as it is to avoid those things that lead to failure. Following is a list of six key steps that can lead to a winning strategy:

1. Organize for success.

2. Create a project plan.

3. Develop a means to track performance to plan.

4. Implement the plan.

5. Train and retrain if necessary.

6. Anticipate and prepare for the culture shock associated with change (Zimmer 1999).

Getting organized and having a well-conceived plan in place is key. In order to know if the project is progressing satisfactorily, the team needs to put a monitoring system in place that measures success criteria against outcomes.

The last step of the six keys cannot be ignored. It is a natural force in humans and organizations to initially resist change. Leadership must be prepared to effectively counter resistance and continue the process of implementation and improvement.

As mentioned earlier in this chapter, some organizations are more receptive to change than others. The organization's leadership, project sponsors, and team members must champion the project and be persistent in seeing it through to full implementation.

Case Study

A community health clinic located in a city of 250,000 provides the bulk of Medicaid and homeless care to the community. The clinic maintains approximately 27,000 active medical records and has a staff of ninety employees. The professional clinical staff includes five family practice physicians, seven physician's assistants, and four clinical nurse practitioners.

The clinic management recently received notice that it is eligible to receive funding to convert its existing computer systems to a new integrated system that will include electronic medical records. Until this point, all records have been hard copy and records tracking throughout the clinic has been manual. Billing and financial management systems have been maintained on separate systems. Clinic appointments were maintained on another system. The office applications software consisted of outdated word-processing and spreadsheet systems. The database software was not integrated with other software.

The proposal for the new integrated system includes office management applications, such as word processing, spreadsheets, database, presentation, email, patient appointment scheduling, computer-based medical records, billing, and financial management software. Laboratory specimens are sent out to a reference lab located at the county hospital. The lab has expressed interest in linking its lab results software to the new integrated system via its intranet. The lab manager indicates that the results can be transmitted in a format compatible with any relational database.

The administrative and clinical support employees are fairly stable, with an average personnel turnover rate of eight percent and an average longevity of five years. The physicians on staff have only been with the clinic from six months to three years. The chief executive officer (CEO) has given the director of health information management the task of managing a project to migrate existing systems to the new information management system.

Money available for the transition must be spent by the end of the fiscal year. There are ten months left. The CEO has also indicated that all existing medical records need to be scanned and loaded into the new system, and expects to see this completed within the next two years.

Case Study Questions

1. Up to seven representatives from different departments can serve on the multidisciplinary project team. Prepare a list of types of individuals you would want to serve on this project.

2. What departmental areas of the healthcare organization should be represented?

3. What specific skills, knowledge, and expertise should team members possess?

4. Draft a time line for the project that includes the major phases of the project and key tasks for each phase.

Project Application

Microsoft Office (http://www.microsoft.com/office/project/default.asp) provides information about Microsoft Project software and demonstrates how to set up a project schedule and track its progress.

There are other Web resources available through searches using keywords such as time-line or project management that make tools such as Gannt and PERT charts available.

Go to the Internet and enter search words for project management software or timeline into a search engine. Explore and develop a list of sources that provide software for managing a project.

Review how you may apply project management methods to group projects in class or the workplace.

Summary

Effective project management requires an understanding of the four phases a project team will go through in the life cycle of a project: initiation, planning, execution (or implementation), and closure. During each of the four phases, the team completes a logical series of processes that are similar to the steps of the performance improvement team process presented in chapter 1. Organizational behavior and group dynamics play an important role in the success or failure of a project. Commitment and support from organizational leadership is also critical to project completion and success.

References

Ash, Tim. 1998. Seven reasons why Internet projects fail. *UNIX Review's Performance Computing* 16(11):15.

Center for Project Management. http://www.center4pm.com.

Glacel, Barbara Pate. 1997. Teamwork's top ten lead to quality. *Journal for Quality and Participation* 20(1):12–16.

Globerson, Schlomo, and Ofer Zweifael. 2002. The impact of the project manager on the project management planning process. *Project Management Journal* 33(3):58–64.

Hersey, Paul, and Kenneth H. Blanchard. 1988. *Management and Organizational Behavior.* Engelwood Cliffs, N.J.: Prentice-Hall.

Keeling, Ralph. 2002. *Project Management: An International Perspective.* New York City: St. Martin's Press.

Kerzner, Harold. 1998. *Project Management: A Systems Approach to Planning, Scheduling and Controlling,* sixth ed. New York City: John Wiley and Sons.

Montebello, A. R., and V. R. Buzzotta. 1993. Work teams that work. *Training and Development* 47(3):59.

PMI Standards Committee. 2000. *A Guide to the Project Management Body of Knowledge.* Newtown Square, Penn.: Project Management Institute.

Seidl, Patricia. 2002. Project management. Chapter 26 in *Health Information Management: Concepts, Principles, and Practice,* Kathleen M. LaTour and Shirley Eichenwald, editors. Chicago: American Health Information Management Association.

Zimmer, Brian T. 1999. Project management: a methodology for success. *Hospital Material Management Quarterly* 21(2):83–89.

Chapter 18
Managing the Human Side of Change

Learning Objectives

- To apply change management techniques to implement performance improvements

- To describe the three phases of change

- To identify key steps in change management

Background and Significance

In today's world, there is no such thing as permanent stability. The processes and structures that worked last year may be ineffective this year. The products and services that are considered cutting edge quickly become obsolete. Similarly, healthcare delivery in the United States has been in a state of rapid and unpredictable evolution ever since the Medicare and Medicaid programs were implemented in the 1970s and the prospective payment system (PPS) was instituted in the 1980s.

The overarching reason for change in healthcare organizations today is the need to improve the quality of care while at the same time controlling the cost of services. Hospitals and other healthcare organizations have institutionalized performance improvement (PI) programs to meet this need. Systems, processes, and staff competencies undergo a circular cycle of change as improvements are made in the clinical, administrative, and governance areas of the organization.

Performance improvement is based on the quantitative analysis of data, processes, and structures, but PI efforts also have a very human side. After all, healthcare is provided by individuals working in extremely complex organizations, not by robots that can be reprogrammed or replaced when change is needed. Failure to consider the human side of PI and the impact of change can derail even the best-conceived improvement efforts.

Healthcare professionals have always understood the importance of what they do. Today, during an era of dramatic and ongoing change, most clinical and allied health professionals strive to improve patient care services and outcomes. In this era of rapid change in healthcare, we see a proliferation of "disruptive technologies and business models that may threaten the status quo but will ultimately raise the quality of care for everyone"

(Christensen, Bohmer, and Kenagy 2000, p. 104). Still, most employees find it difficult to alter their work habits, and healthcare professionals are no exception.

The Three Phases of Change

Every change that affects individuals is experienced in three phases. Whether changing their dietary habits or the way they write a patient care plan, all individuals go through a similar process. In his research on human reaction to change, Kurt Lewin (1951) calls the three phases *unfreezing, changing,* and *refreezing.* Lewin observed that during each of these basic phases, forces in favor of change and forces that resist change work against one another. The analysis of such competing forces in the face of a particular planned change is often called a force-field analysis. These phases have also been referred to as ending, transition, and beginning. The ending phase is characterized by grief and letting go. The transitional phase is characterized by confusion and creativity. The beginning phase is characterized by acceptance and hope for the future.

The three phases of change are not clear-cut steps. Rather, they overlap one another. At any particular point, one of the three phases is likely to predominate, while the emotions and concerns associated with the other two phases fall to the background. The movement is gradual, as one phase gives way to the next.

Grief is a natural human reaction to loss of any kind. Simply defined, grief is the conflicting feelings that come along with the end of something familiar or a change in an accepted pattern of behavior. Before individuals can go on to a new beginning, they need to let go of their old identity or their old way of doing things. According to William Bridges, "the failure to identify and be ready for the endings and losses that change produces is the largest single problem that organizations [and individuals] in transition encounter. The organization institutes a quality improvement program, and no one foresees how many people will experience the 'improvements' as a loss of something related to their job" (Bridges 1991, p. 5).

Like change, grief occurs in stages. Some propose that the process consists of three stages: shock, despair, and recovery. Still others posit that the process includes four stages—denial, resistance, adaptation, and recovery (Kreitzer 1998). Whatever the number and names of the different stages, it is clear that a transformation in thinking and perception must occur.

Between the ending of the old and the beginning of the new lies a transitional zone. This transitional period is experienced regardless of the desire to change or whether the perception of the change is good or bad. The transitional phase is unsettling and uncomfortable. Individuals often report feeling confusion, anxiety, and unsteadiness in the midst of change. The old way of doing things is gone, but the new way of doing things still feels untried and uncomfortable.

Left to their own thoughts and feelings and without sufficient information during this phase of organizational change, individuals may decide to escape their discomfort and confusion by leaving the organization. When the change process is understood, however, the transitional period can be a time of renewal and creativity. Often certain individuals within a larger group are optimistic about change. Their enthusiasm is contagious, and a wise organization will recognize and capitalize on such individuals to infuse their optimistic attitude into groups and departments in order to build positive momentum for change (Blancett and Flarey 1998).

During the final phase of change, the beginning of the new way of doing things, the people who make up the organization settle into a more comfortable state. The new processes or staff structures become familiar, and individuals come to understand and accept their new roles. Some may still worry that the new way of doing things may not work or that it may even make things worse. Some may feel last regrets about the ideas or coworkers they had to leave behind. Some may even miss the freedom of the transitional zone and find settling back into a routine rather boring. A beginning can also be a disappointing time when changes seem to have been made for no discernible reason. Still, eventually, the new way of doing things becomes frozen or the accepted way. An unfamiliar task often becomes automatic within the cycle of change:

> Whenever people learn something sufficiently well, they cease to be aware of it. When you look at a street sign, for example, you absorb its information without consciously performing the act of reading. Computer scientist, economist, and Nobelist Herb Simon calls this phenomenon "compiling"; philosopher Michael Polanyi calls it the "tacit dimension"; psychologist T.K. Gibson calls it "visual invariants"; philosophers Georg Gadamer and Martin Heidegger call it "the horizon" and the "ready-to-hand"; John Seely Brown [. . .] calls it the "periphery." All say, in essence, that only when things disappear in this way are we freed to use them without thinking and so to focus beyond them on new goals (Weiser 1991).

Any organization, and especially a large healthcare organization, thrives best when operating with established structures and processes.

Change Management

Like PI, change can be thought of as a process to be understood and managed. For the purposes of this chapter, *change management* can be defined as a group of techniques used to understand the process of change and accept PI in work processes. One or more members of the PI team may become change manager for the project, or a manager in the areas affected by the change may play this role. Many organizations hire consultants to handle the change management process when the planned changes will have a significant impact on employees and medical staff.

The steps in the process of change management include the following:

1. Identifying the losses
2. Acknowledging the losses
3. Providing information and asking for feedback
4. Marking the endings
5. Managing the transition
6. Clarifying and reinforcing the beginning
7. Celebrating the successes

Nothing is as crucial for ensuring the success of a change plan as effective education and communication about change management. Often the best strategy for change is preparation. Training often gets overlooked or is not well planned. It is a good idea to educate your team on the principles of change management. This will better prepare them for the reactions of others to change and equip them with the skills to handle them appropriately (Amatayakul 1999).

At Norton Healthcare in Louisville, Kentucky, chief information officer Marilynn Black (2003) reported that during a time of heavy transition, the organization realized that communication and training were the keys to success. As the organization prepared to unroll an electronic medical record system project titled CareLink at two hospitals, it appropriately managed the change process in the following ways:

- Educating leadership regarding the dependency of information technology (IT) project success on appropriate change management

- Establishing ownership and accountability of organizationwide operational leadership for the project early in the process

- Creating a steering committee composed of operational leaders from functional areas impacted by Carelink to govern implementation and identify and discuss operational risks

- Creating a vision and project management structure to support the change

- Creating a go-live plan to support the late adopters

- Building a training program to support the change

Furthermore, leaders at Norton Healthcare wholeheartedly embraced the task of training staff by doing the following:

- Implementing a variety of communication techniques to prepare staff and managers in a countdown to the day of transition

- Creating a job-specific training curriculum

- Constructing a learning center

- Constructing a training morale package of materials for staff to review

By all accounts, the transition was successful and went smoothly, thanks to the forethought and planning of leaders and to their attention to change management best practices.

Identifying the Losses

When a PI project is still in the planning stage, identifying the losses that will result from changes that need to be made may be difficult, especially if the change affects more than one area of the organization. The PI team should describe the proposed improvement in as much detail as possible. Using flowcharts to map out processes may help the team to identify all of the areas that will be involved. Changes made in one area may create the need for secondary changes in other areas. The task of the PI team is to identify all of the people who will need to let go of a current way of doing things before an improvement can be implemented. The team also needs to determine exactly what will need to come to an end for the project to be successful.

Identifying the losses after the improvement has been implemented is much easier. The PI team or change manager for the project needs only to ask affected people questions such as "What is different for you now?" or "What don't you do anymore now that . . . ?" It is important to remember that no new process or structure can be implemented until its predecessor has come to an end. Ideally, identifying and acknowledging the losses necessitated by a change should come before the change is implemented, but the ending phase will happen whether it is planned for or not.

Acknowledging the Losses

Depending on the extent of the changes to be made, representatives of the PI team, department managers, or senior executives should explain the planned changes to the people who will be affected. The change manager should be prepared to accept and acknowledge the reactions that result, even when they seem like overreactions.

According to Susan Helbig, "how you frame your message is key to garnering interest, support or a firm commitment from various stakeholders" (Helbig 2003, p. 67). Even among those communicating in this way with staff, there are different roles for different purposes. Messengers simply provide information about the change. Advocates garner enthusiasm for change. Change agents manage the process and are available to answer more detailed questions. Each of these roles should be used at the appropriate time in the change process (Helbig 2003).

When endings take place, people may feel angry, sad, anxious, confused, and depressed. All of these feelings are normal reactions to loss. Allowing people to express their emotions openly is difficult, but critical to success. Sometimes even minor changes may become symbols of much more comprehensive changes that took place in the past but were never fully acknowledged. Minor changes may also be treated as harbingers of more drastic changes in the future. For example, a process redesign that will result in the elimination of one staff position may be seen by employees as the first of many layoffs to come. The key to effectively handling this step in the change management process is active listening: ask questions about how people feel, and acknowledge the legitimacy of those emotions.

Providing Information and Asking for Feedback

The timing and content of communications should be carefully planned in advance as an element of project design. It is crucial that the proposed changes be described in specific detail early in the PI process. If people do not understand the purpose of a change, they will have difficulty accepting it. If they are not sure what the change will entail, they will come to their own conclusions about which processes will end and which will continue. If they are not told how the change will affect them, they may assume the worst.

The PI team or its representative must provide as much information as possible to the people who will be affected directly by the proposed change. Withholding information may lead to intense speculation about the changes to come, and such speculations often create feelings of helplessness and anger. Although some information may need to remain confidential, the change manager should never fabricate answers to questions that he or she cannot answer fully. Rather, the manager should acknowledge the questions and provide as much information as possible. Glossing over the potentially negative aspects of the change only creates mistrust.

Information about the change project should be communicated consistently and often. Repeating the information in a variety of ways will help people accept the change. Newsletters, special announcements, staff meetings, and other forms of communication can all be used to convey the message. The people affected by the change should also be kept up-to-date as proposed changes are developed, and they should be given the opportunity to provide feedback to the PI team.

The PI team should be sure to seek feedback during every critical stage of the project. They should ask for information about the concerns of stakeholders early in the project design phase. During testing, stakeholders should be asked for feedback on what is working well and what is not, and they should be directly involved in creating a detailed plan for implementation. During implementation, the people who apply the change should be asked to suggest refinements and improvements. And after the project is complete, they should be asked to provide feedback on whether the goals of the project were met.

Marking the Endings

Actions always speak louder than words. The change manager should find some method to mark the ending of the old way of doing things. Removing old paper chart storage equipment in an organization that has transitioned to an electronic health record can mark a symbolic end to the old way.

Managers sometimes make the mistake of criticizing the way things were done in the past as a way of introducing improvements. Creating negative pictures of the past is not effective. Instead, the past should be honored for the positive things accomplished and for foundations laid for the future.

Managing the Transition

The length of the transitional period between old and new depends on the extent of the changes to be implemented. Obviously, the restructuring of a whole organization would require a much longer transition than the installation of a document imaging system. However, both would require that people become accustomed to a new way of doing things. In the first case, everyone in the organization would be affected, clinical staff as well as administrative staff, nurses as well as health information managers. The installation of a new piece of diagnostic equipment, however, might affect only a small number of people. The technicians using the equipment would be affected most directly, but the change might also affect the nursing staff, the patient transport staff, and the medical staff working in that specialty area.

The transitional period is a difficult time for everyone. Productivity is likely to diminish as energy levels fall and people feel unsure of themselves and of the systems on which they depend to accomplish their work. Old resentments may resurface, and staff turnover may increase. During any transition, people tend to oppose the change, endorse the change, or reserve judgment. Interpersonal or interdepartmental conflict may result. Things can feel out of control and chaotic, and people may dread going to work every day.

The change manager's task is to help people to understand that chaos is a necessary and normal part of change. It is during this period that people learn new skills, redefine their roles, and work through their questions about the new processes or structures to be implemented. Creative solutions to unforeseen problems are devised, and new relationships are forged. Special training in creative problem solving and team building may be helpful during this period.

The change manager also needs to ensure that nothing falls between the cracks. Temporary systems may be needed to maintain operations. For example, interim team leaders may be assigned to handle staff scheduling, or temporary record-handling procedures may be instituted during the transition between a paper-based and a computer-based health record system.

Above all, the change manager must keep the channels of communication open. Information about the progress of the project and the problems being encountered during implementation must be shared among the areas affected, the PI team, and senior management. New policies and procedures should be developed, and position descriptions should be revised. Change managers may find it helpful to use storyboards to document and explain the improvements being made.

The purpose of the change or the problem that required resolution should be explained and repeated in every communication during the transitional period. The people affected by the change should also be involved in the development of a step-by-step plan for phasing in the new process. The plan should spell out the role that each individual is to play during the transition and after the new system has been fully implemented. The plan should also establish and communicate realistic target dates. As progress is made toward target goals, public acknowledgment is helpful for building positive momentum. Newsletters, group announcements, and posters all work well in this regard.

Clarifying and Reinforcing the Beginning

The arrival of a new manager, the installation of new equipment, or the move to new offices marks the beginning of long transitional processes. Before a beginning can be successful, the people affected by the change must have gone through both an ending and a transition. A true beginning confirms the end of the old way of doing things.

The PI team, the change manager, and other decision makers in the organization need to reinforce new beginnings on every level. Nothing will doom a change effort more quickly than conflicting messages. For example, an improvement that involved making the change from traditional directive management to a self-managed team would not succeed unless the employees were delegated the authority to make decisions. If the former department manager continued to make every important decision, other people on the team would revert to silence. Similarly, customer service initiatives would not survive if senior managers emphasized cost-cutting over quality in their communications to staff.

Celebrating the Successes

Humans use ceremonies to mark beginnings as well as endings. The change manager should find a way to help people celebrate their successes in making meaningful changes in the way they perform their work. A ribbon-cutting ceremony at a new facility, an open house for a reorganized department, and a demonstration of new equipment for colleagues outside the department are some examples of celebrations that mark new beginnings. Even the accomplishment of a minor procedural improvement should be acknowledged with a sincere thank-you to all people involved in conceiving, planning, and implementing the change.

Case Study

Faced with rising costs and declining revenues, the board of directors for a community hospital located in a large East Coast city decided to combine its obstetrics and pediatrics units. The obstetrics service was located in the oldest part of the facility, and the labor and delivery rooms were cramped and inefficient. A large, nearby medical center offered comfortable, family-centered accommodations and state-of-the-art equipment.

As a result, several obstetricians had recently moved their practices from the community hospital to the medical center. The board was reluctant to discontinue obstetrics services because the hospital had a long history of providing maternity care to the surrounding community, but overhead costs for the underutilized and obsolete unit were skyrocketing and the cost of replacing the unit was prohibitive.

The pediatrics unit, in contrast, was housed in the newest wing of the hospital. The hospital's emergency department was one of only two in the city equipped to provide pediatric trauma services, and a number of patients were admitted to the service through emergency. In addition, a large physician group that specialized in pediatric oncology admitted hundreds of patients to the facility each year. The physician group was recognized nationwide and handled referrals from pediatricians throughout a four-state area.

The consolidation of services made sense to the board on several levels. From a cost-control perspective, closing the obsolete facility would save the hospital millions of dollars in renovation expenses. The consolidation also made sense from a patient care point of view. Expectant mothers who were known to have high-risk pregnancies could plan to deliver their babies at a facility that specialized in treating pediatric patients. In addition, the head of the obstetrics unit was scheduled to retire soon, and the board saw an opportunity to make significant changes in the clinical area. The board believed that the obstetrics service should change its focus from handling routine deliveries to handling high-risk pregnancies and thus overcome the competitive disadvantage the hospital now faced in comparison to the medical center. The board voted to seek funding for a new center of excellence for the care of high-risk mothers and newborns.

Combining the two units will require fundamental changes. The two units are managed in very different ways. The department head of the obstetrics unit is a traditionalist, and the obstetrics staff is accustomed to deferring to his judgment when problems arise. By contrast, the department head for the pediatrics unit believes in self-managed work teams, and her unit is structured into cross-functional teams with independent decision-making authority. There are other obvious differences as well: one unit treats adults, while the other treats children. One unit treats women who stay in the unit for a day or two at most, while the other treats infants, children, and adolescents for conditions that require complex treatment regimens and long hospital stays. Parents often sleep in their children's rooms for weeks at a time. Each of the units has a separate identity, and a certain animosity has developed between the units over time.

Case Study Questions

1. How do you think the nurses who work on the two units will feel about the change? The department heads? Who will lose what? How could those losses be acknowledged?

2. Create a communications plan for the project. How would you describe the purpose of the change? What tools would you use to communicate information about the project?

3. Who do you think should act as the change manager for the consolidation?

Summary

Performance improvement initiatives sometimes fail when human factors in change are ignored or mismanaged. By recognizing the steps people go through to accomplish change and managing the transition from old to new, change managers can ensure the success of performance improvement efforts.

References and Suggested Readings

Amatayakul, M. 1999. *The Role of Health Information Managers in CPR Projects.* Chicago: American Health Information Management Association.

Black, Marilynn. 2003. CARELink Project Report: CareLINK Critical Success Factor—Integrating Change Management into the IT Implementation Process. *Norton Healthcare Internal Finance Committee Memorandum.* April 24, Louisville, Ky.

Blancett, Susan Smith, and Dominick L. Flarey. 1998. *Health Care Outcomes, Collaborative, Path-Based Approaches.* Gaithersburg, Md.: Aspen.

Bridges, William. 1991. *Managing Transitions: Making the Most of Change.* New York City: Perseus Books.

Buckley, D. S. 1999. A practitioner's view on managing change. F*rontiers of Health Services Management,* Fall, pp. 38–43.

Christensen, C. M, R. Bohmer, and J. Kenagy. 2000. Will disruptive innovations cure health care? *Harvard Business Review,* September-October, pp. 102–110.

Galpin, Timothy J. 1996. *The Human Side of Change.* San Francisco: Jossey-Bass.

Helbig, Susan. 2003. Communicating change with style. *Journal of the American Health Information Management Association* 74(5):66–67.

Lewin, Kurt. 1951. *Field Theory in Social Science: Selected Theoretical Papers,* D. Cartwright, editor. New York City: Harper and Row.

Kohles, Mary K., William G. Baker, and Barbara A. Donaho. 1995. *Transformational Leadership: Renewing Fundamental Values and Achieving New Relationships in Health Care.* Chicago: American Hospital Publishing.

Kotter, J. P. 1995. Leading change: why transformation efforts fail. *Harvard Business Review,* March–April, pp. 59–67.

Kreitzer, Donald J. 1998. What I learned about change, I learned in practice, not from the literature. *Proceedings of the 17th Annual Midwest Research-to-Practice Conference in Adult, Continuing, and Community Education,* October 8–10, Ball State University, Muncie, Ind.

Senge, Peter, et al. 1994. *The Fifth Discipline Fieldbook: Strategies and Tools for Building a Learning Organization.* New York City: Doubleday.

Weiser, Mark. 1991. UBICOMP paper: the computer for the 21st century. *Scientific American,* September. Accessed online at http://www.ubiq.com/hypertext/weiser/SciAmDraft3.html.

Chapter 19
Developing the Performance Improvement Plan

Learning Objectives

- To describe the areas that should be addressed in the development of a healthcare organization's performance improvement plan

- To identify how performance improvement activities are implemented and findings communicated throughout the organization

Background and Significance

Performance improvement (PI) in healthcare is most effective when it is planned, systematic, and organizationwide, and all appropriate individuals and professions work collaboratively to plan and implement activities. When individuals from different departments representing the scope and care across the organization are included in PI activities, complex problems and processes can be improved. Collaboration on PI activities enables an organization to create a culture that focuses on PI and to plan and provide improvement that endures. When planning PI activities, the organization should identify those areas needing improvement and the desired changes that will lead to sustained improvements, as well as reduce the risk of sentinel events. This chapter will introduce students to the process of developing a PI plan.

Strategic Planning

Planning for performance improvement activities in healthcare organizations should be an outgrowth of the organization's overall strategic planning process. The **strategic plan** is developed by the organization's senior leaders and board of directors, whom, as discussed in chapter 13, the public holds accountable for the quality of the organization's products and services. Commonly, strategic planning includes a process called SWOT analysis where the leaders complete an assessment of the organization's **S**trengths, **W**eaknesses, **O**pportunities, and **T**hreats. Findings from SWOT analyses are used to validate the mission of the organization as a whole and determine the direction the organization is going as a business entity during the coming year. In addition, the leaders carefully consider input from the community the organization serves, its scope of services, the available technologies, staff expertise, the needs and expectations of its customers, and outcome information from PI activities of the past year.

Performance Improvement Plan Design

Key to the implementation of an effective PI program is a written plan that systematically describes the structure and approach the organization will follow in the continuous assessment and improvement of its important functions and processes. Figure 19.1 shows an example of an organizationwide PI plan. The Joint Commission on Accreditation of Healthcare Organizations (JCAHO 2003) recommends that the following activities be included in an organization's PI plan:

- The organization's leaders decide the scope and focus of performance monitoring and data collection activities to evaluate all existing, new, modified, and high-risk processes
- These activities are planned in a collaborative and interdisciplinary manner
- The organization uses criteria to define performance measures or selects performance measures from an external source
- Data are systematically collected, aggregated, and analyzed on an ongoing basis
- Improvement opportunities are identified and changes made that will lead to and sustain improvement

The leaders are expected to select an organizationwide approach to PI and clearly define how all levels of the organization will address improvement issues. As discussed in part I, there are many valid PI methodologies. The approach taken should reflect the organization's leadership style.

Because most organizations identify more improvement opportunities than they can act on, priorities must be set. Criteria are helpful in setting priorities and can include:

- The expected impact on performance
- High-risk, high-volume, or problem-prone processes
- The organization's available resources and staff

For example, at Community Hospital of the West, the managers and employees participate in organizationwide strategic planning. As part of strategic planning one year, they identified a list of their organization's important functions and processes where they thought improvement opportunities existed. (See figure 19.2.) The participants used brainstorming techniques and nominal group techniques to identify and prioritize improvement opportunities. The prioritized ranking shows the number of points that the various opportunities for improvement earned upon review by the entire organization. The prioritized opportunities were then reevaluated and regrouped by the hospital's board of directors and senior leaders in light of other available survey and outcome data. Using preestablished prioritization criteria, the leaders finalized the hospital's improvement goals, the results of which can be found in appendix A of figure 19.1.

**Figure 19.1.   Performance Improvement and Patient Safety Plan,
Community Hospital of the West**

The Community Hospital of the West is a 350-bed tertiary medical center with an inpatient and outpatient continuum of care that provides medical and surgical services to a community of approximately 250,000. Our specialties include: cardiology, orthopedics, obstetrics/gynecology, oncology, and psychiatry. The Community Hospital of the West board of directors is committed to continually improving the delivery and effectiveness of the care and services provided and proactively monitoring and assessing care delivery, patient safety, and the satisfaction of its customers. The board supports an environment that encourages the identification of improvement opportunities from all sources throughout the organization and community, and the provision of care and service that is reflective of the organization's mission, vision, and values.

Mission

Community Hospital of the West is the preeminent regionally integrated healthcare delivery system in the Intermountain West dedicated to providing compassionate, quality, high-value healthcare services to the residents of our communities.

Vision

To provide leadership in patient-centered care, built on a foundation of knowledge, innovation, and human values

Values

Continuous improvement, compassionate and caring, innovative, patient advocate, respect for human dignity, collaborative, and ethical

Performance Improvement Approach and Model

An interdisciplinary, continuous performance improvement approach is recognized across our continuum of care and service areas utilizing a Plan-Do-Check-Act (PDCA) model. Patient care and safety, and all other important organizational functions, are continually monitored, analyzed, and improved.

Organizational Performance Improvement Structure & Expectations

The leaders of Community Hospital of the West [the board of directors, the medical staff officers, and the senior hospital administrators] are committed to the integration of performance improvement activities. All staff are educated in the principles of performance improvement and participate in: identifying opportunities for improvement; data collection and reporting activities; performance improvement team activities; and ongoing education. The board of directors has overall responsibility for ensuring the quality of care and services provided to the community. The board has delegated implementation responsibility for the organizationwide continuous performance improvement activities to the Performance Improvement and Patient Safety Council.

Performance Improvement and Patient Safety Council

The Performance Improvement and Patient Safety Council is an interdisciplinary senior-leadership committee that provides oversight and direction for the design and implementation of the organizationwide, continuous performance improvement and patient safety program. The council annually reviews outcome data and survey information as part of its strategic planning and prioritization processes. The council reports monthly to the medical executive committee and quarterly to the board of directors any adverse outcomes, significant process variations, and actions taken to improve care and address patient safety issues, both proactively and reactively. Standing committees of the medical staff, clinical and department discipline meetings, and this council are responsible for managing and improving patient care and safety issues within their particular high-risk areas. Prioritized measures that include high-risk and problem-prone areas identified throughout the organization are trended, analyzed, and reported to the council by assigned committees/staff on a preestablished schedule. The performance improvement resource department coordinates the implementation of the performance improvement and patient safety plan. The department

(Continued on next page)

Figure 19.1.   *(Continued)*

provides organizationwide support in the design of data collection tools, data display, statistical analysis, benchmark data research, and the preparation of council reports. The council is responsible for receiving findings and acting on recommendations from the board, all committees, departments, performance improvement teams, customer survey data, sentinel events, near misses, and other identified trends in areas such as risk management and infection control. The council is also responsible for the design of the organizationwide staff development program related to continuous performance improvement and patient safety, and for the assessment and assignment of an annual proactive risk reduction activity. At least annually, the council reviews the activities of the performance improvement program and makes recommendations for the continuous improvement of the performance improvement and patient safety plan to the board of directors. The council membership is composed of a physician chairperson, the chief executive officer, medical director, clinical and administrative service directors, performance improvement team members, and other invited staff and guests as appropriate. The council meets at least every other month and as needs indicate.

Standing Committees of the Medical Staff

All standing committees of the medical staff are chaired by a physician with representation [as appropriate] from hospital leadership, department directors, and frontline staff. Reports are submitted to the Performance Improvement and Patient Safety Council, which in turn forwards critical events and findings to the executive committee and to the board of directors. Communication throughout the organization among the board, committees, councils, hospital departments, medical staff, employed and contract staff, and its patients/families is open and flows in all directions, as appropriate and as allowed by regulations.

Medical Executive Committee

An elected official from the medical staff, medical staff committee chairs, medical director, chief executive officer, clinical service director, compliance officer, and the chief financial officer are standing members of this committee. This committee meets monthly and coordinates the business of the medical staff [recommending changes to their bylaws, rules and regulations, reviewing appointment/reappointment recommendations, election of officers, etc.] and the integration of patient care and hospital support services. Significant performance improvement and patient safety–related issues forwarded from the Performance Improvement and Patient Safety Council are reviewed, discussed, acted upon, and forwarded to the board and, as appropriate, to other departments, committees, and staff.

Medical Staff and Specialty Department Meetings

Each clinical staff specialty meets at least quarterly to review and discuss performance improvement activities, staff development issues, and other related planning and directing activities. The medical staff [at large] meets at least annually for the election of officers, bylaws review, and general staff education.

Ethics Committee

This committee is responsible for serving as a resource regarding medical/ethical issues that surface for patients, their families, and the organization's clinical care providers.

Credentials Committee

This committee is responsible for the design and implementation of the organization's credentialing process and includes reviewing applications for appointment and reappointment, defining privilege delineation criteria, evaluating physical health issues, etc. Recommendations on all credentialing-related issues are reported to the executive committee and forwarded to the board of directors for final approval.

Utilization & Documentation Standards Committee

This committee is responsible for the review of findings from the monitoring activities on patient-specific data and information, timeliness of clinical record entries, and appropriateness of admissions and continued stays. Significant findings and recommendations are reported to the Performance Improvement and Patient Safety Council, the executive committee, provider quality profiles, other committees, and departments and individuals, as appropriate.

Figure 19.1.   *(Continued)*

Pharmacy & Therapeutics Committee

This committee is responsible for formulary review and development, policy setting, procedure development, medication-related safety education, and monitoring the safety and efficacy of medication use throughout the organization. Medication monitoring includes a systematic, ongoing process of reviewing prescribing/ordering, procurement and storage, preparation and dispensing, administration, and adverse drug reactions. Data collection, analysis of aggregate data for patterns and trends, recommendations for process/system changes, and reporting of significant findings and actions to the Performance Improvement and Patient Safety Council is performed by this committee.

Environmental Safety Committee

This committee is responsible for planning and directing environmental services within all seven environments of care. Education of staff on environmental safety issues and performance monitoring, data analysis, and continuous improvement efforts are this committee's responsibility. This committee meets monthly and reports data collection, analysis, and improvement initiatives to the Performance Improvement and Patient Safety Council at least quarterly. An environmental safety report identifying and reviewing improvement goals is submitted to the board of directors annually. Committee representation includes hospital and medical staff leaders, engineering and maintenance, housekeeping, central processing, security, and employee health.

Strategic Planning Process

The strategic planning process occurs annually prior to the start of the fiscal year and coincides with organizationwide plan/program reviews and the budgeting process. Performance improvement and patient safety program review is initiated by the council using findings from the leaders' strategic goals, the council's self-assessment, and staff survey data on the program's effectiveness. Additional information such as aggregate outcome data from performance measures, effectiveness of corrective actions implemented as a result of process variations and adverse outcomes, input from customer surveys, status on past year's goals, findings and actions from the annual proactive risk assessment/reduction activity, regulatory and hospital process changes are all reviewed and considered in the planning process and in the prioritization of performance goals and measures for the upcoming year.

Criteria for Prioritization of Improvement Goals, Performance Measures, and Data Collection

- Does the improvement opportunity/measure support the organization's mission, scope of care, and service provided and/or population(s) served?
- Is the performance measure a required [regulatory] measure, and does it provide performance information on an important function?
- Does the opportunity improve patient safety?
- Does the opportunity relate to an event that resulted in a sentinel event or near miss?
- Does the opportunity reflect patient feedback on needs or expectations?
- The degree of adverse impact on patient care that can be expected if the improvement opportunity remains unresolved.
- Does the opportunity reflect a high-volume, problem-prone or high-risk process?
- Are resources available to conduct the improvement process?
- Does the opportunity involve changing regulatory requirement?

Performance Measurement

This organization collects data on important processes in order to monitor its performance. Data collection is prioritized based on this organization's mission, scope of care, services provided, and populations served. Data collection is systematic and may be used to establish a performance baseline, describe process performance stability, identify areas for more focused data collection, and/or to determine if improvement has been sustained. Available benchmark information for established performance measures is drawn from internal and external databases. Data collection, responsibilities, and reporting

(Continued on next page)

Figure 19.1.   *(Continued)*

schedules are defined in an appendix to this plan. Data that are collected to monitor performance include the following:

- Performance measures related to accreditation requirements (ORYX)
- Patient safety issues including the following error-prone areas: medication events, falls, blood events, procedure/treatment/surgical events, behavioral events, equipment events, and laboratory events
- High-risk areas that may have the potential to result in a sentinel event including: operative procedures that place patients at risk, medication use, restraint use, seclusion when it is part of the care or services provided, care or services provided to high-risk populations, outcomes related to resuscitation, AWOLs, and attempted suicides
- Adverse drug events (ADEs)
- Needs, safety concerns, expectations, and satisfaction of patients and their families
- Failed processes related to the JCAHO's National Patient Safety Goals
- Utilization management activities
- The performance of new and modified processes
- Quality control activities in the clinical laboratory, diagnostic radiology, nutritional services, nuclear medicine, radiation oncology, and pharmacy
- Infection control surveillance and reporting
- Medical record documentation for quality of care and timeliness
- Risk management information including sentinel events, near misses, complaints, findings from inspections by regulatory agencies, compensable events
- Environmental safety
- Efficacy of services provided through contract or written agreement
- Processes related to medication use
- The appropriateness and effectiveness of pain management
- Processes related to blood and blood components
- Appropriateness of behavior management procedures
- Autopsy results, when performed
- Customer demographics and diagnoses
- Financial data
- Staff opinions and needs
- Measures established when performance improvement and patient safety teams are chartered to design/redesign a process
- Other measures that may warrant targeted study

Measurement Process and Tools

When clinical conditions or systems are evaluated, measurement includes the following components:

- Design and assessment of new processes
- Assessment of the dimensions of performance, utilizing data from customer satisfactions, surveys, financial analysis, clinical outcomes of care, and functional outcomes of care
- Development of indicators of care or service that are measurable and focus on processes or outcomes
- Utilization of benchmarks or thresholds for performance
- Identification of data sources
- Development of a method of data collection and organization of data measures
- Measurement of the level of performance and stability of important existing processes
- Aggregation and trending of data
- Use of established clinical practice guidelines as a framework for standards of care and practice, when applicable
- Evaluation of individual cases that have potential or actual risk to the patient (adverse event/sentinel event review)

Benchmarks/thresholds are based on current professional literature, national standards, clinical practice guidelines, or internal benchmarks for improvement. Thresholds are derived from retrospective data

Figure 19.1.   *(Continued)*

relative to previous measurements within the organization or from comparable organization data. Measures that have no retrospective data or comparable information can establish rate-based measurements utilizing calculated means. A benchmark is a quantitative goal embraced by the organization and is reflective of best practices within the internal and/or external environment. These goals serve as a mechanism for acceleration of performance curves through the process of continuous improvement.

Data Sources and Sampling

Data sources include medical records, encounter data, satisfaction surveys, complaint information, internal clinical databases (e.g. information from order-entry system, diagnosis and procedural coding system, departmental logs, observation, surveys, and interviews, etc.). Sampling methodology shall be relevant to the performance measures or study being conducted. For general review studies, a sample size of 5% or 30 cases, whichever is greater, may be utilized. When the statistical significance of a study is critical, a scientific methodology is recommended. Control charts are used to measure key indicators on an ongoing basis to assist in determining sustained improvement(s). Statistical process control methods are utilized to identify whether an indicator is in control.

Aggregation and Analysis of Performance Data

The results of systematic, ongoing measures are aggregated and analyzed for identification of trends, variances, and opportunities to improve patient care and safety. Data analysis should answer the following questions:

- What is our current level of performance?
- How stable are current processes?
- Do any steps in the process have undesirable variation(s)?
- Have strategies to stabilize or improve performance been effective?
- Are there areas that could be improved?
- What should our improvement priorities be?
- Was there sustained improvement in the processes that were changed?

Data Review

Trended data are reviewed when:

- Trended performance measures significantly and undesirably vary from those of other organizations, requiring a more detailed review.
- Trended performance measures significantly and undesirably vary from recognized standards, benchmarks, or statistical process controls.
- The occurrence of an event is questionable or too infrequent to make judgments about "patterns" in care or to perform an analysis to detect statistical significance.

Event Definitions

Near miss is defined as an opportunity to improve patient safety–related practices based on a condition or incident with potential for more serious consequences. A root-cause analysis may be performed when a near miss occurs.

Reportable event is defined as an unintended act, either of omission or commission, or an act that does not achieve its intended outcome. An incident report is completed by staff and forwarded to the Performance Improvement & Patient Safety Resource Department. Reportable events are trended quarterly. These events are reviewed by their respective committees and/or service area directors and recommendations for corrective actions reported to the Performance Improvement and Patient Safety Council.

Sentinel event is defined as an unexpected occurrence involving death or serious physical or psychological injury or the risk thereof. Serious injury specifically includes loss of limb or function. A root-cause analysis is performed when a sentinel event occurs. All *Sentinel Event Alert* publications from the JCAHO will be reviewed for relevance to our organization.

(Continued on next page)

Figure 19.1.   *(Continued)*

Intensive review of an incident requires the review of medical records or other data elements to determine if process problems exist and if an ongoing performance measure should be established to monitor process stability. An intensive review is undertaken when:

- A significant adverse drug reaction or medication error occurs.
- An external regulatory agency requests the review.
- At the request of the Performance Improvement and Patient Safety Council.
- Performing proactive risk reduction activities.

Root-cause analysis is conducted when a significant negative deviation from expected outcomes occurs or when a near miss occurs and further study is recommended by the council.

Peer Review Process

Cases are referred to peer review when they meet criteria as defined in the physician peer review plan. Findings are referred to appropriate committees for review and action, as warranted, and to individual physician practice profiles and reviewed as part of the reappointment process.

Performance Improvement Model

The model for performance improvement is Plan-Do-Check-Act (PDCA) and is defined as:

- **Plan** is based on the results of data collection or the assessment of a process. The plan should include how the process will be improved and what will be measured to evaluate the effectiveness of the proposed process change(s).
- **Do** includes the implementation of process changes. These changes may be tested before changing policy or procedure or conducting extensive education.
- **Check** evaluates the effect of the action taken at a given point in time.
- **Act** is to hold the gain and to continue to improve the process.

Evaluation

The measurement, assessment, and evaluation processes will continue to provide the necessary information as to the effectiveness of the improvement. If the identified problem continues to persist despite the planned improvements, the PDCA model will continue until sustained improvement is achieved. Any findings, conclusions, recommendations, actions taken, and results of the actions taken as a result of the performance improvement process are documented and reported to the appropriate individuals, departments, or committees. This information shall be used in the reappointment of providers, recontracting with contract agencies providing outsourced patient care services, and the employee evaluation process.

Patient Safety Risk Reduction Model

The model used to conduct the annual proactive risk assessment and reduction activity is the failure-mode-effects-analysis (FMEA).

Communication

Performance improvement and patient safety activities are communicated through the established committee structure as well as through regular clinical, discipline and staff meetings, e-mails, the annual storyboard fair, and the intranet. Members of the council are responsible for maintaining communication related to performance improvement and patient safety initiatives. The treating physician is responsible for informing patients and their family (when appropriate) about the outcomes of their care, including unanticipated outcomes such as sentinel events, and for documenting such communication in the patient's clinical record.

Education

The organization's leaders and council members are responsible for ongoing educational activities related to the performance improvement and patient safety program. This includes orientation of new employees at hire and new board members, and annual education of all employees at the annual employee fair, and through participation in performance improvement teams.

Figure 19.1.   *(Continued)*

Staff Support

The Employee Assistant Program (EAP) is a resource to support staff involved in a sentinel event and other work-related performance issues. The EAP is a confidential employee service. The clinical leadership is also responsible for meeting with staff involved in sentinel events to provide a means for communication and support.

Confidentiality

All performance improvement and patient safety activities set forth in this plan, including minutes, reports, and associated work products, are confidential and may not be released or discussed with any person or agency except those mandated by hospital policy or state or federal law.

Annual Review

The performance improvement and patient safety plan is reviewed annually as part of the organization-wide strategic planning process. The plan review is based on the organization's mission, evaluation of goals from the previous year, data collection results, and external regulatory changes.

APPENDIX A: PERFORMANCE IMPROVEMENT GOALS

Goal 1: To improve patient, physician, and employee satisfaction

A. Patient Satisfaction: Improve patient satisfaction as measured by the Gallup Survey.

Action Plan:
- Implement the caring model of nursing.
- Refine and improve the centralized scheduling process.
- Improve patient education and communication in the area of advanced directives.

Measurement:
- Patient responses on the Gallup Survey will shift from satisfied or dissatisfied to increase the very satisfied by 5%.
- The number of positive comments will increase by at least 5%.
- The number of billing complaints will decrease by at least 5%.

B. Physician Satisfaction: Improve physician satisfaction as measured by the biannual medical staff survey.

Action Plan:
- Implement the caring model of nursing.
- Refine/improve the centralized scheduling process.
- Evaluate and downsize committee structure as appropriate.

Measurement: The results of the biannual medical staff survey will shift from satisfied or dissatisfied to increase the very satisfied by 5%.

C. Employee Satisfaction: Improve employee satisfaction.

Action Plan:
- Implement the caring model of nursing.
- Award/recognize employees for years of service.
- Review/improve the employee evaluation process.
- Implement a system of merit raises.
- Develop department-specific action plans based on employee survey results and exit interview feedback.

Measurement:
- The annual employee turnover rate will decrease by 5%.
- All employees will be surveyed in June/July at department meetings using Gallup Survey questions.
- A system of performing exit interviews will be implemented.

(Continued on next page)

Figure 19.1.   *(Continued)*

Goal 2: To improve the infrastructure and systems used to collect, measure, and assess information so that information will be secure, accurate, appropriately accessible, useful, timely, and effective

Action Plan:
- Provide education in basic information management principles and provide tools, including software/hardware training, for leaders and other staff as needed.
- Improve/develop point-of-service data documentation with associated monitoring and evaluation of outcomes.
- Develop standing agendas for meetings to support appropriate flow of information throughout the organization.
- Develop standardization in documentation measurement, including expansion of the organizational data dictionary.
- Improve communication pathways to provide information to, and encourage feedback from, patients, trustees, physicians, employees, and other customers.

Measurement:
- Performance improvement/risk and safety reports will be complete, accurate, and timely.
- Compliance with documentation requirements as measured by data quality monitoring program will increase.
- Communication as measured by employee, physician, and patient survey results and through unsolicited comments will show improvement.

Goal 3: To improve leadership orientation, education, and performance

Action Plan:
- Revise/develop executive team and manager orientation and reference manual.
- Provide/attend at least two leadership education programs focusing on assessed needs for trustees, medical executive staff, executive team, and managers.
- Establish protocols for development, implementation, and review of standing physician orders.

Measurement:
- Educational program feedback surveys from participants will demonstrate program effectiveness.
- Trustee self-evaluation results will improve.
- Trustee evaluations of the CEO will improve.
- Survey results and self-evaluations will improve.
- Use and annual review of standing physician order protocols will be monitored.

Goal 4: To develop and improve employee competence and performance

Action Plan:
- Develop and administer an organizationwide educational needs assessment program, including feedback from physicians and employees as well as patient satisfaction surveys.
- Develop and implement educational programming to address prioritized needs, for example, mandatory staff education requirements, restraint protocols, confidentiality issues, universal precautions, and patient and family education process, etc.

Measurement:
- Results of mandatory staff education posttests and demonstrated competencies.
- Employee perception that needs have been met will be measured by participant evaluations of the educational offerings and employee satisfaction surveys.

Figure 19.2   **Example of a Strategic Planning Document Showing Potential Performance Improvement Opportunities**

Community Hospital of the West
Important Functions and Opportunities
Annual Strategic Planning

Functions and Opportunities	Priority Points
Patient care	
• Define restraint protocol	22
• Provide physical therapy services on weekends	1
• Change menu service	4
• Improve patient transport process	10
Patient education	
• Develop community awareness program	12
• Expand patient and family education	48
• Improve discharge instruction and documentation procedures	25
• Improve education for surgical patients	16
• Expand blood donor program	9
Patient rights	
• Include discussion of patient rights as part of admissions process	23
• Educate staff, families, and patients about the function of the ethics committee	21
• Respect patient's right to privacy and treatment with dignity	22
Patient assessment	
• Define assessment process	34
• Code status on admission	55
• Do more complete assessment on preop patients	11
Infection control	
• Enforce universal precautions	58
• Develop infection control program for home care	14
• Ensure that patient rooms are clean before assigning new patients to rooms	19
• Improve traffic control in patient care areas	18
Continuum of care	
• Address regional psychiatric services support	12
• Target high-risk patients for preventive care	17
• Respond to changing regulations on authorized procedures	15
• Define proper follow-up call from hospital to patient	10
• Define proper protocol for standing orders	27
Management of environment of care	
• Develop master plan for remodeling patient care areas	22
• Refine role of housekeeping	26
• Look at complaints about waiting areas	22
• Develop a system for providing hazardous spill carts	2
• Address after-hours/weekend security issues	25
• Remodel the operating room transitional area	3
• Upgrade operating room furniture and equipment	8

(Continued on next page)

Figure 19.2   *(Continued)*

Organizational improvements
- Develop plan to reduce medication errors   21
- Explore concurrent data collection and reporting processes   13
- Implement supply chain management   4
- Improve radiology and operating room scheduling process   18
- Support process improvement activities through development of teamwork   16

Leadership
- Develop a physician–hospital organization to work with managed care   13
- Conduct a community needs assessment   7
- Build feedback from employee and physician satisfaction surveys
  into the strategic planning process   7
- Develop a vision and goals for each department/service   5
- Downsize the number of committees   9
- Clarify leadership's role in all important organizationwide functions   8
- Introduce staff to board of trustees and define board's expectations of staff   2
- Define protocol for charity care   1
- Define hospital's system for acknowledging patients' deaths   2
- Clarify role of administrator on call   1
- Update and maintain all departmental policies and procedures   8

Management of human resources
- Provide identity badges for all physicians   4
- Improve communications with key customers (patients, employees, physicians)   14
- Improve mandatory staff education process   20
- Develop and implement competencies/skills checklists for every department   18
- Improve system for designating PRN staff—who to call, how many, etc.   1
- Update physician directory   1
- Utilize intranet training   3
- Develop policy on lab testing for employees and physicians   2
- Develop staff cross-training program   10
- Decrease staff turnover   1

Management of information
- Standardize organization of policies and procedures among departments   18
- Raise awareness of confidentiality issues   41
- Inventory the information the organization collects and determine
  what is necessary for quality control and leadership/governance needs   9
- Provide Internet access in the library   6
- Provide training and policy development on CMS coding rules   6

Many of the items on the strategic planning process document were related to patient care (see the section on assessment of patients, care of patients, and others in figure 19.2). Before strategic brainstorming, the organization had been collecting data using the Gallup Patient Satisfaction Survey. (See chapter 6.) An in-depth analysis of the data revealed a negative trend in multiple indicators related to nursing care. The indicators included the following:

- Overall nursing care
- Staff showed concern
- Nurses anticipated needs
- Nurses explained procedures
- Nurses demonstrated skill in providing care
- Nurses helped calm fears
- Staff communicated effectively
- Nurses/staff responded to requests

After considerable discussion of the brainstormed opportunities and of the Gallup survey data, the leadership decided to implement a new approach to nursing care in response to issues affecting nursing care.

Following considerable literature research, the leadership decided to implement Jean Watson's theory of human caring. The theory of human caring "recognizes the dignity and worth of individuals and that their responses to illness are unique; acknowledges the individual's right to continuous autonomy; helps individuals reach maximum capacity; recognizes that nursing takes place within a human-to-human caring relationship; and supports caring as the core of nursing practice, recognizing that caring is effectively demonstrated and practiced interpersonally" (Watson 1985). In Watson's theory, caring nursing is exhibited by five behaviors:

- Nurse introduces self to patients and explains role in care that day.
- Nurse calls the patient by preferred name.
- Nurse sits at the patient's bedside for at least five minutes per shift to plan and review care.
- Nurse uses a handshake or a touch on the arm.
- Nurse uses the mission, vision, and value statements of the organization in planning care (Dingman, et al. 1999).

Implementation of this care model was made part of the hospital's strategic plan for improving patients' perceptions of nursing care. All other opportunities were aligned to this major goal. It was intended that changing the strategy for nursing care in this dramatic way would be evidenced in subsequent measures of nursing effectiveness by the Gallup Patient Satisfaction Survey.

The first part of the PI goals document was organized to reflect the needs of the organization's most important customers: patients, physicians, and employees. Opportunities were then listed beneath the customers to whom they pertained. Maintaining visibility of its customers in the plan and focusing on the customer is a component of this organization's PI approach.

The rest of the goals document focuses on the important processes and functions of the organization most likely to affect high-risk, high-volume, and problem-prone outcomes. Each section identified an action plan and the means by which the efforts at improvement would be measured. These measurements should provide good data for the organization to assess itself and plan the following year's PI goals, thus maintaining the continuous PI philosophy and cycle in the organization.

Other important areas that should be addressed in an organization's PI plan include the following:

- The process for initial and ongoing education of the board and medical and employed staff members on its PI process and annual improvement goals

- The expectations of all members of the organization in PI activities

- The process for annual review of the effectiveness of the organization's PI program

Implementing the Performance Improvement Plan

The organization's leaders have a central role in initiating, performing, and maintaining the organization's PI priorities.

Implementation of the organizationwide plan for PI is a challenge at best. The plan should clearly delineate how members of the organization are educated on the PI process and what their roles and responsibilities are in carrying out the organization's plan. For example, the Community Hospital of the West's PI and patient safety plan (figure 19.1) describes how the board of directors delegates plan implementation responsibility to the PI and Patient Safety Council. The council in turn defines which committees, departments, and staff are responsible for data collection, assessing and reporting findings, as well as other performance improvement activities.

The scope and focus of what will be measured, which data will be collected and by whom, and the frequency and intensity with which the data will be collected and reported should be clearly defined by the healthcare organization. Table 7.1 (page 100) in chapter 7 shows an example of one organization's data collection and reporting schedule. While healthcare organizations do have some control over what types of data they collect, healthcare regulations mandate much of what must be collected. For example, recent initiatives around national patient safety goals require that all accredited healthcare organizations collect data on patients' views on how to improve patient safety, staff opinions and suggestions for improving patient safety, and staff willingness to report errors.

Performance improvement priorities should be data driven. That is, the data the organization collects about its own performance should be analyzed and considered when setting improvement priorities. Data should be aggregated and displayed in a way that provides for easy assessment of the findings. Establishing benchmarks on each measure the organization collects data on and displaying the benchmark information alongside the aggregate data measure is one way to quickly identify less-than-desirable outcomes in performance. Figure 15.1 (page 261) in chapter 15 provides one example of how an organization displays its outcome data. Historically, healthcare organizations have been referred to as "data rich": they spent too much time collecting data and failed to turn the data into meaningful information or use it in setting performance priorities. Setting measurement and data collection priorities are newer standards that have evolved over the past few years to help organizations better identify, manage, and act on collected data.

In demonstrating the effectiveness of its PI model, a healthcare organization must show evidence that the entire cycle has been completed on each of its prioritized improvements. Often enough, organizations fail to complete the cycle. Data can be collected, aggregated, and analyzed, and an improvement priority can be established; but without providing evidence of an effective improvement, the cycle remains incomplete. It is not unusual for a process change(s) to result in variations in performance initially or that a process change is the correct "fix" to an identified problem. The ongoing, systematic review of measurements identified to confirm process stability is the "check" step in the improvement cycle. Monitoring a process change for up to six months with ongoing data checks (sampling at regular intervals) once stability is achieved validates the improvement or the need for continued improvements (the "act" step), and completes the improvement cycle.

Case Study

Students should look at the list of functions and opportunities (figure 19.2) and the PI plan (figure 19.1) for Community Hospital of the West and answer the questions listed below.

For student experience at school or work, they should draft a PI plan. The plan should identify specific PI priorities, measures selected to monitor improvement priorities, possible corrective actions, and any other important information that would describe the PI process.

Case Study Questions

1. How is the performance improvement plan linked to the list of functions and opportunities identified during strategic planning?

2. Can you identify items in the prioritized strategic planning process document that are related to goals in the performance improvement plan?

3. Note that the items in the strategic planning process document are very specific. Were related items from the list grouped into a more general category for the final performance improvement goals?

4. Are the measurements identified for the goals truly quantifiable? That is, will the measurements actually lead to objective data that can be evaluated for evidence of improvement?

Summary

Planning the direction of a healthcare organization's performance improvement (PI) program is a complex activity. The program must be created in concert with the organization's overall strategic plan. From the many potential improvement opportunities, the organization must prioritize which issues will be the focus of improvement efforts during the coming year.

Planning the PI program includes reviews of the results from the strategic planning process, the organization's mission, community needs, customer expectations, and other related outcome data on the organization's performance. The document may also include a discussion of the educational initiatives to be developed in support of the PI program, descriptions of the components of the program, methodology, and other pertinent information.

References

Baker, Susan Keane. 1998. *Managing Patient Expectations: The Art of Finding and Keeping Loyal Patients.* San Francisco: Jossey-Bass.

Dingman, Sharon K., et al. 1999. Implementing a caring model to improve patient satisfaction. *Journal of Nursing Administration* 29(12):30–37.

Joint Commission on Accreditation of Healthcare Organizations. 2003. *Comprehensive Accreditation Manual for Hospitals.* Oakbrook Terrace, Ill.: JCAHO.

Joint Commission on Accreditation of Healthcare Organizations. 2000. *Improving the Care Experience.* Oakbrook Terrace, Ill.: JCAHO.

Performance Improvement Consultative Group (PICG). www.pihealthcare.org.

Plisek, P. E. 1995. Techniques for managing quality. *Hospital and Health Services Administration* 40(1):50–79.

Watson, Jean. 1985. *Nursing: Human Science and Human Care, A Theory of Nursing.* Norwalk, Conn.: Appleton-Century-Crofts.

Chapter 20
Evaluating the Performance Improvement Program

Learning Objectives

- To explain why performance improvement programs are evaluated

- To identify the aspects of the performance improvement program that should be evaluated

- To describe what organizations should do with the information gathered through evaluating the performance improvement program

Background and Significance

The processes of planning and evaluating a performance improvement (PI) program should mirror one another. Taken together, they are a cyclical activity; planning leads to evaluation, and evaluation provides the impetus for new planning. The task of appraising the PI program is generally completed on an annual basis, and the results should be reported to the healthcare organization's board, management, employees, and medical staff.

Performance improvement programs are evaluated for four reasons:

1. *To determine whether the organization's approach to designing, measuring, assessing, and improving its performance is planned, systematic, and organizationwide.* Forethought and deliberation help to focus the organization on important issues, and lead to better results in program activities. Systematizing the PI program makes it possible for participants to understand what is expected and enables them to anticipate program requirements. Committing to the organizationwide nature of the program ensures that everyone in the organization is in concert with the program's objectives and understands what they are expected to contribute. The PI evaluation process includes the participation of all employees and organizational leaders, from frontline staff up to the level of the board. Involvement of the organization's staff in the evaluation process can be demonstrated through team/committee input, questionnaires and

surveys, self-evaluation, suggestion boxes, and so forth. For example, each department in the organization can assess the educational competency of its individual staff as it relates to their PI knowledge. Department competency is then compared to an established organizationwide PI competency goal.

2. *To determine whether the organization's approach and its activities are carried out collaboratively.* Performance improvement activities should be multi-disciplinary. They should improve performance across department lines; in other words, they should be cross-functional or intradepartmental. Another expectation is that all factors that contribute to a problem will be remedied. For example, PI teams should be assessed in an annual evaluation of frontline staff or multidisciplinary team participation through documentation of regular attendance at meetings, data collection, and overall success of the team's improvement efforts.

3. *To determine whether the organization's approach needs redesign in light of changes in the strategic plan or organizational objectives.* If the organization's mission, vision, values, organizational structure, or strategic initiatives have changed since the PI program was planned, then new measures and assessment activities may have to be undertaken. For example, if an organization has a strategic initiative to add a new service, all departments, organization leaders, and the board of directors would be included in defining measures and assess-ment activities for the new service. Modifications should be implemented as soon as possible after major changes in organization objectives have been made.

4. *To determine whether the program was effective in the improvement of overall organizational performance.* It is important to identify whether the PI program was responsible for important improvements in organizational performance or whether those improvements were due to other factors. It is also important to identify whether improvements were achieved and maintained. Review of program performance should identify whether program processes are efficient, effective, timely, and appropriately supported with personnel, budget, and other resources. Figure 20.1 shows survey questions used in one organization to assess the effectiveness of the different departments' PI program. The results from the survey can be used to assess and then recommend changes to the organization's PI program.

Components of Program Review

During program review, each of the areas in which clinical and nonclinical services are pro-vided should be examined for continued focus and relevance. Each area should document the opportunities for improvement that were identified and which PI goals were met. Each area should document which problems remain unresolved from prior evaluation periods or have not shown significant improvement. Each area should identify issues in the PI program that, if changed, would better support the organization's overall mission and PI efforts. Although the evaluation of the organization's PI program can be structured in any number of ways, the following outline is typical.

> **Figure 20.1.   Performance Improvement Staff Survey Questions**
>
> * How is PI used in your department/service?
> * Are you asked for suggestions/ideas for processes that need to be improved in your area?
> * Are you asked for suggestions/ideas for processes that need to be improved organizationwide?
> * Have you been involved in any PI activities in the past year?
> * Have the results of PI activities been communicated to you on a regular basis? If yes, how communicated?
> * Have you used the results of any PI activities in the delivery of patient care/service in the past year?
> * Have PI activities improved patient care or service in your area in the past year?
> * What performance improvements need to be made to improve the safety of patient care delivery?
> * Are PI activities understood by the staff in your work area?
> * What process related to patient care is strongest in your work area?
> * What process related to patient care is weakest in your work area?
> * If you could improve only one patient process or outcome, what would you choose?

Executive Summary

This section of the program evaluation report summarizes the main points in the annual review. The summary should provide the reader with a snapshot of the year's best and most significant issues and should answer the question "If readers looked at only one or two pages of the report, what would you want them to be aware of and remember?"

Overview

Information on collaboration, strategic planning, and operating goals related to PI should be reviewed in this section. This might include the organization's mission statement. In general, the mission statement should continue to reflect the organization's key customers and long-term direction. Establishing direction and planning from organization leaders is essential in illustrating how PI is focused and prioritized from year to year. How and why teams or processes were identified for improvement should be discussed to provide a rationale for how organization leaders determine and prioritize areas for improvement and development. This section links the organizational goals identified by the board with the way the goals were operationalized.

Performance Improvement Structure

The organizational structure that supports, directs, and coordinates all PI activities should be described, and any changes should be illustrated in this section. A key element is how findings from PI activities are shared with the board, hospital and medical staff leaders, and other staff. The reporting relationships should be reviewed to remind how the organization routes, communicates, and shares performance-related information. Other questions that may be addressed in this section include the following:

* Does the performance improvement methodology currently used in the organization continue to support its PI activities and management style?

- Could the model be made more efficient or formalities lessened in some activities?

- Do all members of the organization understand the model and the ways it can helpfully structure PI activities?

Improvement Opportunities

Trends from aggregate data analysis, interventions and corrective actions or improvements, or changes in policies and procedures may be highlighted in this section using graphic representation and storyboard findings rather than detailed reports. System improvements may be broken into four areas.

Patient-Focused Improvements

This section focuses on clinical performance improvements that have affected patient care. Patient care activities, such as patient rights, assessment and treatment of patients, continuum of care, infectious disease management, and so forth, may be the outline used in addressing these improvements and/or prioritized opportunities. The efforts of patient care improvement activities are reviewed, as are team processes and areas of measurement. Examples may include the results of monitoring pain management on postsurgical patients, waiting times for emergency patients, or antibiotic administration times.

Organizational Improvements

The focus of this area is nonclinical and may deal with functions such as environment-of-care issues, staff training needs, and leadership development goals. Organizational improvements may include activities such as reengineering the admitting process, reducing the suspense days on unbilled accounts, or reducing staff injuries. The efforts of all organizational improvement activities are reviewed, as are process changes and areas of measurement.

Ongoing Measurements

These measurements and results are related to important functions and processes that are monitored on an ongoing basis. They include measurements that are required by regulatory agencies, such as staffing effectiveness, disease-specific monitoring, medication use, blood and blood component use, and customer satisfaction. Major results and impacts should be included for each function. Sentinel events and resulting root-cause analyses should also be presented, along with information related to near misses. The results of risk reduction assessments and strategies should be discussed here as well. Any ongoing variations noted in aggregate data findings would become a target priority area for ongoing improvement in the coming year's plan.

Comparative Summary Measurements

If comparative databases have been used in the organization's improvement activities to assess outcomes or determine areas for improvement, a synopsis of critical measures and improvement results should be included and reviewed here. Information on key patient care issues related to the JCAHO ORYX initiatives is an example of one external comparative database used by healthcare organizations.

Performance Improvement Team Activities

This part of the report should highlight the work of PI teams sanctioned by the organization's PI council and should provide answers to the following questions:

- Has the work of the team resulted in measurable and sustained improvements?

- What process changes and training occurred to support/sustain the improvement(s)?

- Has the organization's staff assigned to performance improvement team projects been effectively trained to work on performance improvement teams?

- Are staff willing and able to take on the important roles in team activities?

- Do they participate and interact well at team meetings to work through PI processes?

- Can they document important team milestones with appropriate tools?

- Have they implemented the organization's PI model appropriately?

- Have they learned to listen and question effectively in interpersonal communication?

- Can they communicate the team's process and outcomes effectively to the rest of the organization?

Other PI Review Topics

Other topics that may be addressed in the organization's performance improvement plan review may include areas of focused change or improvement.

Customer Satisfaction

Have internal and external customers been identified for all PI projects? Have customers' requirements been identified in detail? Has customer satisfaction been measured objectively and with appropriate tools?

Risk Exposure

Is adherence to procedures monitored and appropriate education undertaken when necessary? Are the right services always rendered to the right patient, at the right time, with the right procedure? Are unforeseen occurrences always reported per policy and procedure? Are care processes modified when necessary to prevent injury to patients, visitors, and employees?

Human Resources Development

Are appropriate hiring, staffing, development, and assessment activities occurring in all departments? Are employees being screened for disease as required by state health regulations? Are required documents being maintained on all employees? Are required credentials and licenses being verified on all employed staff? Are required credentials and licenses being verified on all independent practitioners? Are appropriate competency reviews being performed on all staff?

Accreditation and Licensure

Is the organization continuously ready for review by accrediting and licensing agencies? Have adverse outcomes from regulatory surveys in the past year been corrected? Is the organization's presurvey self-assessment process effective? Have new standards and changes to standards been implemented?

PI Program Effectiveness and Recommendations

This section of the PI program assessment outlines the areas identified and prioritized for improvement, measurement, and data collection during the next year. In large part, these recommendations are based on the findings outlined in the annual report, as well as goals identified in the organization's strategic planning process.

Case Study

Look at the case study for chapter 3 (pp. 28–29). The team at Western States University Medical Center continued to work on the issue. During the ensuing months, the team elected a leader and found a facilitator. Its vision statement was constructed: "Patient location is correctly identified in XYZ and ABC systems at all times." It tentatively identified process customers as nursing staff, patient accounting staff, admitting staff, and patients.

The team decided early in the process that it did not have sufficient data to make decisions about system capabilities, nor to recommend solutions to the problem. It developed data collection activities to get a handle on the realities of patient location within the institution, then spent the first four months collecting data. It collected data on the number of patients admitted, discharged, and transferred within the system; examined the user procedure for entering the admissions, discharges, and transfers in the system; and finally summarized the amount of admitting, discharging, and transferring that each user performed. It did observational studies of the patient transfer procedure, identifying how long the typical patient transfer took from one unit to another, the exact times of initiation and completion, and the time of transfer input into the information system. It identified the number and type of errors that occurred concerning patient admission, discharge, and transfer, and the effects those errors had on the census and in other departments.

At the end of the first four months of activity, the quality council of Western States University Medical Center began its annual review of the important aspects of the PI program. The team leader of the patient transfer team was asked to submit a one-page synopsis of the team's activities, which included a brief description of team activities and a complete set of the data collected by the team.

Case Study Questions

1. What would be your assessment of the patient transfer team's accomplishments thus far if you were a member of the quality council?

2. What recommendations would you make with reference to the team's future activities?

Summary

The annual evaluation of a performance improvement (PI) program ensures that the program focuses on opportunities for improvement that are truly important to the organization. The foundation of a PI program assessment includes how well the organization's leaders set expectations; developed plans; and managed processes to measure, assess, and improve the quality of the organization's operations (such as governance, management, clinical, and support activities).

Evaluation should encompass all areas of the program, including the organization's strategic goals, PI structure, team training and functioning, as well as ongoing and comparative measurement monitoring of the results of established improvement priorities. The findings of the evaluation should be incorporated into the following year's PI plan.

References

Bassett-Lathrop, Carolyn, et al. 1995. A structure for organizing a facility's annual performance review. *Journal of HealthCare Quality* 17(4).

Dingman, Sharon K., et al. 1999. Implementing a caring model to improve patient satisfaction. *Journal of Nursing Administration* 29(12):30–37.

Watson, Jean. 1985. *Nursing: Human Science and Human Care: A Theory of Nursing.* Norwalk, Conn.: Appleton-Canterbury-Crofts.

Chapter 21

Understanding the Legal Implications of Performance Improvement

Learning Objective

- To describe the legal aspects of performance improvement activities conducted in healthcare organizations

Background and Significance

In healthcare organizations, performance improvement (PI) activities are affected by a number of laws, rules, and regulations. Because PI projects and processes can be complex and at times controversial, understanding the legal context in which the activities are carried out is critical. This chapter applies familiar legal theories to PI in healthcare and addresses the legal implications of PI in relation to the following areas:

- Copyright law and the ownership of PI and quality management methodologies, data, and studies

- Contract law

- Tort law and the standard of care

- Privacy and confidentiality issues

- Peer review, immunity, and sentinel event reporting

Copyright Law

The Copyright Act of 1976 guarantees the creators of intellectual works the right to control how their works will be used; in other words, it establishes ownership rights for intellectual property. Copyright generally extends through the life of the author plus fifty years. When copyright is held by an organization, the term of the copyright is seventy-five years from the date of publication or one hundred years from the date of creation, whichever is longer.

Copyright law protects the original expression contained in an intellectual work. In this context, *expression* means the words, sounds, and images used to express an idea or describe a process. It is important to understand that copyright law protects only the expression and not the actual ideas or facts that make up the content of the work. Two descriptions of the same facts may be copyrighted by two individual authors, as long as the second version is not a copy of the first. For example, any number of books could be written about the Second World War, and the author of each book could own the copyright on the book he or she wrote.

All original work is automatically covered by copyright law from the time it first appears in recognizable form. Copyrightable materials include magazine articles, books, databases, audio and video recordings, illustrations, photographs, reports, and lectures. Copyright law also applies to intellectual property such as computer programs and software applications.

Original intellectual work can be created by an individual, a group, or an organization, and any of these individuals or groups can hold the copyright on an original work. Federal copyright law also recognizes works "made for hire." When employees prepare copyrightable materials as part of their employment responsibilities, the employer is considered the author for copyright purposes. Similarly, when one person or organization hires another person or organization to prepare an original work, the copyright is owned by the person or organization that requested the work.

Copyright law treats intellectual property as personal or physical property. Therefore, the ownership of intellectual property can be transferred to another party in whole or in part. For example, writers often transfer their copyrights on original written works to the organizations that publish their articles and books. In addition, the copyright holder can permit another person or organization to use his or her property. In such cases, a **permissions** agreement gives a second party the legal right to use previously copyrighted material. For example, an author may enter into an agreement that allows another party to reprint her entire book, or a specific chapter or chart from her book.

Software licenses are another example of limited rights of use for copyrighted intellectual property. A software license generally allows a person to use the software, but the user does not own the code that forms the basis of the software. The owner of the copyright is the original software developer or another party to whom the copyright was transferred or sold.

One notable exception to copyright law is any work that is created through the workings of the federal government. Such works are considered to be in the **public domain;** they can be used freely, but they cannot be copyrighted by anyone. Works on which the copyright has expired also fall into the public domain.

Copying the original work of another author and claiming it as one's own is a serious violation of copyright law known as **plagiarism.** Similarly, using previously copyrighted work without the owner's permission may be illegal even if the copyright holder is given credit as the author.

Under some very limited circumstances, copyrighted materials may be used without the owner's written permission. Such exceptions to copyright law fall under the **doctrine of fair use.** The intent of the fair-use exceptions to copyright law is to allow limited use for purposes usually associated with nonprofit activities. Fair use of copyrighted works includes making photocopies for purposes of criticism, comment, news reporting, education, scholarship, and research. Copyrighted works may also be quoted in materials published for similar purposes.

In determining whether the use made of a work in any particular case is a fair use, each of the following factors must be considered:

- The purpose and character of the use, including whether such use is of a commercial nature or is for nonprofit educational purposes
- The nature of the copyrighted work
- The amount of material used in relation to the whole copyrighted work
- The effect of the use on the potential market for, or value of, a copyrighted work

In practical terms, the provisions of copyright law are relevant to PI projects in two areas: first, in protecting the results of PI efforts as intellectual property and, second, in respecting the ownership rights of other copyright holders.

The results of most PI initiatives are documented in written reports. Some projects require the development of unique research methodologies. Because the work is performed by employees as part of their regular jobs or by consultants under contract, the copyright on such materials is usually owned by the healthcare organization. Any databases or process tools developed during PI projects are also subject to copyright ownership.

To protect the organization's copyright on such materials, a copyright notice should be included at the beginning of every written report and computer database. Materials distributed as single pages should carry a copyright notice on the bottom of each page. Formal registration is not necessary to protect the copyright on such original work. The author of the material needs only to place a notice of copyright on the work using the following language:

Copyright *(year)*, by *(name of organization)*.
All rights reserved.

When a project requires a significant investment of resources or results in products that will be sold or licensed to other organizations, it may be advantageous to formally register the copyright. The federal government has established an office for copyright registration. Registration of a copyright provides enhanced legal protection by documenting a clear public notice of the copyright's existence and establishing the owner as the first party to copyright the material. Instructions on registering a copyright can be obtained from the Copyright Office in Washington, D.C., or from its Web site, http://lcweb.loc.gov/copyright/.

Protecting an organization's ownership rights by copyrighting its intellectual property will not ensure the confidentiality of clinical and business-related information. Privacy and confidentiality issues are discussed later in this chapter.

Participants in PI projects often use published and unpublished materials from sources outside the organization. Such sources may include publishers, other healthcare organizations, standards organizations, consulting firms, and software developers. When copyrighted written materials are used for internal research and education, photocopying is usually allowable under the doctrine of fair use. When photocopies of copyrighted materials are to be distributed to a number of participants, a reference to the original source of the material and a notice of the source's copyright should be placed on every copy.

It should be remembered, however, that photocopying is not an acceptable alternative to buying books or publishers' reprints when they are available. The same is true of materials developed for clinical improvement projects. Organizations such as third-party payers often invest enormous amounts of time and money in developing criteria to be used in the evaluation of patient care. Such criteria are often sold or licensed by the copyright owner in an effort to recoup investments in the project, and such materials cannot be used or photocopied without permission.

The need to obtain a license to use copyrighted software is more familiar. Copyright owners of software products often grant site licenses to healthcare organizations. Many educational products available on the Internet are also sold through site licenses. It is important to remember that materials made available on Web sites are protected under the provisions of the copyright law and should be used with the same consideration as print materials.

To avoid allegations of copyright violation, a PI team should take the following steps:

- Clarify ownership rights when dealing with consultants by executing written work-for-hire agreements.

- Execute confidentiality agreements when materials will be reviewed by parties outside the organization.

- Attach copyright notices to works developed in the healthcare organization, and register the copyrights with the federal copyright office when appropriate.

- Purchase multiple copies of products to which the doctrine of fair use does not apply.

Contract Law

A **contract** is a mutual promise between two or more parties that the law recognizes and considers enforceable. Promises are made every day by all sorts of people under all sorts of circumstances. What distinguishes an enforceable promise, or contract, from those that are not considered enforceable promises?

In order for the law to recognize a contract as valid and enforceable, four conditions must be met. First, there must be **mutual assent** between or among the parties. This assent (or agreement) should be made in writing. Without a written agreement, genuine issues related to proof may arise. The courts are reluctant to take into consideration any contract terms that are not expressly written down. Therefore, it is critical that all the items the parties want to be binding are reflected in the language of the contract.

The test of whether mutual assent exists is determined by the answers to two questions: was a valid offer communicated by one party, and was a valid acceptance of the offer communicated by the other party? A common example of a contract is the agreement entered into when a house is sold. Did the buyer make an offer that the seller understood? Did the seller communicate acceptance of the offer to the buyer in a way that the buyer understood? When the answer to both questions is yes, the condition of mutual assent has been met.

The second legal consideration in the validity of a contract is whether the contract has a **legal purpose.** For example, the courts will not enforce agreements or contracts between individuals if the transaction involved the transport or selling of illegal drugs.

The third consideration is competency, that is, whether the parties are competent to enter into a binding agreement. Individuals may be judged to be incompetent owing to immature age or mental impairment. For example, most states require that parties to contracts must be at least eighteen years old.

Finally, a contract must be supported by what is referred to as legal consideration. All parties to an agreement must promise to do something that they are not otherwise legally obligated to do. Generally, consideration involves making a monetary payment on the part of one party and doing some specified task on the part of the other party. For example, a software vendor might agree to provide software with certain characteristics necessary to perform specific tasks, and the buyer might agree to pay for the software as well as not to violate the copyright or license granted for the software. If one party failed to perform its obligation, the other party could sue for breach of contract.

A contract should contain the following information and provisions:

- *Identification of the parties:* The contract should include the legal names and addresses of the persons or organizations that are party to the contract. Post office box numbers should not be used as legal addresses because they would not be sufficient for serving legal notices such as subpoenas.

- *Terms of the agreement:* The contract should include a clear statement of the purpose of the agreement, as well as the beginning and ending dates of the agreement. This is particularly important when work is to be performed within rigid deadlines. Provisions related to penalties for not performing the work in a timely manner may also be included.

- *Provisions for termination:* The contract should include termination provisions. Termination of the agreement "for cause" should be allowed. That is, the contract could be terminated if one of the parties failed to perform or performed inadequately, or one of the parties otherwise breached the agreement. A termination-without-cause clause would allow termination of the agreement with sixty or ninety days' written notice. Agreements that terminate for cause usually include provisions that assign costs to the party that does not fulfill the contract.

Contract issues often arise in the context of PI when a healthcare organization engages a vendor or consultant to perform a quality study or to lead the organization in developing its PI processes. Contracts are also involved when the organization agrees to buy a certain type of software or to use a certain copyrighted data-gathering methodology. Confidentiality agreements are another example of contracts that are commonly used in the field of healthcare PI.

When vendors or consultants are engaged to provide support for PI projects, several considerations should be covered during the contracting process. First, the organization's representative should make sure that the vendor or consultant holds adequate professional liability and workers' compensation insurance. When an agreement covers certain goods and services with a value of $10,000 or more and the organization is a Medicare contractor, the vendor or consultant must agree to provide access to certain records. The books and records needed to verify the costs related to the contracted services must be made available to appropriate state and federal agencies for four years after the services are rendered.

The parties to contracts should also warranty that they will not participate in any activity that violates state or federal law prohibiting fraud and abuse, kickbacks, or other prohibited activity. Vendors and consultants should also provide assurance that they have not been suspended from participation in any federal program. In addition, the parties should agree to comply with other laws, rules, and regulations, such as the Americans with Disabilities Act, the Civil Rights Act, and the Drug Free Workplace Act.

The contract should also make it clear that the work to be performed by the vendor or consultant is to be considered "work for hire" under the copyright law. The contract should indicate who owns copyright to any product or report developed, including, but not limited to, any graphic art, photographs, videos, and computer images. The agreement should also make it clear which party will be responsible for obtaining permissions and licenses or for filing copyright applications.

Contracts with vendors and consultants should also state that the vendor or consultant is not allowed to use the organization's name in any promotional materials without the express written permission of the organization. Every contract should contain a clause that disallows subcontracting of the work to another organization not party to the contract. Amendment of the agreement should not be allowed without the written consent of all parties.

The agreement should contain confidentiality language. The terms of the contract must acknowledge that, if during the period of the agreement the vendor comes into contact with proprietary, copyrighted, or trade secret information using organizational resources, such information may not be disclosed to others by the vendor without consent. In healthcare, this is particularly relevant to patient-identifiable information.

During the contracting process, the vendor or consultant should be informed of the organization's corporate compliance program and should receive and agree to comply with the organization's code of conduct and business standards. The vendor should also be required to report first to the party that hired it before consulting with any outside entity or government official regarding any alleged violation of law or regulation that the vendor may have reason to believe has occurred.

For example, problems may arise when a vendor supplies a proposal for a quality management information system, which a healthcare organization accepts. However, these kinds of proposals are often so detailed that the entirety of the terms may not always be included in the contract. In that instance, a contract may include the proposal's terms with simple language that the proposal is "attached and its terms and obligations are incorporated by reference." In this way, the vendor can be held to promises it made in the marketing and sales of its product.

Business Associates

The compliance deadline for title II of the Health Insurance Portability and Accountability Act (HIPAA) of 1996 went into effect on April 13, 2003. As a result, healthcare organizations need to be more vigilant and certainly more compliant than before in their dealings with business associates.

Business associates can be defined as entities that perform functions on behalf of a healthcare organization. Business associates create, receive, or have access to individually identifiable health information maintained by the healthcare organization. Typically, healthcare organizations have current contracts with business associates called business partner agreements.

These contracts need to be reevaluated to ensure that the business partner is upholding the privacy and security regulations of HIPAA. Once the healthcare organization has defined who its business associates are, it needs to define the privacy and security expectations of its business associates in regard to HIPAA compliance.

The security rule includes four categories of requirements: administrative safeguards, physical safeguards, technical security services, and technical security mechanisms. These categories dictate how patient health information should be accessed and utilized by the business associate.

Administrative safeguards include measures to protect data such as forming and managing security policies and procedures. They also include controlled access to patient health information and written policies and procedures to support irregular situations that could result in a security or privacy breach.

Physical safeguards are designed to protect the physical control of patient health information. Physical safeguards include media control, physical access control, and training regarding security policies and procedures. Security standards apply to the healthcare organization as well as its business associates. The technical securities and mechanisms are put in place to protect electronic data, including limiting access to, and creating integrity controls for, private and secured information.

Contracts should include a clause outlining the retention, return, or destruction method for all patient health information that has been transported to the business associate. Patient health information that was sent to the business associate should be destroyed and not returned. The method of destruction should be specified within the contract, along with proof of destruction. A termination clause should also be included in the contract, and both the business associate and the healthcare organization must sign the contract in order for it to be a legally binding agreement.

Internal policies and procedures should be created to ensure privacy and security of health information. Healthcare organizations should have a policy and procedure for sending patient health information to business associates that includes information on the method and form of delivering the patient health information. They should also set up secured environments for business associates to access patient health information from outside the organization.

Policies and procedures should be created and implemented for business associates that will need access to the organization's network. Organizations can utilize firewalls, routers, and switches to assist with network-based filtering, as well as user IDs and passwords to assist with identification and authentication controls.

HIPAA requirements include staff training, as well as the significance of the privacy and security regulations. Staff members must be held accountable for violating privacy or security policies. Training should be conducted at orientation and on an annual basis, given that regulations are subject to change over time. Policies and procedures should also be documented and updated each time they are modified and reviewed at least on an annual basis.

Tort Law and the Standard of Care

A **tort** is a wrongful act committed against a person or a piece of property. A tort is a civil wrong and, as such, is considered separate from criminal acts and breaches of contract. A tort can be either intentional or unintentional. Intentional torts are committed with the intent to do something wrong, and unintentional torts are committed without such intent.

Tort law in the United States is based on English law. It provides a way for individuals, groups, businesses, corporations, and other nongovernmental organizations to resolve disputes. Many kinds of actions, or lawsuits, can be brought under the general heading of tort law in civil courts, as one private party sues another private party. A patient, for example, can sue a physician for battery.

In contrast, wrongful acts that violate criminal laws are prosecuted in criminal courts. In criminal cases, a unit of government is the party pursuing a claim against an individual accused of committing a crime. Under some circumstances, both a criminal claim and a civil claim can be made for the same wrongful act. Criminal cases result in fines and jail sentences, while civil cases result in fines.

In healthcare, the most notable tort is the tort of negligence. *Negligence* is generally defined as a failure to perform an act that a reasonably prudent person would perform (omission) or the performance of an act that a reasonably prudent person would not perform (commission) under a given set of circumstances. Negligent conduct represents a departure from the way a reasonable person would act or a departure from the usual or normal standard of behavior. In healthcare, negligence is a departure from the usual standard of care.

Four elements must be present to prove that a tort of negligence occurred. These items must be present whether the plaintiff is suing a physician for providing negligent care or a CEO is suing his lawyer for negligent advice in handling of his company's affairs. All four of the following elements must be present:

1. *Duty to use due care:* A relationship must have been established between the parties in which one party has an obligation to act as a reasonably prudent person would act toward the other. The relationship between a physician and a patient is an example of such a relationship. The requirement that the physician must conform to a specified standard of care is necessary in order to establish liability for a breach of that standard of care.

2. *Breach of duty:* One party failed to conform to a specific duty owed a party.

3. *Injury:* The party to whom the duty was owed was harmed because the duty was breached.

4. *Cause:* A connection between the breach of the duty and the injury or harm can be established. This element is known as a causal connection. To establish a causal connection, there must be a reasonably close relationship between the breach of the duty and the harm, and the harm must have been foreseeable.

A major goal of all PI initiatives is to develop the best possible clinical practices. All PI efforts should be fair and reasonable because they establish a standard of care in the eyes of the law. For example, a professional organization might establish a standard that requires that the organization assign four nurses for every ten patients in a certain patient care unit. If fewer than four nurses were working on the unit at a particular time, the staffing level could be deemed to be below the standard of care.

In determining whether a person or an organization breached an established standard of care, the law looks to several sources to determine just what the standard should be. First, a court or jury would look at what the law sets out as a standard of care. It would review what, if anything, state or federal laws and regulations say about appropriate behavior. Evidence of the established standard of care for certain professional activities might also be found in regulations and guidelines promulgated by governmental agencies or in guidelines, codes of ethics, and other treatises published by professional associations or societies. For example, the Joint Commission on Accreditation of Healthcare Organizations (JCAHO) identifies a standard of care for hospitals and other healthcare organizations such that failing to meet the standard could be said to be negligent if injuries occurred that could have been prevented had the standard been followed.

Changes in the standard of care occur over time with advances in technology and medical science. Changes in any established medical practice must be supported by reliable research. Studies can result in a new standard of care, and restudying an issue to validate a suggested change in the accepted course of action may be the most prudent PI approach.

Privacy and Confidentiality Issues

Performance improvement initiatives involve reviewing the medical care provided in the healthcare organization. To accomplish this review, healthcare professionals look at multiple episodes of care for particular types of patients. Most often, the review is accomplished by referring to the health records of specific, name-identified patients. Healthcare professionals involved in peer review activities occupy a unique position. They must understand the rights of all of the parties involved: patients, providers, and healthcare organizations. In addition, they must mediate among the parties in a way consistent with the law and yet serving the legitimate needs of patients and providers, the public good, the requirements of healthcare research, and the responsibilities of healthcare organizations.

Healthcare organizations are responsible for developing, implementing, and enforcing strict policies on the privacy of health information. They are also responsible for educating all of the parties involved in handling patient-identifiable information. Where health record information is available online, patient privacy is mandated by HIPAA and other regulations. The act and the regulations cover every use of patient data, even use by the entity that is deemed to be the legal owner of patient records, the healthcare organization. The act contains provisions specifically related to removing all patient identifiers from health records to be used for purposes other than patient care. The federal regulation singles out quality studies as one purpose for which patient identity should be protected.

Using patient-identifiable health records and other identifiable health information is common practice in PI initiatives and quality studies. Therefore, a working knowledge of HIPAA regulations is essential. Some states have also enacted legislation or have legislation pending that require that patient-identifiable information be removed from a health record before that record can be used in any study. Privacy advocates suggest that patient-identifiable information should be used only with the express written consent of the patient or the patient's legal representative.

In general, healthcare entities provide medical staff, employees, and agents of the organization access to patient-identifiable information to perform quality studies. At a minimum, all individuals who have access to patient-identifiable information should be required to undergo initial and annual training in the patient's right to privacy and the confidentiality of health records as well relevant laws and regulations related to patient privacy and release of information. They should be provided copies of relevant policies and procedures, and they should certify that they understand the organization's polices and agree to abide by them.

A **confidentiality agreement** or **confidentiality statement** similar to the ones shown in figures 21.1 and 21.2 should be imposed on the organization's employees, independent practitioners, agents, contractors, vendors, and others.

Figure 21.1.   Sample Confidentiality Agreement for Employees, Contractors, Vendors, and Agents

HEALTHCARE ENTITY, INC.
CONFIDENTIALITY AGREEMENT

Healthcare Entity, Inc. ("the Entity"), has a legal and ethical responsibility to safeguard the privacy of all patients and to protect the confidentiality of health, business, and proprietary information. In the course of its business relationship with the Entity, _____ ("Contractor") may come into possession of confidential patient information, even though it may not be directly involved in providing patient care services, or confidential and proprietary business and legal information.

In consideration of and as a condition to its business relationship between the parties, Contractor, its employees and agents, will hold the following information ("confidential information") in strictest confidence, and in accordance with the Entity's policies and procedures:

(1) Any information supplied by the Entity;
(2) Any information that is the direct result of services provided to the Entity by Consultant; and
(3) Any information about the Entity's business operations, services, products, or patients.

Contractor further agrees, on behalf of itself, its employees and agents, as follows:

(1) To maintain the confidentiality of patient information in accordance with the Entity's confidentiality policies;
(2) To make a copy of the Entity's policies available to its agent and employees for reference and to adopt and observe policies and procedures that meet the Entity's standards with regard to maintaining patient information in a secure manner and properly disposing of information which is no longer needed or is converted to another medium;
(3) To review the Entity's confidentiality policies with its employees and agents and instruct them that:
   • they are to access patient information only as necessary to carry out the responsibilities of their employment or agency;
   • they are to maintain confidentiality of patient information in accordance with the Entity's policies;
   • violation of patient confidentiality may be grounds for disciplinary action up to and including termination of employment or contract.
(4) To obtain from each employee or agent who uses or has access to patient information on Contractor's behalf a written statement that he or she has been informed about and understands the obligation to protect confidentiality and agrees to do so. This signed statement shall be maintained in the employee or agent's personal or contract files and shall be renewed annually.
(5) That the Entity may at its sole discretion revoke access to patient information at any time, if it has a good faith belief that patient confidentiality may be breached by Contractor, its agents or employees. Contractor further agrees to immediately suspend or terminate access to patient information by any agent or employee so requested by the Entity.
(6) That the Entity may audit access to and use of its patient information by Contractor, its agents or employees at any time on an ongoing basis and ask for and receive copies of the agent and employee statements required pursuant to paragraph (4) above.

This agreement is effective as of the date signed and shall remain in effect until revoked by either party. Provisions related to maintaining confidentiality of patient information shall survive the termination of this agreement.

For Contractor: For the Entity:

_____   _____
Signed   Signed
Type Name: Type Name:
Type Title: Type Title:

_____   _____
Date   Date

Figure 21.2.   Sample Confidentiality Statement

1. I _____ (agent or employee name) do hereby acknowledge my
obligation to maintain patient confidentiality and agree to not divulge, discuss, or otherwise disclose
any information relating to a patient or any aspect of his or her care unless otherwise expressly
allowed by the Entity. I further acknowledge and understand that (1) I will access patient information
only as necessary to carry out the responsibilities of my employment or agency; (2) I will maintain
confidentiality of patient information in accordance with the Entity's policies; and (3) violation of
patient confidentiality may be grounds for disciplinary action up to and including termination of
employment or contract.

2. I have read the above statement and understand the consequences for breach of patient confidentiality.

_____

Signature of employee or agent

Name:

Title:

Department:

Date:

It is important to keep in mind that independent practitioners have the same rights to
privacy in the course of performance improvement review and study. Study findings
regarding the practice of a particular provider deserve the same level of confidentiality as
do the records of patients used to perform the review. Selection of cases for review in committee must be performed by predefined policy and procedure and must adhere to review
criteria applied uniformly across all providers. Access to the patient records of a provider
must be accorded only for the purposes of the review. Information developed for review
activities and known to review committee members must remain confidential and within
the confines of committee discussion and records only. Release of information may only
occur for the purposes of further review by higher-level committees in the performance
improvement organizational structure or for review by state licensing entities. Quality management policy and procedure must make this necessity clear to all persons involved in
performance improvement review activities.

Peer Review, Immunity, and Sentinel Event Reporting

Performance improvement activities and quality assurance studies in healthcare organizations are conducted under the control and direction of peer review committees. A **peer
review committee** is a group of like professionals established according to an organization's medical staff bylaws, the organization's policy and procedure, or the requirements
of state law. The peer review system allows medical professionals to critique and criticize
the work of their colleagues candidly and without fear of reprisal. Organizations that use
peer review to assess treatment protocols and PI efforts benefit greatly from the feedback
that neutral peer reviewers can provide.

The process of establishing peer review committees varies. There are two levels of
peer review to which an independent practitioner may be subject. The first level is the peer
review undertaken within healthcare organizations to improve practitioner performance or

to validate that the practitioner's standard of care is commensurate with that of the community of practitioners to which he belongs. Generally, this review is undertaken routinely during the credentialing process (as discussed in chapter 12). Occasionally, when a practitioner sustains negative outcomes in the treatment of a patient or patients, a special review will be undertaken by peers in the same department or specialty to ensure that the care provided by the practitioner was appropriate.

For example, one neurosurgeon at Community Hospital of the West routinely got his laminectomy patients up and walking the day after surgery. At this time, the early 1970s, his neurosurgical peers believed that allowing patients to get out of bed and to walk so soon after surgery exposed the patient and the healthcare organization to excessive risk of a negative outcome. After extensive peer review over a series of years, the neurosurgeon's privileges were withdrawn at the institution because he refused to comply with the standard of care of his community of fellow neurosurgeons. Interestingly, getting patients up and walking on the day following surgery became the standard of care in the late 1990s. Thus, this neurosurgeon was ahead of his time.

When a practitioner is deemed by organizational peer review to practice in a dangerous or completely unacceptable manner, he or she may then be referred to the second level of peer review processes, that of the professional licensing bodies of a state government. Referral to state licensing bodies may also occur when the public at large complains to the licensing body about the practitioner. Usually, the licensing of independent practitioners is undertaken in close cooperation with medical and other societies in a particular state, and so peers are again used to review and validate practice at the state level. Determinations made at this level could result in disciplinary action or even revocation of the practitioner's license.

In general, institutional peer review committees perform the following activities:

- Evaluating the quality of care rendered by providers of healthcare services
- Determining whether the services provided were performed in compliance with applicable standards of care
- Determining whether the cost of the services rendered was reasonable in a given geographic area
- Determining fitness to practice

Institutional level I review most commonly considers the first two issues on the list, while state level II review would more often consider the last three issues on the list. The medical professionals involved in evaluating the work of their peers are provided certain legal protections, or peer review privileges. Peer review privileges are established by medical staff bylaws at level I and by state law at level II, and may vary from state to state. The protections have two goals: to ensure the confidentiality of peer review activities and records, and to protect committee members from civil liability for good-faith participation in the peer review process.

Level II peer review committees are organized under strict legal guidelines. They must adhere to the rules established by state law or regulation in order to be eligible for peer review privileges. At level I, the medical staff bylaws are more concerned with ensuring objectivity and appropriate due process for each practitioner undergoing routine or special review.

Generally, the proceedings and records of peer review committees are not subject to legal **discovery.** (That is, any written documentation of a committee's activities is not admissible as evidence in a legal proceeding.) Original documents or records available from sources outside the peer review committee are not immune from discovery or use in civil proceedings merely because they were presented during the proceedings of a peer review committee. For example, if a patient filed a malpractice action against a physician, the patient's medical record would still be admissible in court. However, the minutes of peer review discussion of that patient's case would generally not be discoverable. In addition, any peer review processes authorized by state law and carried out in good faith are exempt from state antitrust law.

Sentinel Events

The mandate on sentinel events from the JCAHO, effective April 2, 1998, expects healthcare organizations to report certain events or occurrences that the JCAHO calls sentinel events. A **sentinel event** is an unexpected occurrence involving death or serious physical or psychological injury or the risk thereof to either patients or employees. Serious injury specifically includes loss of limb or function. The phrase "or the risk thereof" includes any process variation for which a recurrence would carry a significant chance of a serious adverse outcome. Such events are termed *sentinel* because they signal the need for immediate investigation and response. The purpose of the JCAHO's self-reporting program is to identify aggregate patterns in healthcare quality and to contribute to the overarching goal of preventing harm and promoting safe healthcare. The standard requires that a root-cause analysis or a detailed investigation be performed when a sentinel event occurs. The JCAHO stated four purposes for requiring the reporting of sentinel events:

- To have a positive impact on improving patient care

- To focus the attention of an organization that has experienced a sentinel event on understanding the causes that underlie the event and on making changes in the organization's systems and processes to reduce the probability of such an event recurring in the future

- To increase the general knowledge about sentinel events, their causes, and strategies for prevention

- To maintain the confidence of the public in the accreditation process

Root-cause analysis is a process for identifying the underlying problems that result in variations in care and outcome or the sentinel event. The analyses focus on systems and processes, with the goal of identifying opportunities for improvement. The product of the root-cause analysis is an action plan that identifies the strategies that the organization intends to implement to reduce the risk of similar events occurring in the future.

The subset of sentinel events that is subject to review by the JCAHO includes any occurrence that meets either of two basic criteria. Only those sentinel events that affect the recipients of care (patients, clients, or residents) and that meet the following criteria are reportable to the JCAHO:

- The event resulted in an unanticipated death or a significant and permanent loss of function not related to the natural course of the patient's illness or underlying condition.

- The event was one of the following (even if the outcome was not death or major permanent loss of function):

    —Suicide of a patient in a setting where the patient received around-the-clock care (for example, a hospital, residential treatment center, or crisis stabilization center)

    —Unanticipated death of a full-term infant

    —Infant abduction or discharge to the wrong family

    —Rape

    —Hemolytic transfusion reaction involving administration of blood or blood products with major blood group incompatibilities

    —Surgery on the wrong patient or the wrong body part

Obviously, these kinds of events are likely to trigger investigations and quality studies to improve patient care.

Since 1998, the JCAHO has also published a newsletter called *Sentinel Event Alert*. The aim of *Sentinel Event Alert* is to disseminate information and lessons learned from root-cause analyses of actual sentinel events. These periodic newsletters identify specific sentinel events, describe their common underlying causes, and suggest steps to prevent occurrences in the future (JCAHO 2003b).

Past issues of *Sentinel Event Alert* have covered topics such as treatment delays, medical errors, blood transfusion errors, wrong-site surgery, and inpatient suicides. *Sentinel Event Alert* is available for free on the JCAHO's Web site.

The JCAHO's policy on sentinel event reporting raises issues related to the access of outside accrediting bodies to PI information. Sentinel event information includes reports and investigative materials that traditionally have been created as part of a healthcare organization's peer review process. As such, the information has been deemed confidential and was not available to outside third parties. Although the information related to any studies or investigations that occur after a sentinel event contributes to an understanding of quality concerns or breakdowns in systems that affect delivery of high-quality care, vigorous and candid discourse might be discouraged if the parties did not believe that the information was protected from disclosure. Oftentimes, such occurrences trigger a quality study to determine whether certain problems or systems flaws are chronic quality problems.

What should an organization do when faced with this dilemma? If the facility decided to report a sentinel event to the JCAHO, it would essentially waive its peer review privilege. If, however, it decided not to report the event, it would face sanction from the accreditation body.

The JCAHO has indicated that the reporting policy is voluntary. Healthcare organizations will not be automatically placed under greater accreditation scrutiny for choosing not to self-report sentinel events as long as a credible root-cause analysis is conducted and appropriate actions are taken. Healthcare organizations, however, remain concerned about third-party access to peer review information.

Although organizations are granted limited confidentiality when they report sentinel events and follow up with credible root-cause analysis and an appropriate action plan, the potential for discovery of the facts of the situation is real. Healthcare organizations must weigh carefully their decision and should consider at a minimum conducting thorough root-cause analyses through existing peer review mechanisms. The decision to report a sentinel event should be made by the organization after careful review of the risks and benefits of reporting. Certainly, any quality study that is ordered following a sentinel event should be done through the auspices of the peer review committee process to maximize confidentiality and immunity.

Case Study

One evening, Helen James, director of quality management at the Community Hospital of the West, was at home watching television when the telephone rang. On the other end of the line was one of the clerks from Health Information Services at the hospital. The clerk had received a request from Dr. Cooper, a neurosurgeon, to pull all of the records of spinal anesthesia for Dr. Johnson, an anesthesiologist. Dr. Cooper had informed the clerk that the records were needed for a peer review for quality improvement. The clerk already had a full workload and did not know how to begin looking up one doctor's records if they're not incomplete.

Ms. James pondered the situation for a moment. She did not recall that Dr. Johnson, the anesthesiologist, was on the list for credentials review this year, and none of his cases had come up before the surgical case review committee. Why would Dr. Cooper need Dr. Johnson's records?

Case Study Questions

1. What should Ms. James say to Dr. Cooper?

2. Because it is unlikely that Dr. Johnson is eligible for peer review, what should her policy be regarding access to his records?

Summary

Performance improvement activities in healthcare organizations are affected by a number of laws, rules, and regulations. Understanding the legal context in which PI activities are carried out is critical. The following areas of law have practical application in PI projects: copyright law, contract law, tort law and the standard of care, and privacy and confidentiality issues. Legal issues must also be considered in peer review activities and sentinel event reporting.

References

Joint Commission on Accreditation of Healthcare Organizations. 2003a. Facts about Patient Safety. Accessed online at http://www.jcaho.org/accredited+organizations/patient+safety/index.htm#3.

Joint Commission on Accreditation of Healthcare Organizations. 2003b. Sentinel Event Alert. Accessed online at http://www.jcaho.org/about+us/news+letters/sentinel+event+alert/sentinel+event+alert+index.htm.

Kozak, Ellen M. 1996. *Every Writer's Guide to Copyright and Publishing Law.* New York City: Henry Holt.

McWay, Dana C. 1997. Risk management and quality assurance. In *Legal Aspects of Health Information Management.* New York City: Delmar Publishers.

Sullivan, Tori. 2002. Mind Your Business Associate Access: Six Steps. *Journal of the AHIMA* 73(9):92ff.

U.S. Copyright Office. 2000. Copyright Law of the United States of America. Title 17 of the *United States Code.* Circular 92, April.

Chapter 22
Predicting the Future of Performance Improvement in Healthcare

Learning Objectives

- To describe new theories and models of performance improvement
- To determine the feasibility of the new theories and models and their potential for long-term impact on the healthcare system

Background and Significance

The introduction of this book provides a historical account of events and performance improvement (PI) models that brought the healthcare system to its current level of quality improvement (QI) at the beginning of the twenty-first century. Although the healthcare system continues to improve, it still faces criticism regarding lapses in quality and patient safety. Health services managers must deal with environmental changes, challenge paradigms that may no longer be effective, and look for trends and new paradigms that continue to improve quality in the future.

Differing Views on PI in Healthcare

Some experts are very positive about their view and expectations of the healthcare industry, as reflected in the following statement:

> Healthcare organizations are continuously looking for ways to improve quality, increase patient satisfaction, and reduce costs. By adopting practices used by centers of excellence—including a specialty-focused, coordinated, and multidisciplinary approach to care as well as a commitment to innovation—organizations can enhance their clinical and financial performance (Wolf 2002, p. 23).

A counteropinion in a 1999 report from the Office of the Inspector General (OIG) regarding accreditation of healthcare organizations states that "while they matter enormously to hospitals, Joint Commission survey results fail to make meaningful distinctions among hospitals" (OIG 1999, p. 2). A recent article expressed similar sentiments addressing the arguments and concerns raised in an Institute of Medicine Medical Practice Study that declared, "between 44,000 and 98,000 people die in US hospitals annually as a result of medical errors" (Leape 2000, p. 95).

Emerging Trends in Healthcare PI

The following discussion addresses a number of emerging trends, any of which may become the next paradigm for healthcare PI.

On March 1, 2001, the Institute of Medicine published a two-part report, *Crossing the Quality Chasm,* that presented medical error as a chronic health threat that compares with breast cancer, motor vehicle accidents and AIDS in its lethality. The IOM committee's report makes the following charge: "In its current form, habits, and environment, American health care is incapable of providing the public with the quality health care it expects and deserves" (Berwick 2002, p. 85). Research published in a previous report titled, *To Err Is Human,* served as a foundation for more comprehensive research that made a compelling case for the concern about the health and safety of patients entering the U.S. healthcare system. The report asserts that old systems are unreliable and mere attempts to fix what is broken would not do enough to correct the problem. The situation requires a major change in the entire healthcare system (Berwick 2002).

According to Thomas C. Dolan, president of the American College of Healthcare Executives, the report "has attracted a great deal of attention" from healthcare administrators and the public. The second part of the report provides "a blueprint for redesigning the healthcare system to improve care" (Dolan 2002, p. 4).

Time will tell whether changes in the healthcare system will result from modification of existing models or emerge from creative innovations. The Institute of Medicine committee recommended six major aims on which healthcare organizations and professionals should be focused and aligned. The six aims for providing care should be the following:

- Safe
- Effective
- Patient centered
- Timely
- Efficient
- Equitable

Ideally, the healthcare system will continue to seek out new areas of focus for PI. At the beginning of the twenty-first century, five primary areas are getting attention. These include information systems/information technology, payment systems, Six Sigma, systems thinking/learning organizations, and evidence-based medicine/evidence-based management.

These areas of focus have emerged or are emerging in the healthcare system and questions remain regarding how effectively they will diffuse into the industry.

It may be helpful to consider new QI trends from the same perspective that Henry Mintzberg approaches strategic planning (Mintzberg 1994). The model in figure 22.1 depicts how managers plan to pursue an intended strategy, and as their intended strategy fails to be realized, how they adopt a more deliberate strategy. Meanwhile, conditions in the environment emerge that evolve into a realized strategy. The emergent strategy is neither planned nor intended but evolves out of trial and error.

Mintzberg's emergent strategy model can be adapted to formulate a model for managed care. Originally, managed care was implemented with the intent of reducing costs and

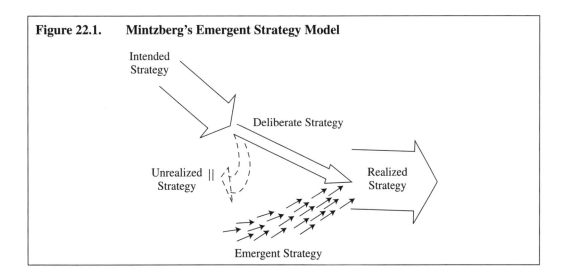

Figure 22.1.   Mintzberg's Emergent Strategy Model

increasing profits. The sequence of events following the emergence of managed care can be diagrammed using Mintzberg's emergent strategy model.

Cost reduction and profit increase were partially met using supply-side control, but efforts to decrease demand for healthcare services were unrealized. The financial incentives to employ error prevention programs or seek least costly alternatives for healthcare services were not effectively built into the design of managed care plans. The shift in managed care focus began to narrow with greater attention to supply-side cost containment as an interim strategy.

As other strategic forces have emerged in the economy, a new paradigm for demand management strategy has evolved. Some of these forces include pressure for greater preventative measures, nurse advice, decision support technology, health promotion, and continuous quality improvement (CQI). These emerging forces have combined into a new paradigm that is becoming known as demand management. Emergent strategy offers a model for understanding the way new ideas or approaches to improvement may actually become a reality in the healthcare industry.

Information Systems/Information Technology

In the last two decades of the twentieth century, information technology capability has expanded exponentially. Computers have the capability to store and retrieve incredibly large amounts of data. However, in many cases they are merely used to perform functions that were done manually in the past. For example, Donald Berwick (1999), in his presentation, "Escape Fire," at the Eleventh Annual National Forum of Quality Improvement in Healthcare held in New Orleans, notes the following reality in spite of recent technological advances in the handling of healthcare information:

> In the current model, information is treated generally as a tool for retrospection, a record of what has happened, a stable asset that sometime we might or might not use to see the past, or to defend or prosecute a lawsuit.

In chapter 15, it was explained that PI is an information-intensive activity. Information management continues to migrate from paper systems to paperless systems that rely on computer technology for storage and retrieval of data. In the past, data were recorded into hard-copy records by hand or through the use of computer keyboards. For years, the only document regarding a patient's diagnosis and treatment that was legally admissible in a court of law was the paper medical record.

There was a major shift in legal precedent when a physician's electronic signature became admissible in court. Some state governments were quick to embrace the idea. The Joint Commission on Accreditation of Healthcare Organizations (JCAHO) was more reluctant but has come to accept this as status quo.

The current trend in documentation is to move to a paperless environment. This trend has both pros and cons. Computer technology can speed up the transmission and retrieval of large amounts of information. It also makes information easily accessible to multiple individuals from great distances. Healthcare providers have instant access to patient information and can literally consult with specialists around the world. Research and evaluation of healthcare practices has been expanded because researchers can retrieve information that was previously inaccessible from data repositories.

Data Confidentiality and Overload

At least two issues of concern have been raised with advances in information technology: confidentiality and data overload. Confidentiality-sensitive documents and the privacy of the individual patient are important issues that have far-reaching legal implications if not handled appropriately. As microcomputer technology becomes more widespread, more individuals will be walking around with devices in their possession that contain patient and proprietary information. Although system administrators have put firewalls and security measures in place, each system is only secure as long as individuals with access comply with security measures and protect confidential information.

Chapter 15 discusses the ways in which computer technology improves one's ability to transform data into information but also makes the point that larger amounts of data can be involved in the process. As the use of monitoring and diagnostic equipment has advanced with the use of digital technology, large amounts of data generated by equipment can be recorded automatically and stored data repositories. Without a well-developed database architecture or individuals who are knowledgeable in query language retrieving data in a meaningful way, these large amount of information can be overwhelming.

Standardization of software systems has always been problematic and will continue to be a problem. For the past few decades, healthcare organizations have purchased a wide variety of information systems that are not compatible with one another. Already, billions of dollars have been spent on these systems, and many organizations are reluctant to overhaul their systems in order to have a standardized and integrated system because the cost is so prohibitive. Currently, most organizations continue to move their information systems toward total integration over the years as their budgets allow.

Handheld Technologies

Are we looking at the end of handwritten data? Technology is already in place that converts handwriting into a digitized document. In a matter of years, the new generation of students entering universities will be accustomed to taking notes on Palm Pilots and beaming data between one another and their instructors. As they complete their college degrees, they will enter the workforce expecting their coworkers to be conversant in the same technology.

The use of handheld devices is becoming more and more prevalent in healthcare organizations. Most executives and professionals own devices such as Palm Pilots and use them for both personal and professional purposes. Using a handheld device that incorporates cellular telephone technology, Internet access, and personal computing is becoming widely available and affordable to the average individual.

Cyborg Technology

Cyborg technology is also becoming a reality in everyday life. No one thinks twice about cardiac pacemakers and defibrillators, which are implanted to collect data about the heart and can be read by passing a wireless device over the patient's chest. Two million people have already received implants of this kind (Schonberg 2002). Cochlear implants have restored hearing to the deaf. Implanted chips for storing patient information and physically locating individuals are also being developed and tested. Health information management (HIM) professionals continue to wrestle with the challenge of managing high volumes of patient information data generated by cyborg technology as well as the legal and ethical implications associated with use of that technology.

Artificial Intelligence

Another area of information technology that continues to grow is artificial intelligence. Software using artificial intelligence that employs algorithm-directed decision making for healthcare professionals is emerging, especially in the area of triage. The implications for using artificial intelligence to make critical diagnostic and treatment decisions with or without human intervention have far-reaching legal and ethical consequences. Health information professionals will be heavily involved in this development because of the need to document the decision-making process.

Internet Use in Healthcare

With the development of electronic medical records, the use of the Internet expands. Currently, there are four groups of barriers to adoption of Internet-based systems. First is the patients, who do not have universal access. Second, some physicians are reluctant to accept Internet-based systems, primarily because of time constraints and malpractice exposure. Third, for administration, the tempo of change in information technology increases cost, and there are concerns about allowing proprietary information to be available on the Internet. Finally, there are concerns about privacy and confidentiality. Until the Health Insurance Portability and Accountability Act (HIPAA) regulations are more clearly understood, the consequences of failed security or inappropriate disclosure will continue to be a barrier to the adoption of Internet-based systems (Kerwin 2002). Donald Berwick observes, "total access 24-7-365 begins to be achievable only when we agree—scientists, professionals, patients, payers, and healthcare workforce—that the product we choose to make is not a visit. Our product is healing relationships, and these can be fashioned in many new and wonderful forms, if we suspend the old ways of making sense of care. . . . The healthcare encounter as a face-to-face act is a dinosaur" (Berwick 1999).

Payment Systems

Payment systems are an integral part of the healthcare delivery system. Healthcare expenditures make up a significant portion of the U.S. economy. The payment system has evolved from a time when a country doctor may have been reimbursed through bartering to the development of fee-for-service reimbursement and managed care plans. Eggleston (2000, p. 173) notes:

> It has been recognized for a long time that how we pay healthcare providers and how we insure consumers against the risks of medical expenditures—that is, how we design health insurance payment systems—can have significant consequences for the equity and efficiency of a healthcare system.

Managed Care and Prospective Payment Systems

Payment systems may also have a positive or negative impact on the quality of healthcare delivery. The implementation of managed care and prospective payment systems (PPSs) was hailed as a new way to cut costs and encourage better demand-side decision making on the part of healthcare consumers. Based on more than twenty-five years' experience as a healthcare administrator, the author has observed a phenomenon in healthcare that the farther an individual is removed from actual payment for a service, the less he or she values the service provided. In community health programs in particular, individuals who pay the least in out-of-pocket expenses for healthcare services tend to be the most vocal in their criticism of the system. One can make the logical assumption that this trend carries through when comparing fee-for-service to PPSs. With the shift from fee-for-service as the predominant payment system to greater prevalence of small copayments in managed care systems such as HMOs, the perception of the value and quality of healthcare also shifts.

Managed care systems also created new limits on access to care. Under the fee-for-service system, a larger number of patients could easily find access to a specialist rather than a primary care provider. Under managed care, an individual with a sore knee or twisted ankle cannot go directly to an orthopedist but must first see a primary care physician. Regardless of the quality of care delivered and a positive outcome, there is a tendency for patients to perceive the care of an orthopedist as more specialized, and thus of higher quality, than that of a primary care provider.

Electronic Billing in Healthcare

An element of the prospective payment system that requires CQI is the reimbursement system. In the past, the billing office of a clinic or hospital prepared a paper bill that was mailed to the patient or the insurance company. In today's system, many patients pay a copayment at the time of the encounter, and the bill to the insurance company is sent through an electronic data submission system via Internet or phone.

Although the new billing system software has edit checks to catch errors, mistakes and rework still occur. Previously, accounts receivable were tracked over 30, 60, and 90 days, as providers traded mail with insurance companies to correct errors. Now bills are reimbursed within hours or days if no errors are found. Errors result in no reimbursement until there is a reconciliation of the data submitted. This has caused healthcare organizations to focus on the quality of information coming to the individuals who input the billing information into the electronic billing system. This will continue to be a critical PI area as organizations evaluate the system of collecting billing information from the point of service to inputting it into the electronic system. Regardless of how payment mechanisms evolve in the future, the quality of data collected and the way data are processed will require some level of PI.

Six Sigma

Benchmarking has been long practiced within the domain of the healthcare system. However, some healthcare organizations have begun to benchmark against other industries and are selecting models that may be adapted to the healthcare industry. For example, some healthcare organizations have begun applying the Six Sigma philosophy to their PI programs. Six Sigma is practiced widely in business sectors outside healthcare, and this philosophy is gaining acceptance in the healthcare industry (Shortell and Selberg 2002).

Six Sigma uses statistics for measuring variation in a process with the intent of producing error-free results. Sigma refers to the standard deviation used in descriptive statistics to determine how much an event or observation varies from the estimated average of the population sample. For example, a student who scores 130 on an IQ test would be considered to have a higher IQ than 97.5 percent of the population. The average IQ is considered to be 100 and the standard deviation for IQ is 15 points. Thus, a score of 130 is a variation of two standard deviations above the average. Only 2.5 percent of the population is estimated to have scores above two standard deviations.

Using this kind of statistic to measure variation in quality required some refinement of Six Sigma in healthcare. The intent of any PI program is to minimize error; and by extension, to decrease the number of errors occurring in any observation of a number of encounters or activities is the ultimate goal of PI.

Six Sigma was chosen as a target statistic because even two or three standard deviations would not be acceptable in certain scenarios. A 2.5 percent error rate for making correct change at a movie theater may be acceptable, but that error rate in airlines avoiding fatal crashes is completely unacceptable because airlines have hundreds of flights in the air on any given day. Even if there were only one hundred flights per day, two to three fatal crashes per day would be devastating to the airline industry, not to mention the population as a whole. Therefore, it is important to keep this PI approach in proper perspective if it is to be applied to healthcare.

The Six Sigma measure indicates no more than 3.4 errors per one million encounters. Consider the challenge of achieving no more than 3.4 errors per one million prescriptions, surgeries, or diagnoses. In certain areas, this standard may seem unattainable, and in others, it may not be rigorous enough. However, incorporating Six Sigma into PI requires considerable organizational change.

Systems Thinking

Systems thinking continues to emerge in continued efforts to improve quality. Approaching quality from a systems orientation rather than a process orientation requires development of individual and collective skill sets. Peter Senge (1990) outlines a suite of five disciplines that individuals should master. They are:

- Building a shared vision
- Team learning
- Personal mastery
- Mental models
- Systems thinking

A critical element of systems thinking is viewing an organization as an open system of interdependencies and connectedness rather than a collection of individual parts and professional enclaves. This approach sees interrelatedness as a whole and looks for patterns rather than snapshots of organizational activities and processes.

Process orientation has traditionally relied on measurements of cost, productivity, quality, and time to improve processes. There is a tendency to approach processes in a linear fashion, fixing one process at a time. In some cases, organizations have begun to incorporate organizational learning and human capital into measurement (Forsberg, et al. 1999). The difference in systems thinking is that rather than relegating PI to teams, the five disciplines are implemented to help the total organization learn and make the changes in processes the way things are done throughout the entire organization.

Traditional education in America has approached learning from a viewpoint heavily influenced by Sir Isaac Newton, which views the world with a focus on single material objects rather than relationships between objects or concepts (Wheatley 1994). By contrast, systems thinkers consider the world around them from a holistic point of view. They look at relationships rather than the parts and pieces in isolation.

Anyone who has worked in a hospital or an integrated healthcare system should intuitively understand how systems thinking is more appropriate for managing complex organizations with very diverse professional staffs. However, many PI approaches in the past have relied on Newtonian models. Performance improvement efforts such as quality control and quality assurance focused heavily on structure and compliance standards. As PI efforts have begun to focus on process and outcomes, the need to understand and embrace systems thinking has become more apparent. Focusing on the outcome of a patient encounter forces all members of the healthcare team to consider a shared vision and to learn as a cohesive unit rather than operating as individuals who only contribute their expertise and energies to patient care in an episodic and isolated manner.

The physician's role has always been crucial to healthcare delivery. Two important characteristics of the healthcare system have shaped the role of physicians. "The first is the legal system: only physicians are permitted to provide certain services. Second, both patients and insurers lack the necessary information to make many medically related decisions" (Feldstein 2003, p. 37). In the past, health services organizations operated under the assumption that the physicians possessed all the medical knowledge necessary and everyone responded to their instructions.

Then schools of nursing began to specialize, and nursing began to develop a body of knowledge and skills unique to its profession. Other areas of specialization such as medical laboratories, pharmacy, radiology, physical therapy, and health information have also developed their own unique bodies of knowledge. Few individuals, including physicians, are capable of retaining the extensive knowledge and understanding of all specialized areas in health service organizations. Out of necessity, a greater degree of interdependence has emerged between different health professions. There is a need for "instruction" to flow in many new directions other than from the physician to all others in the organization. Some leaders in healthcare now consider Peter Senge's model for learning organizations a viable PI model.

DRGs and Systems Thinking

Senge (1990, p. 13) uses the Greek word *metanoia* to describe a shift of the mind that is more than just memorizing and recalling information. For example, health information managers do more than just take in information about DRGs and coding. They experience a shift of the mind that allows them to look at documentation in a medical record and know that it is sufficient to ensure reimbursement from a third-party payer and for audit purposes. When DRGs were introduced, this approach resulted in interdepartmental confrontations. Coding personnel became well versed in DRGs because appropriate reimbursement was dependent on proper documentation, but some physicians who had become accustomed to older systems of reimbursement were not receptive to properly documenting in a way that would facilitate reimbursement from third-party payers.

Until physicians, nurses, and health information personnel began to learn collectively about the DRG process and how their respective roles were interdependent, there were major staff conflicts, and significant retooling was needed to successfully receive reimbursement. Becoming a learning organization requires a commitment to improved communication. By experience, most healthcare organizations have discovered that collective learning with regard to common terms and understanding of emerging models such as DRGs is necessary to improved performance.

Evidence-Based Management

Evidence-based medicine was addressed earlier in the text in chapter 10, and there has been an offshoot of this concept that now addresses management. The Institute of Medicine report generated research that goes beyond outcomes assessment to evidence-based medicine. "Evidence-based medicine is based on the principle that providers and patients should adopt those medical treatments that scientific clinical trials find to be safe and efficacious. It advocates that clinical practice should be grounded in scientific evidence and continually improved by the results of robust research" (Van de Ven and Schomaker 2002, p. 89). This a movement that aims to "introduce rationality into the adoption and diffusion of innovations" (Denis and Langley 2002, p. 32).

Professors of healthcare administration have been encouraged by Shortell to merge evidence-based medicine and evidence-based management into their curricula. This includes addressing the six aims that the Institute of Medicine report recommends to healthcare organizations to make their services safe, effective, efficient, timely, personalized, and equitable. Three overarching principles must also be applied. They are patient-centeredness, system-mindedness, and evidence based. In order to apply these principles, ten key issues (Shortell 2002) are raised:

- Care based on continuous healing relationships
- Customize based on patient needs and values
- The patient as the source of control

- Shared knowledge and free flow of information

- Evidence-based decision making

- Safety as a system property

- The need for transparency

- Anticipation of needs

- Continuous decrease in waste

- Cooperation among clinicians

Where evidence-based medicine has focused on collecting data from large patient populations, looked for indicators, and assessed outcomes that indicate the need from clinical guidelines, evidence-based management will require a broader systems approach to identifying interdependencies between administrative and clinician activities. For example, precertification for certain procedures was implemented by insurance companies as an administrative procedure to reduce cost. A growing body of evidence has been identified that indicates the precertification process may adversely affect the quality of care provided to patients and, in some cases, actually increase costs in the long run. Consider the cases where precertification for a procedure that may have allowed intervention to early diagnosis and treatment of an aggressive cancer was delayed by the administrative process and the patient was subjected to more expensive and invasive treatment with a prognosis of an early death.

Shortell also presents six challenges facing leaders in healthcare:

- Redesign of care processes

- Effective use of information technology

- Knowledge and skills management

- Development of effective teams

- Coordinated care longitudinally

- Performance and outcome measurement

In the same manner that clinicians have used scientific investigation, managers must also use scientific inquiry to produce evidence that supports shifts in policy and practice in health services management. He also makes the point that the focus of evidence-based management must be organizationwide and applied to at least four levels of change to effectively improve quality. These four levels include four considerations: individual, group or team, organization, and larger system or environment.

Fiction, Fad, or Fundamental?

The challenge facing health information professionals is determining which new ideas regarding quality programs and models are a trend that will last over time and which ones will fade into the background when a new trend emerges. For example, Avedis Donabedian is often credited with the development of the three classes of quality assessment: structure, process, and outcome. The first two were more readily embraced by the health system; it was only in the latter years of the twentieth century that outcomes have become a major focus of PI programs. The health system continues to shift toward a broader, systems-oriented PI approach.

ISO 9000 Certification

If healthcare organizations expand into global entities, they will be required to deal with the same issues that other industries face when doing business outside the United States. ISO 9000 certification is part of a PI system that is required in order to conduct business in certain foreign countries. The International Organization for Standards in Geneva, Switzerland, first published ISO 9000 standards in 1987. Into the 1990s, it was much more prevalent in European countries than in the United States. ISO 9000 sets specification standards for quality management with regard to process management and product control. In the healthcare setting, product control is quality control of patient care activities. Companies that document and demonstrate compliance with ISO 9000 standards receive certification.

In 1998, the Ministry of Public Health in Thailand announced that it planned to require state hospitals to meet ISO 9000 standards (*Bangkok Post* 1998). At the same time, there was a move on the part of a few hospitals in the United States to adopt ISO 9002 standards, an enhancement of the original ISO 9000 standards. The American Legion Hospital in Crowley, Louisiana, dropped JCAHO accreditation and adopted ISO 9002. The Pulaski Community Hospital in Virginia's New River Valley continued with JCAHO accreditation but also adopted ISO 9002 standards (Babwin 1998).

Whether or not other hospitals will follow the lead of these two hospitals is yet to be seen. Although the American Legion Hospital must undergo a state audit to continue receiving Medicare payments, adopting ISO 9002 certification and dropping JCAHO accreditation reportedly reduced costs by $100,000 a year. Although it seems unlikely that many will follow the lead of the American Legion Hospital, many may choose to continue JCAHO accreditation and adapt ISO 9002 standards. If this becomes a trend, healthcare professionals will need to become conversant in ISO 9002 standards and learn how they can be applied to direct care clinical processes (Schyve 2000).

Theory of Constraints

Eliyahu Goldratt may someday emerge to become a leader widely recognized for his development of the theory of constraints (TOC) as a process of ongoing improvement. The theory of constraints has been accepted in some sectors of the industry, more so in manufacturing than in service. The service industries have been less accepting of TOC. The concept of throughput and identifying constraints in healthcare organizations is complicated by the complexity of the system and its accounting procedures:

> Throughput is the rate at which a system generates money through sales after reduction for material costs, commissions, and distribution cost. Under TOC, the objective is to maximize throughput while minimizing operating expenses for labor, sales, and administration and simultaneously minimizing investment outlays for inventory, plant, and equipment. The first step in applying TOC is to identify the constraining factor. For manufacturing concerns, the constraining factor is often but not always the time available on a certain machine or process. For companies that employ skilled workers, and for many service organizations, the constraint is often the time of one or a few key employees (Bushong 1999, p. 53).

In healthcare, there is little published on the TOC, and what has been published are a "few single-institution case studies" (Breen, et al. 2002, p. 46). This PI approach may have been an idea ahead of its time within the healthcare industry. Only in recent years, with a shift toward systems thinking and surveying chains of interdependencies, have healthcare managers become more accepting of the TOC model.

A constraint within the context of TOC is thought of as a bottleneck in a process. In the healthcare setting, an example of a constraint might be found in a clinic where a limit on a resource slows down the overall ability of the clinic to move patients in and out of the clinic during the normal workday. If there is a shortage of medical assistants, there can be delays in turning over exam rooms and prepping that backs up patients in the waiting room until an exam room is ready. This delay would then also impact physicians' time because they are also waiting for patients to be placed in exam rooms before they can see them.

The human response to this kind of constraint is frustration, and there is then a tendency to rush things at the end of the clinic schedule. Patients wait longer than necessary to be seen, and the risk of errors when the physician is rushed increases. Thus, the constraint in a health clinic can be a function of the health professional's appointment schedule and efficient use of exam rooms or procedure rooms. However, the patient is an important factor in that he or she must arrive at appointments on time and at the same time view each experience as a high-quality encounter.

In order to apply TOC, the following five focusing steps (Breen 2002) must be completed:

1. *Identify the system's constraint.*

   The challenge here is to conduct an effective root-cause analysis. When team members are skilled in systems thinking, they are better able to see relationships and interdependencies between organizational structures and processes. The flowcharting of processes is a useful tool for focusing on a system constraint.

2. *Decide how to exploit it.*

   Exploit a constraint by deciding how best to change structure, policies, or procedures to open the flow of goods and/or services through the constraint. In the example cited above, the shortage of medical assistants may be exploited in a number of ways. Although the solution many would select is to hire more medical assistants, other ways to exploit the constraint might include shifting some tasks and responsibilities currently assigned to medical assistants to other staff members. Another solution may be found in determining if certain policies are creating inefficiencies by requiring tasks that might be eliminated or combined if the policy were changed.

3. *Subordinate and synchronize everything else to above decisions.*

   It is important to remember that a decision to change the conditions of a constraint is likely to be met with resistance. If the decision is made to hire more medical assistants, the entire organization must be dedicated to recruiting, selecting, orienting, and training new medical assistants. This means other hiring actions must not take priority over hiring medical assistants. At the same time, this decision must be synchronized with other activities in the organization. For example, work schedules would need to be coordinated to accommodate training schedules for new hires.

4. *Elevate the system's constraint.*

This means investing in the constraint. If more funding is needed for new hires, facilities, or equipment, the organization must commit the needed funds. Although the focus in step two is to find ways to exploit the constraint without significant increases in spending, step four may require the commitment of more funding.

5. *If in the above steps the constraint has shifted, go back to step one. Do not allow inertia to become the system's constraint.*

If during the process of implementing steps one through four the constraint shifts, go back to step one and address the new constraint. This step-by-step process of TOC is like another PI program, an ongoing process of continuous improvement.

Theory of constraints requires commitment from the entire organization but offers promise as a PI tool for healthcare organizations. "Of particular benefit is the method's focusing power by targeting the weakest link in a system or process. Improvement efforts can be costly, but TOC makes the most of any effort by aiming at the single improvement that will significantly affect the overall system"(Womack and Flowers 1999, p. 405). Whether or not it will ultimately be embraced as a new paradigm in healthcare is not yet clear. Any organization that chooses to implement TOC should assess the readiness of the leadership team to accept this new PI approach before doing so.

Case Study

In the mid-1990s, six members of the executive team of the Community Health Center were trained in TOC by a consultant. The purpose was to explore the potential for applying TOC to management of the health center. Much effort was focused on step one, but the different levels of sophistication in systems thinking among the team members was problematic and made it difficult to get through step two. Without a better understanding of the chain of interdependencies, deciding how to exploit a constraint was difficult. Until the team and the organization learned to embrace and understand systems thinking, it was clear that effectively using TOC was futile. Some members of the team were still operating under the mental model of classic healthcare that responds to occurrences as episodic and isolated. Furthermore, a lack of understanding of the chain of interdependencies makes it virtually impossible for an organization to subordinate and synchronize everything to exploitation of an identified constraint.

Case Study Questions

1. What opportunity for performance improvement can you identify in this case?

2. What was the fundamental problem preventing implementation of a new model for quality improvement at the health center?

3. If you had been the chief executive officer of the health center, what steps would you have taken to better prepare the executive team for implementing TOC?

Summary

The healthcare system continues to evolve, and HIM professionals must embrace new approaches and ideas for improving performance. As electronic medical records become the norm, HIM professionals must learn to shift from knowledge and management of paper records to the world of computers and data management if they are to stay relevant. Information systems and information technology have become the next locus of power within healthcare organizations. The capability of information systems to seamlessly interface between medical records and payment systems is powerful. This makes information systems a key point of leverage with healthcare organizations. Systems administrators have become chief information officers and now are evolving into business strategists. In the past, they have been perceived as midlevel technical managers of hardware and software, but as organizations have made huge capital investments in information systems and become dependent on information systems to conduct day-to-day business, they have taken on a more powerful role. Health information managers are inexorably linked to the corporate information system. Shifts in payment systems will continue and require adaptation by the organization.

Six Sigma may become a new paradigm in health organizations as they are challenged to reduce errors significantly. How the concept is implemented will determine its acceptance. The models for systems thinking/learning organizations continue to be adopted throughout the healthcare system. Leaders will have to become literate and skilled in systems thinking. Acceptance of evidence-based medicine and management will depend on management's capability to access research and to conduct research.

The theory of constraints has been applied in a number of areas in healthcare in the United States and England (Breen, et al. 2002). Whether or not the concept will be widely accepted has yet to be determined. The innovative manager must be able to distinguish between new trends that are efficacious and those that are merely fads that will fade away when the excitement about them wanes. It is imperative that managers who lead the way continually seek to improve their knowledge and skills by informing themselves about new ideas and approaches to performance improvement.

References

Babwin, Don. 1998. Move over, JCAHO. *Hospitals and Health Networks* 72(10):57–58.

Berwick, Donald. 1999. Escape fire. *Eleventh Annual National Forum of Quality Improvement in Healthcare,* New Orleans.

Berwick, Donald M. 2002. A user's manual for the IOM's 'Quality Chasm' report. *Health Affairs* 21(3):80–90.

Breen, Anne M., Tracey Burton-Houle, and David C. Aron. 2002. Applying the theory of constraints in healthcare, part 1: the philosophy. *Quality Management in Healthcare* 10(3):40–46.

Burton, L. R. 1998. Implementing medical call centers for demand management in healthcare organizations: a model for internal installation or outsourcing. Unpublished.

Bushong, J. Gregory, and John C. Talbott. 1999. An application of the theory of constraints. *CPA Journal* 69(4):53–55.

Denis, Jean-Louis, and Ann Langley. 2002. Forum: Jean-Louis Denis and Ann Langley. *Healthcare Management Review* 27(3):32–34.

Dolan, Thomas, C. 2002. Crossing the quality chasm with ACHE. *Healthcare Executive* 17(1):4.

Donabedian, Avedis. 1966. Evaluating the quality of medical care. *Milbank Memorial Fund Quarterly* 44 (3):166–206.

Eggleston, Karen. 2000. Risk selection and optimal health insurance–provider payment systems. *Journal of Risk & Insurance* 67(2):173.

Feldstein, Paul J. 2003. *Health Policy Issues: An Economic Perspective, 3rd Edition.* Washington, D.C.: AUPHA Press.

Forsberg, Torbjorn, Lars Nilsson, and Marc Antoni. 1999. Process orientation: the Swedish experience. *TQM Magazine* 10(4/5):S540–S547.

In brief: state hospitals aim for ISO 9000, *Bangkok Post,* 26 November 1998.

Institute of Medicine. 2001. *Crossing the Quality Chasm: A New Health System for the Twenty-first Century.* Washington D.C.: National Academy of Sciences.

Investing in the new economy: the convergence of the healthcare, biotechnology, electronics, computer, and telecommunications industries. *PR Newswire,* 8 April 1999.

Kerwin, Kathryn, E. 2002. The role of the internet in improving healthcare quality. *Journal of Healthcare Management* 47(4):225–36.

Leape, Lucian L. 2000. Institute of Medicine medical error figures are not exaggerated. *Journal of the American Medical Association* 284(1):95–97.

Lipshitz, Raanan. 2000. Chic, mystique, and misconception: Argyris and Schon and the rhetoric of organizational learning. *Journal of Applied Behavioral Science* 36(4):456.

Mintzberg, Henry. 1994. *The Rise and Fall of Strategic Planning.* New York City: The Free Press.

Office of the Inspector General, Department of Health and Human Services. 1999. *The External Review of Hospital Quality: The Role of Accreditation.* OEI-01-97-00051.

Schonberg, Erick. 2002. The cyborg known as you. *Business 2.0 Media Inc.* 3(8):33.

Schyve, Paul M. 2000. A trio for quality: accreditation, Baldridge and ISO 9000 can play a role in reducing medical errors. *Quality Progress* 33(6):53–55.

Senge, Peter. 1990. *The Fifth Discipline.* New York City: Doubleday Currency.

Shortell, Stephen M. 2002. Crossing the quality chasm: what it will take to make the leap. *Presentation at Association of University Programs in Health Administration (AUPHA) 2002 annual meeting.* June 23, Washington, D.C.

Shortell, Stephen M., and Jeff Selberg. 2002. Working differently: the IOM's call to action. *Healthcare Executive* 17(1):6–10.

Van de Ven, Andrew H., and Margaret S. Schomaker. 2002. Commentary: the rhetoric of evidence-based medicine. *Healthcare Management Review* 27(3):89–91.

Wheatley, Margaret. 1994. *Leadership and the New Science: Learning about Organization from an Orderly Universe.* San Francisco: Berrett-Koehler.

Wolf, Emily J. 2002. Employing practices used by centers of excellence. *Healthcare Executive* 17(1):20–23.

Womack, David E., and Steve Flowers. 1999. Improving system performance: a case study in the application of the theory of constraints. *Journal of Healthcare Management* 44(5):397–407.

Glossary and Index

Glossary

Absolute frequency: The number of times that a score or value occurs in a data set.

Accreditation: The act of granting approval to a healthcare organization based on whether the organization has met a set of voluntary standards developed by an accreditation agency.

Accreditation standards: An accrediting agency's published rules, which serve as the basis for comparative assessment during the review or survey process.

Action plan: A set of initiatives that are to be undertaken to achieve a performance improvement goal.

Adverse drug reaction: A patient's undesired, unintended, or unexpected detrimental response to a medication given in dosages that are recognized in accepted medical practice.

Affinity diagram: A graphic tool used to organize and prioritize ideas after a brain-storming session.

Agenda: A list of the tasks to be accomplished during a meeting.

Bar graph: A graphic data display tool used to show discrete categories of information.

Benchmarking: The systematic comparison of the products, services, and outcomes of one organization with those of a similar organization; or the systematic comparison of one organization's outcomes with regional or national standards.

Blitz team: A type of PI team that constructs relatively simple and quick "fixes" to improve work process without going through the complete PI cycle.

Brainstorming: An idea-generation technique in which a team leader solicits creative input from team members.

Case management: The principal process by which healthcare organizations optimize the continuum of care for their patients.

Cause-and-effect diagram: An investigational technique that facilitates the identification of the various factors that contribute to a problem; also called a fishbone diagram.

Certification: The act of granting approval for a healthcare organization to provide services to a specific group of beneficiaries; also, the act of granting a healthcare professional approval to practice. *See also* Credential.

Change management: A group of interpersonal and communication techniques used to help people understand the process of change and accept improvements in the way they perform their work.

Check sheet: A data collection tool used to identify patterns in sample observations.

Clinical guidelines: The descriptions of medical interventions for specific diagnoses in which treatment regimens and the patients' progress are evaluated on the basis of nationally accepted standards of care for each diagnosis.

Clinical Laboratory Improvement Amendments (CLIA):   The 1988 reenactment of the 1967 Clinical Laboratory Improvement Act, the federal regulations outlining the quality assurance activities required of laboratories that provide clinical services.

Clinical path: A graphic tool used to communicate established standards of patient care for specific diagnoses; also called clinical pathway, care map, and critical path.

Clinical practice standards: The established criteria against which the decisions and actions of healthcare practitioners and other representatives of healthcare organizations are assessed in accordance with state and federal laws, regulations, and guidelines; the codes of ethics published by professional associations or societies; the criteria for accreditation published by accreditation agencies; or the usual and common practice of similar clinicians or organizations in a geographical region.

Clinical privileges: The accordance of permission by a healthcare organization to a licensed, independent healthcare practitioner (physician, nurse practitioner, or another professional) to practice in a specific area of specialty within that organization.

Closed record review: The examination of patient records assumed to be complete with respect to all necessary and appropriate documentation by surveyors from accreditation organizations.

Community-acquired infection: An infection that was present in a patient before he or she was admitted to the hospital.

Compliance: The process of meeting a prescribed set of standards or regulations in order to maintain active accreditation, licensure, or certification status.

Compulsory review: The examination of a healthcare facility and its processes and infrastructures as required by state laws and regulations.

Conditions of Participation: A set of regulations published by the Centers for Medicare and Medicaid Services (CMS) to outline requirements of approved programs providing healthcare services to beneficiaries of Medicare and Medicaid programs.

Confidentiality agreement or statement: A document that outlines the responsibility of healthcare workers, contractors, vendors, and agents for complying with their employer's or client's policies and procedures for protecting patient and corporate information and information systems.

Continuous data: Data that may have an infinite number of possible values in measurements that can be expressed in decimal values.

Continuous monitoring: The regular and frequent assessment of healthcare processes and their outcomes and related costs.

Continuous quality improvement (CQI): A component of total quality management (TQM) that emphasizes ongoing performance assessment and improvement planning.

Continuum of care: The totality of healthcare services provided in all settings, from the least intensive to the most intensive; the emphasis is on treating individual patients at the level of care required at any given time by their course of treatment.

Contract: A mutual promise between two or more parties that the law recognizes and considers enforceable.

Control areas: Physical facilities within an organization where utilities are turned on or off or are otherwise monitored.

Control chart: A data display tool used to show variation in key processes over time.

Copyright: A legal principle that protects the original expression contained in an intellectual work, that is, the words, sounds, and/or images used by an author or authors to express an idea or to describe a process.

Cost: The amount of financial resources consumed in the provision of healthcare services.

Credential: A formal agreement granting an individual permission to practice in a profession, usually conferred by a national professional organization dedicated to a specific area of healthcare practice; or the accordance of permission by a healthcare organization to a licensed, independent practitioner (physician, nurse practitioner, and another professional) to practice in a specific area of specialty within that organization.

Credentialing process: The examination of an independent healthcare practitioner's licenses, specialty credentials, and professional performance upon which a healthcare organization bases its decision to confer or withhold permission to practice (privileges) in the organization.

Criterion: *See* Indicator.

Critical path: *See* Clinical path.

Critical performance measures: Those outputs by which the quality of an organization's services will be measured by patients, clients, visitors, and the community.

Cross-functional: A term used to describe an entity or activity that involves more than one healthcare department, discipline, or profession.

Customers: Those individuals who receive a product or a service from a process; *internal* customers are those individuals within the organization who receive products or services from an organizational unit or department, and *external* customers are those individuals from outside the organization who receive products or services from an organizational unit or department.

Deemed status: The term used for the assumption by the Centers for Medicare and Medicaid Services (CMS) that an organization meets the Medicare and Medicaid *Conditions of Participation* as a result of prior accreditation by the JCAHO or CARF.

Direct observation: A means of gathering data about a process in which participants in the process are observed.

Discovery: The pretrial disclosure on the part of one or both litigants of any facts or documents considered germane and admissible in a legal proceeding.

Discrete or count data: Numerical values that represent whole numbers, for example, the number of children in a family or the number of unbillable patient accounts; discrete data can be displayed in bar graphs.

Doctrine of fair use: A convention that permits the limited use of copyrighted intellectual property for purposes usually associated with nonprofit activities, including making photocopies for purposes of criticism, comment, news reporting, education, scholarship, and research.

Document review: An in-depth study performed by accreditation surveyors of an organization's policies and procedures, administrative records, human resources records, performance improvement documentation, and other similar documents, as well as a review of closed patient records.

Due process: The right of individuals to fair treatment under the law.

Effectiveness: In the language of the Joint Commission on Accreditation of Healthcare Organizations (JCAHO), the degree to which a healthcare intervention is provided in the correct manner, given the current state of knowledge, with the goal of achieving the desired/projected outcome for the patient.

Efficacy: In the language of the Joint Commission on Accreditation of Healthcare Organizations (JCAHO), the degree to which the treatment intervention used for a patient has been shown to accomplish the desired/projected outcomes.

Efficiency: In the language of the Joint Commission on Accreditation of Healthcare Organizations (JCAHO), the ratio of the outcomes for a patient to the resources consumed in delivering the care.

End product: The final result(s) of healthcare services in terms of the patient's expectations, needs, and quality of life, which may be positive and appropriate or negative and diminishing.

Evidence-based medicine: The care processes or treatment interventions that researchers performing large, population-based studies have found to achieve the best outcomes in various types of medical practice.

Exit conference: A meeting that closes a site visit during which the surveyors representing an accrediting organization summarize their findings and explain any deficiencies that have been identified; at this time, the leadership of the organization is also allowed an opportunity to discuss the surveyors' perspectives or provide additional information related to any deficiencies the surveyors intend to cite in their final reports.

Expectations: The characteristics that customers want to be evident in a healthcare product, service, or outcome.

Facility quality indicator profile: A report based on the data gathered using the Long-Term Care Minimum Data Set that indicates what proportion of the facility's residents have deficits in each area of assessment during the reporting period and, specifically, which residents have which deficits; the profile also provides data comparing the facility's current status with a preestablished comparison group.

Fishbone diagram: *See* Cause-and-effect diagram.

Flowchart: An analytical tool used to illustrate the sequence of activities in a complex process.

Food and drug interactions: Unexpected conditions that result from the physiologic incompatibility of therapeutic drugs and food consumed by a patient.

Formulary: A list of drugs approved for use in a healthcare organization; the selection of items to be included in the formulary is based on objective evaluations of their relative therapeutic merits, safety, and cost.

Gantt chart: A type of data display tool used to schedule a process and track its progress over time.

Ground rules: An agreement concerning attendance, time management, participation, communication, decision making, documentation, room arrangements and cleanup, and so forth that has been developed by PI team members at the initiation of the team's work.

Hard issues: The processes and products upon which customers base their perceptions of quality.

Healthcare Integrity and Protection Data Bank (HIPDB):   A national database that maintains reports on civil judgments and criminal convictions issued against licensed healthcare providers.

Histogram: A type of bar graph used to display data proportionally.

Icon: A graphic symbol used to represent a critical event in a process flowchart.

Incident report: *See* Occurrence report.

Independent practitioner: Any individual permitted by law to provide healthcare services without direction or supervision, within the scope of the individual's license as conferred by state regulatory agencies.

Indicator: A performance measure used to monitor the outcomes of a process; also called a criterion.

Information management standards: One chapter of the *Comprehensive Accreditation Manual for Hospitals (CAMH)* of the Joint Commission on Accreditation of Healthcare Organizations (JCAHO) that promulgates the JCAHO's requirements regarding the data and information used for various purposes in hospital organizations.

Interview: A discussion of the qualifications and experiences of a job applicant with respect to the employment process; or a discussion about an organization with its leadership during the accreditation or licensure survey process.

Leadership: The senior governing, administrative, and management groups of a healthcare organization that are responsible for setting the mission and overall strategic direction of the organization.

Legal purpose: An element of a legal contract that states the reason the contract is being created.

License: The legal authorization granted by a state to an entity that allows the entity to provide healthcare services within a specific scope of services and geographical location; states license both individual healthcare professionals and healthcare facilities.

Licensure: The process of granting a facility or healthcare professional a license to practice.

Line chart: A type of data display tool used to plot information on the progress of a process over time.

Material safety data sheet (MSDS):   Documentation maintained on the hazardous materials used in a healthcare organization; the documentation outlines such information as common and chemical names, family name, and product codes; risks associated with the material, including overall health risk, flammability, reactivity with other chemicals, and effects at the site of contact; descriptions of the protective equipment and clothing that should be used to handle the material; and other similar information.

Mean: The average value in a range of values that is calculated by summing the values and dividing the total by the number of values in the range.

Median: A measure of central tendency that shows the midpoint of a frequency distribution when the observations have been arranged in order from lowest to highest.

Medication error: A mistake that involves an accidental drug overdose, an administration of an incorrect substance, an accidental consumption of a drug, or a misuse of a drug or biological during a medical or surgical procedure.

Minimum Data Set (MDS):   The data set that the Centers for Medicare and Medicaid Services (CMS) requires long-term care facilities to collect on all residents who are federal program beneficiaries.

Minutes: The written record of key events in a formal meeting.

Mission: A broad statement describing what a healthcare organization does; or a statement of the goals and purpose of a performance improvement initiative.

Mutual assent: A written agreement describing the promises made by each party to a contract.

National Practitioner Data Bank (NPDB):   A federally sponsored national database that maintains reports on medical malpractice settlements, clinical privilege actions, and professional society membership actions against licensed healthcare professionals.

Nominal data: Values assigned to specific categories; also called categorical data.

Nominal group technique: A QI technique that allows groups to narrow the focus of discussion or to make decisions without becoming involved in extended, circular discussions.

Nosocomial infection: An infection acquired as a result of an exposure that occurred in a healthcare facility after the patient was admitted.

Occurrence report: A structured data collection tool that risk managers use to gather information about potentially compensable events; also called an incident report.

Opening conference: A meeting conducted at the beginning of the accreditation site visit during which the surveyors outline the schedule of activities and list any individuals whom they would like to interview.

Opportunity for improvement: A healthcare structure, product, service, process, or outcome that does not meet its customers' expectations and, therefore, could be improved.

Ordinal data: Values assigned to rank the comparative characteristics of something according to a given set of criteria; also called ranked data.

Outcomes: The end results of healthcare services in terms of the patient's expectations, needs, and quality of life; may be positive and appropriate or negative and diminishing.

Outputs: The measurable products of an organization's work.

Pareto chart: A type of bar graph used to determine priorities in problem solving.

Peer review committee: A group of like professionals, or peers, established according to an organization's medical staff bylaws, organizational policy and procedure, or the requirements of state law; the peer review system allows medical professionals to candidly critique and criticize the work of their colleagues without fear of reprisal.

Peer review organization (PRO):   *See* Quality improvement organization (QIO).

Performance improvement council: The leadership group that oversees performance improvement activities in some healthcare organizations.

Performance improvement team: Members of the healthcare organization who have formed a cross-functional group to examine a performance issue and make recommendations with respect to its improvement.

Performance measures: Those outputs by which the quality of the organization and its work units is assessed by patients, clients, visitors, and community leaders.

Permissions: The means by which a copyright holder formally allows another person or organization to use his or her intellectual property.

Pharmacy and therapeutics (P and T) committee: The multidisciplinary committee that oversees and monitors the drugs and therapeutics available for use, the administration of medications and therapeutics, and the positive and negative outcomes of medications and therapeutics used in a healthcare organization.

Pie chart: A data display tool used to show the relationship of individual parts to the whole.

Plagiarism: The act of copying the original work of an author and claiming it as one's own.

Potentially compensable event (PCE):   An occurrence that results in injury to persons in the healthcare organization or to property loss.

Process: The interrelated activities of healthcare organizations—which include governance, managerial support, and clinical services—that affect patient outcomes across departments and disciplines within an integrated environment.

Public domain: Intellectual property that, because of its age, expiration of copyright, or development by a public agency, can be used freely and cannot be copyrighted by anyone.

QI toolbox techniques: Tools that facilitate the collection, display, and analysis of data and information and that help team members stay focused; include cause-and-effect diagrams, graphic presentations, and others.

Quality assurance (QA):   A term commonly used in healthcare to refer to quality-monitoring activities during the 1970s and 1980s, at which time it connoted a retrospective review of care provided with admonishment of providers for substandard care.

Quality improvement organization (QIO):   Private or public agencies contracted by the Centers for Medicare and Medicaid Services (CMS) to undertake examination and evaluation of the quality of healthcare rendered to beneficiaries of federal healthcare programs.

Ranked data: *See* Ordinal data.

Recall logs: Documentation of communications from manufacturers regarding problems with equipment.

Relative frequency: The percentage of times that a characteristic appears in a data set.

Risk: A formal insurance term denoting liability to compensate individuals for injuries sustained in a healthcare facility.

Root-cause analysis: Analysis of a sentinel event from all aspects (human, procedural, machinery, material) to identify how each contributed to the occurrence of the event and to develop new systems that will prevent recurrence.

Sentinel event: An unexpected occurrence involving death or serious injury, or the risk thereof. Serious injury specifically includes loss of limb or function. The phrase "or risk thereof" includes any process variation for which a recurrence would carry a significant chance of serious adverse outcome. Such events are called "sentinel" because they signal the need for immediate investigation and response.

Site visit: An in-person review conducted by an accreditation survey team; the visit involves document reviews, staff interviews, an examination of the organization's physical plant, and other activities.

Soft issues: Staff attitudes upon which customers base their perceptions of quality.

Software license: A formal agreement that allows a person to use a copyrighted software application.

Special treatment procedures (STPs):   A term used by the Joint Commission on Accreditation of Healthcare Organizations (JCAHO) to denote the use of seclusion, restraints, and protective devices during patient care.

Standard deviation: A statistic used to show how the values in a range are distributed around the mean.

Standards of care: An established set of clinical decisions and actions taken by clinicians and other representatives of healthcare organizations in accordance with state and federal laws, regulations, and guidelines; codes of ethics published by professional associations or societies; regulations for accreditation published by accreditation agencies; or usual and common practice of similar clinicians or organizations in a geographical region.

Storyboard: A graphic display tool used to communicate the details of PI activities.

Strategic plan: The document in which the leadership of a healthcare organization identifies the organization's overall mission, vision, goals, and values to help set the long-term direction of the organization as a business entity.

Structures: The foundations of caregiving, which include buildings, equipment, technologies, professional staff, and appropriate policies.

Survey team: A group of individuals sent by an accrediting agency to review a healthcare organization for accreditation purposes.

Survey tools: Research instruments that are used to gather data and information from respondents in a uniform manner through the administration of a predefined and structured set of questions and possible responses.

Team charter: A document that explains the issues the team was implemented to improve, describes the goals and objectives and desired end state (vision), and lists the initial members of the team and their respective departments.

Team facilitator: A PI team role primarily responsible for ensuring that an effective performance improvement process occurs by serving as advisor and consultant to the PI team; remaining a neutral, nonvoting member; suggesting alternative PI methods and techniques to keep the team on target and moving forward; managing group dynamics; acting as coach and motivator for the team; assisting in consensus building when necessary; and recognizing team and individual achievements.

Team leader: A PI team role responsible for championing the effectiveness of PI activities in meeting customers' needs and for the content of a team's work.

Team member: A PI team role responsible for participating in team activities, identifying opportunities for improvement, gathering and analyzing data, sharing knowledge, and planning improvements.

Team recorder/scribe: A PI team role responsible for maintaining the records of a team's work during meetings, including any documentation required by the organization.

Timekeeper: A PI team role responsible for notifying the team during meetings of time remaining on each agenda item in an effort to keep the team moving forward on its PI project.

Tort: A wrongful act committed against a person or a piece of property in a civil rather than a criminal context.

Total quality management (TQM):   A management philosophy developed in the mid-twentieth century by W. Edwards Deming and others that encouraged industrial organizations to focus on the quality of their products as their paramount mission.

Transfusion reaction: Signs, symptoms, or conditions caused by a patient's having been given an incompatible transfusion.

Universal precautions: The application of a set of procedures specifically designed to minimize or eliminate the passage of infectious disease agents from one individual to another during the provision of healthcare services.

Vision: A description of the ideal end state or a description of the best way a process should function.

Voluntary review: An examination of an organization's structures and processes conducted at the request of a healthcare facility seeking accreditation from a reviewing agency.

Index

AAAHC. *See* Accreditation Association for
     Ambulatory Health Care
AAMC. *See* American Association of Medical
     Colleges
Absolute frequency, definition of, 35–36
Accreditation
     definition of, 243
     establishment of U.S. process for, *xix*
     focus on quality as key for, *xxiv*
     program evaluation report, review of, 333
Accreditation and licensing agencies, expecta-
     tions of, 4
Accreditation Association for Ambulatory Health
     Care, 245
Accreditation categories of JCAHO, 249–50
Acute care facilities, JCAHO accreditation of,
     247–50. *See also* Joint Commission on
     Accreditation of Healthcare Organizations
Adjourning to mark dissolution of PI team, 292
Administrator, healthcare. *See* Chief executive
     officer
Adverse drug reaction reports, review of, 158
Affinity diagrams
     case study for, 18
     example of, 18
     to prioritize ideas after brainstorming, 17
Agenda
     to communicate meeting's purpose and goals,
          27
     for next meeting, creating, 28
     sample, 27
     setting, 53–54
Aggregate data
     collection of, 260
     to drive priorities of PI plan, 326–27
     to identify practice patterns, trends in risk
          occurrences, and sentinel events, 139

JCAHO standards for, 273
     needs assessment for, 32
     from training checklists to monitor orientation
          and competency, 206
AHA. *See* American Hospital Association
Allied health professions, formalization of
     credentials for, *xvi*
AMA. *See* American Medical Association
Ambulatory Payment Classification system
     initiated in 2001, *xii*
     PPS for outpatient and ambulatory surgery
          services using, *xxi*
Ambulatory surgery centers, accreditation of, 245
American Association of Medical Colleges
     establishment in 1976 of, *xiv*
     medical college standards for approval by, *xii,
          xv*
American Board of Medical Specialties, 208–9
American College of Surgeons
     hospital improvement movement tied to, *xviii*
     Hospital Standardization Program established
          by, *xii, xviii*
     involvement in forming JCAHO, *xix*
American Hospital Association
     involvement in forming JCAHO, *xix*
     key elements of board's PI responsibilities
          noted by, 235–36
American Medical Association
     Committee on Medical Education of, *xv*
     establishment in 1840 of, *xiv*
     involvement in forming JCAHO of, *xix*
     Physician Master File, 208
     state licensing boards encouraged in 1874 by, *xii*
American Osteopathic Association, 245
American Public Health Association, 112
American Society of Health-System Pharmacists,
     156

Americans with Disabilities Act of 1990, 195
Antibiotics, debate about appropriate use of, 114
AOA. *See* American Osteopathic Association
APC system. *See* Ambulatory Payment Classification system
Artificial intelligence in information technology, 357
Assessment, team approach to patient, 148
Association of Professionals for Infection and Epidemiology, 112
Assurance in care quality, 72
Authoritarian system of care, 88, 89

Bacteria resistant to antibiotics, 114
Bar graphs
    to display discrete categories, 40
    examples of, 4148–49
Bell-shaped curve, 38
Benchmark
    baseline for, 31
    displaying data for, 326–27
    establishing, 6–7
    for indicators in continuum of care, 99
Benchmarking
    comparisons in, 13, 31
    definition of, 13
    external, 31–32
    JCAHO standard for contributing data for, 270
    performing, 12
Best practice. *See* Benchmark
Bio-terrorist attacks, emergency preparedness for, 171
Blitz team, 21–22
Blood products, evaluations of use of, 159–60
Blood-borne pathogens, tracking exposures to, 115
Board of directors
    among leaders for PI activities, 232
    case study of functioning of, 240–41
    managing PI activities of, 235–37
    subcommittees of, 237
Boas, Ernst P., therapeutics for chronically ill promoted by, *xvii*
Brainstorming
    case study for, 18
    structured and unstructured, 16–17
Business associates, definition of, 342

Calderone, Mary Steichen, research in contraceptive methods by, *xvii*
Calibration standards for medical equipment, 187
Callen, Maude E., midwife training by, *xvi*
Canadian Medical Association, involvement in forming JCAHO, *xix*

Care environment, improving, 171–93
    case study for, 189
    project application for, 189
    QI toolbox technique for, 189–92
    real-life example for, 188–89
    steps in, 172–88
Care pathway, 148
    cultural values reflected in, 154
    goal of, 148
    implementation of, 154
    sample, 149–53
Care planning. *See also* Treatment plan
    preadmission, 93, 94
    at time of admission, 94
Caregivers
    direct and indirect, clinical staffing needs measured for, 197
    universal precautions applied by, 112–13
CARF. *See* Commission on Accreditation of Rehabilitation Facilities
Case study
    for change management, 310–11
    for customer satisfaction survey development, 83–84
    for drug abuse of clinician, 145
    for graph preparation, 48–49
    for identifying opportunities for improvement, 18
    for implementing new work process, 28–29
    for improving care environment and life safety, 189
    for information technologies, 274–75
    for investigating urinary tract infections, 123–25
    for JCAHO survey, 253, 256
    for optimizing continuum of care, 107–8
    for optimizing patient care, 168
    for peer review, 351
    for PI plan, 327
    for PI program review, 334
    for PI team meeting, 286–87
    for project to convert to electronic medical records, 300–301
    for storyboard presentation of PI activities, 60–62
    for theory of constraints, 365
Cause-and-effect diagram. *See* Fishbone diagram
Cautious affiliation, as stage of team progression, 292
CDC. *See* Centers for Disease Control
Celebration of PI team's successes, 291, 309
Center for Project Management, causes of project failure listed by, 297
Centers for Disease Control, 112
    involved in collective response to bioterrorism, 295
    laboratory services regulated by, 159
    Web site of, 113

Centers for Medicare and Medicaid Services
    administration of Medicare and Medicaid by, 244
    deemed status of accreditation granted by, 247
    HCFA reorganized in 2002 as, *xii*
    OASIS developed for home health agencies by, 163
    and Quality Improvement Organizations, 245–46
    reimbursement amounts for prospective payment system determined by, *xxi*
CEO. *See* Chief executive officer
Certification, definition of, 243–44
Change
    acknowledging losses during, 307
    building momentum for, 304
    cycle of, 305
    endorsement of, 308
    evidence-based management applied to four levels of, 362
    identifying losses from, 306–7
    opposition to, during transitional period, 308
    providing information about, 307–8
    roles played during, 307
    three phases of, 304–5
Change management, 303–11
    case study for, 310–11
    definition of, 305
    steps in, 305
    techniques for, 306
Change manager, 308–9
Charting by exception
    real-life example of, 253–55
    regulations pertaining to, 254–55
Check sheet for data collection, 32, 34–35
"Checking in" process at PI team meetings, 292
Chief executive officer
    accuracy of JCAHO self-assessment attested by, 248
    among leaders for PI activities, 232
CLIA. *See* Clinical Laboratory Improvement Amendments
Clinical guidelines, 162
Clinical Laboratory Improvement Amendments, 159
    equipment codes of, 187
Clinical laboratory services, PI team example for, 26–27
Clinical practice standards, 161
Clinical privileges
    delineation of, 209
    history of conferring, *xii*
    initial appointment for, 208, 209–10
    reappointment for, 208, 210, 238
    sources of information about applicants for, 209–10

Clinical staffing needs, determining, 197
Clinicians' practice patterns and outcomes, monitoring, 140
Closure in PI project management, 295
CMS. *See* Centers for Medicare and Medicaid Services
Collaborative teamwork, as stage of team progression, 292
College of American Pathologists standards for laboratory analyzers, 187
Commission on Accreditation of Rehabilitation Facilities, 244–45, 247, 252
    project application using regulations of, 256
Common cause (normal) variation, in control chart, 45
Communicable disease, professional specialists trained in, 112
Communication of PI activities and recommendations, 53–63
Community-acquired infection, 113
Comparative data, 260
    JCAHO standards for, 273
Comparative summary measurements, described in PI program evaluation report, 332
Competence assessment process, 206
Competency, to enter contract, 40
Competitiveness, as stage of team progression, 292
Compliance
    continuous review by healthcare facility to ensure, 246
    definition of, 244
    JCAHO review of standards, 248
Compulsory reviews, for licensure requirements, 244
Conditions of Participation, Medicare and Medicaid, 243–44, 245–47
    certification of compliance with, 251, 252–53
    project application using regulations of, 256
Confidentiality agreement, sample, 346
Confidentiality statement, 345
    sample, 347
Constraints, 364–365
Continuous data, uses of, 35
Continuous improvement, steps to success in, 12–15
Continuous monitoring and improvement functions, 65–227
Continuous quality improvement, *xxii,* 355
    for reimbursement systems, 358
Continuum of care
    case study of, 107–8
    definition of, 87
    optimizing, 87–109
    patient-centered, 90
    PI team example for, 25–26
    project application of, 107

Continuum of care (continued)
   QI toolbox techniques for optimizing, 98–106
   real-life example of managing, 96–98
   settings for, 93
   treatment settings along the, 88
Continuum of care model, 89
Contract
   confidentiality language in, 342
   definition of, 340
   information and provisions in, 341
   patient health information clause in, 343
   work to be performed specified in, 342
Contract law, 340–42
Contraindications of medications, patient and
      family education about, 155
Control chart
   creating, 47
   example of, 46
   to measure key processes over time, 45
   upper and lower control limits of, 46–47
Copyright
   registration of, 339
   term of, 337
Copyright Act of 1976, 337
Copyright law, 337–40
   applicable to PI projects, 339
   definition of, 338
   treatment of intellectual property under, 338
Copyright notice, language of, 339
Copyrightable materials, 338
Core measures
   JCAHO data set for, 34, 99, 164, 168, 270
   of performance data, 160–61
   summary reporting of, 270–71
Costs
   fixed and operating, 293
   as one focus of QI philosophy, 11
CPM. *See* Critical Path Method
CPT codes. *See* Current Procedural Terminology
      codes
CQI. *See* Continuous quality improvement
CRAF method for categories of recordable infor-
      mation, 54–56
Credentialing of medical staff, 140, 199. *See also*
      Privileges, clinical
   committee review of, 210
   competency testing for medications as part of,
      156
   process of, 208–10
Credentials, professional organizations as
      conferring, 199
Credentials committee
   organizational relationships of, 235
   QI toolbox technique of summary profiles
      used by, 212, 213
   review of medical staff application and docu-
      mentation by, 210

Credentials verification organization, 208
Critical Path Method, 294
Culture shock of change, planning for, 299
Current Procedural Terminology codes, *xxi*
Customer expectations, 67, 69
   project application in determining, 83
Customer requirements
   identifying, 12, 15
   matching end products of process to, 15
Customer satisfaction, 67–85
   case study in surveying, 83–84
   comparing impact of quality risk program
      with, 140
   monitoring and improving, steps in, 70,
      72–74
   program evaluation report review of, 333
   QI toolbox techniques for measuring, 80–83
   real-life examples of measuring, 74–79
Customers
   external, 68, 70, 74–76, 79, 234
   identifying, 12, 14, 67, 69, 70, 74
   internal, 68, 70, 74, 79, 234
   matched with opportunities for performance
      improvement, 326
   products and services important to,
      identifying, 70, 72–73, 74
   types of, 68–70
CVO. *See* Credentials verification organization
Cyborg technology in healthcare, 357

Data, transformation to knowledge of, 260–68
Data analysis
   for comparative performance data performed
      by vendors, 270–71
   in customer satisfaction assessments, 74
   JCAHO standards for, 271–74
   for PI teams, 285
   to turn data collected into meaningful infor-
      mation, 327
Data collection
   analyzing results of, 7, 327
   categories of, 260
   check sheets for, 32, 34–35
   in customer satisfaction assessment, 73, 74
   for electronic billing, 358–59
   monitoring performance through, 5–9
   person responsible for, 34
   PI plan specifying, 326
   for PI teams, 285
   prioritized problem areas as source of, 9
Data collection tools, 32, 34–35
Data comparisons. *See* Benchmarking
Data confidentiality, 356
Data display tools, 40–47
Data repositories, 268

Data sets. *See also* Minimum Data Set for Long-Term Care
  graphs to compare, 40
  sample, 36, 48, 50–51
  statistical analysis using, 36–37
Data types, 35–36
Day surgery registration as example of identifying improvement opportunities, 15–16
Decubitus ulcers, associated with hospital care, 113
Deemed status, 247
Delinquent medical records, 6
Deming, W. Edwards, total quality management concept developed by, *xxii–xxiii*
Department of Health and Human Services, Centers for Medicare and Medicaid as agency of, 244
Department of Homeland Security, involved in collective response to bioterrorism, 295
Design, PI project, 293
Diagnosis-related groups (DRGs)
  standardized payments for, *xxi,* 92
  systems thinking for, 361
Dickinson, Robert Latou, structured health assessment tool of, *xvii*
Digitized documents, 356
Direct observation in data collection for customer satisfaction, 73
Director of nursing, among leaders for PI activities, 232
Disaster drills, regularly conducted, 186
Discharge planning, 94, 96
Discovery, immunity of documents from legal, 348–49
Discrete (count) data, uses of, 35
Dock, Lavinia Lloyd, epidemic nursing practices developed by, *xvii*
Doctrine of fair use, 338–39
Documentation of care
  paperless, 356
  review of clinical, 159, 161
Documentation of orientation and job competencies, 206
Documentation of PI program, 238
  posted on intranet, 269
Donabedian, Avedis
  classes of quality assessment developed by, 362–63
  TQM applied to healthcare by, *xxii–xxiii*
Drug purchases, review of, 158
Drug usage evaluation in pharmacy and therapeutics program, 157
Due process hearing for medical staff reappointments, 210

Education in performance improvement. *See* Training programs

Electronic billing, healthcare software for, 358–59
Electronic medical records
  case study of conversion to, 300–301
  change management for conversion to, 307–8
  use of Internet expanding with implementation of, 357
Electronic signature of physician, as admissible in court, 356
Emergency management plan, 172
  mitigation activities in, 185
  monitoring and improving, 180–86
  preparedness activities in, 185
  recovery actions in, 186
  response activities in, 185–86
Emergency power source for critical areas of facility, 188
Emergency preparations for disasters, 171
Empathy of staff in relating to customers, 72–73
Employees
  alteration of work habits by, 304
  as most important resource and biggest liability, 196
  orientation and training for new, 199, 206
  potential for legal action by, 195
  procedures following exposure to hazardous materials by, 174
  procedures to handle emergencies by, 186
  security management for, 172–73
  suited to performance improvement, 231
  working with supervisor or manager to develop performance goals, 207
Employees, prospective, background check of, 200
Employment application
  review of, 200
  sample, 201–4
Employment interview, 200
End products, of healthcare services
  definition of, 13
  identifying, 13–14
  matched to customers' requirements, 15
  as one focus of QI philosophy, 11
Environmental safety committee, organizational relationships of, 235
Equal Employment Opportunity Act of 1972, 195
Ethics committee, organizational relationships of, 235
Evaluation of PI project results, 295
Event, as basic unit of recording in healthcare, 259
Evidence-based management, 361–62
Evidence-based medicine, 162, 361–62
Execution (implementation) of PI project, 294
Executive director, hospital. *See* Chief executive officer

Facility quality indicator profile, 163, 251
    example of, 165–67
Failure mode, effects and criticality analysis, 129
Family and Medical Leave Act of 1993, 195
Features of healthcare services, distinguishing, 73
Fee-for-service payments, *xxi, xxii,* 358
Feedback during PI project, 308
Fire drills, regularly conducted, 186
Fire prevention plan, 172
Fire safety training, 186
Fishbone diagram
    4 Ms of, 142–43
    sample, 143
    to structure root-cause analysis, 142
Fixed costs of PI project, 293
Flexner, Abraham, medical curriculum review by, *xv*
Flowchart
    care path, 116–21
    icons used in, 122–23
    policies and procedures supporting, 10
    of redesign process, 10
FMECA. *See* Failure mode, effects and criticality analysis
Food and Drug Administration, equipment testing mandate of, 187
Food services department
    cleanliness of, tracking, 114
    real-life example of external customers of, 75–76
Formulary
    approved drugs specified by, 156
    developing effective, 157
    screening mechanisms in maintaining, 158
Full time equivalent staff, JCAHO standards related to, 196
Funneling process, for interview questions, 83
Future, of performance improvement in healthcare, 353–67

Gantt charts
    for PI projects, 293
    project management software creating, 301
    sample, 298–99
    scheduling activities using, 99
Germs, identification and limitation of, 111
Goldratt, Eliyahu, theory of constraints developed by, 363
Governing board of healthcare facility, oversight of policy and procedure review by, 161
Grief
    definition of, 304
    stages of, 304
Ground rules, for PI team meetings, 281–82

Handheld technologies, 356–57
Harmonious cohesiveness, as stage of team progression, 292
Hazard vulnerability analysis
    assessment tool for, sample, 181–85
    to identify and prioritize potential emergencies, 180
Hazardous material use, 174–79
Hazardous materials and waste management plan, 172, 173–80
HCPCS. *See* Healthcare Common Procedure Coding System
Health information services and business office services, PI team example for, 26
Health information services managers, expertise of, 271
Health Insurance Portability and Accountability Act
    compliance deadline for, 342
    information management standards of JCAHO in conformance with, 274
    regulations of, 115, 342–43, 357
    working knowledge of regulations in, 345
Health maintenance organizations, *xxi–xxii,* 358
Health Plan Employer Data and Information Set, 162, 245
Health records. *See* Medical records
Health Resources and Services Administration involved in collective response to bioterrorism, 295
Healthcare Common Procedure Coding System, *xxi*
Healthcare delivery
    evolution of U.S., *xi–xvi,* 303
    payment systems as integral to, 358
Healthcare expenditures, regulatory approaches to control, 92
Healthcare in United States, 91–93
Healthcare industry, differing views on PI in, 353
Healthcare Integrity and Protection Data Bank, 209, 210
Healthcare organization processes, 11
Healthcare performance improvement projects, 289–302
    case study for, 300–301
    copyright law applied to, 339
    deadlines of, 297
    emerging trends in, 354–55
    feedback during, 308
    implementing, 291
    individuals sponsoring, 290
    initiating, 290, 291
    key steps leading to winning strategy for, 299–300
    life cycle of, 290–91
    organizational structure for, 290
    planning, 291, 293–94
    project application for scheduling, 301
    reasons for failure of, 297, 299–300

Healthcare quality improvements, early U.S., *xi–xix*
Healthcare services
    expanding costs of, *xx*
    PI monitoring of utilization of, 89
    results of, 11, 13
HEDIS. *See* Health Plan Employer Data and Information Set
High-risk processes, identification of, 129
High-volume outputs, identifying, 12
Hill-Burton Act, *xii, xix*
HIPAA. *See* Health Insurance Portability and Accountability Act
HIPDB. *See* Healthcare Integrity and Protection Data Bank
Histogram
    characteristics of, 41
    differences among uses of Pareto chart, pie chart, and, 40
    to display continuous data proportionally, 40
    example of, 42, 43
HMOs. *See* Health maintenance organizations
Holistic care, 154
Home health agencies, OASIS used by, 163–64
Hospitals
    first U.S., *xiii*
    growth of city, *xv*
    standardization and accreditation developed for, *xvii–xix*
Hospital Hazard Vulnerability Assessment tool
    purpose of, 180
    sample, 181–85
Hospital standard measures, sample, 100–106
Hospital Standardization Program
    established by American College of Surgeons, *xii, xviii*
    replaced by Joint Commission on Accreditation of Hospitals, *xix*
Human immunodeficiency virus, tracking exposures to, 115
Human resources, 195–227
    program evaluation report review of, 333
Human resources staff
    position descriptions as providing guidance to, 199
    position requisitions as initiating communication between recruiting department and, 197
HVA. *See* Hazard vulnerability analysis

Icons, for flowcharts, 122–23
Implementation phase of PI projects, tasks in, 294
Incentives for staff retention, 208
Incident reports. *See* Occurrence reports

Indexes, online, 269
Indicators
    definition of, 98
    for patient care situations, 162
    ratio for, 98
Infection, discovery of pathologic organisms as causative agent of, 111
Infection control committee, environmental rounds by, 114
Infection control program, actions of, 115
Infectious disease prevention and control, 111–26
    case study in, 123–25
    educational and screening programs conducted in, 114, 115–16
    multidisciplinary approach to, 113
    as organizationwide performance issue, 115
    project application in, 125
    QI toolbox technique for flowcharting process of, 122–23
    real-life example in, 116–21
    steps in, 112–16
    surveillance procedures in, 113–14
    universal precautions in, 112–13, 114
Information, about change, 307–8
Information management, migration from paper systems to paperless, 356
Information management standards (JCAHO), 271–74
Information management tools, standardizing, 269
Information requirements for healthcare organizations, *xxiv*
Information resources, management for PI purposes of, 260
Information resources managers, 271
Information technologies, case study for, 274–75
Information technology capability, growth in, 355–57
Information warehouses, 270
Institute of Medicine, 128, 155
    Crossing the Quality Chasm report on medical errors by, 354, 361
    Medical Practice Study of, 353
Institutional review board, data-gathering tools preapproved by, 73
Insurance industry, healthcare services payments by, *xix–xx*
Insurance strategy for healthcare organization, development of, 130, 141
Intellectual property, copyright of, 338
International Organization for Standards, ISO 9000 standards published by, 363
Internet
    access to, as source of information for PI teams, 268
    use in healthcare of, 357

Interrelatedness in systems thinking about healthcare organizations, 360
Interviews
in data collection for customer satisfaction, 73, 76
design of customer satisfaction, 82–83
Intranet-based communication technologies, 269
Intranets, access to information warehouses using, 270
IOM. *See* Institute of Medicine
ISO 9000 certification, quality management standards in, 363
ISO 9002 standards, hospitals adopting, 363

JCAHO. *See* Joint Commission on Accreditation of Healthcare Organizations
Johnson, Lyndon, Medicare and Medicaid created during administration of, *xx*
Joint Commission on Accreditation of Healthcare Organizations
as accreditation agency, 243, 247–50, 253–56
accreditation decision categories of, 249–50
causes of sentinel events reported by, 206
Comprehensive Accreditation Manual for Hospitals, 244
Core Measures of, 34, 99, 164, 168, 270–71
costs of maintaining accreditation by, 363
data trend analyses of core measures by, 160–61
formation in 1953 of, *xii, xix*
goal of patient care for, 147
indicators in measuring staff effectiveness by, 196–97
investigation of occurrences required by, 139
laboratory analyzer standards by, 187
leadership expectations of, 234
mission of, 244
new emergency management standards by, 171, 185–86
ORYX external benchmarks. See ORYX external benchmark (JCAHO)
patient advocate, defined by, 140
patient rights standards by, 131
patient safety standards and program of, 128, 129
Performance Improvement standards of, QI initiatives influenced by, *xxiii*
physician health standards of, 210
PI plan recommendations of, 314
plans to respond to bioterrorism required by, 295
as resource to establish benchmark, 7
self-assessments by organizations for, 247–48
"Sentinel Event Alert" publication of, 14, 139, 350

sentinel event defined by, 14, 139, 349
sentinel events reportable to, 349–50
special treatment procedure use and patients' rights as concern of, 159
unannounced surveys by, 247
Joint Commission on Accreditation of Healthcare Organizations Environment of Care standards
for hazardous materials and waste management, 173
for life safety management, 186
for medical equipment management, 187
for security management, 172
Joint Commission on Accreditation of Healthcare Organizations Information Management standards, IM.1–IM.10, 271–74

Knowledge, transformation of data into, 260–68

Laboratory services, evaluations of, 159–60
Leadership
challenges in healthcare, 362
factors contributing to effective, 196
of healthcare organization, 234
of PI project, 292, 293
situational, 293
Leadership group to over*see* PI activities, 8, 239, 326
Legal issues for performance improvement, 337–51
Legal purpose of contract, 340
Liability claims, risk exposure of healthcare organizations to, 127
Licenses
process of conferring, definition and application of, 208–10
state regulatory agencies as conferring, 199
Licensure
definition of, 243
focus on quality as key for, *xxiv*
historical perspectives on state, *xii, xiv*
impetus for, *xiv*
for nurses, *xvi*
program evaluation report review of, 333
Life safety, 171–93
Life Safety Code, 173, 188
Life safety management program, 180, 186
Line chart
creating, 44–45
example of, 46
to show progress of process of time, 44
Listening skills of PI team members, 283
Long-term care certification and survey review process, 163, 250–51

Losses caused by change
    acknowledging, 309
    identifying, 308–9

Malpractice suits, 128
    documents used in, 131
    information about competency or medical
        privileging in, 140
    medication errors in, 155
Managed care
    definition of, *xxi*
    effects of, 358
    emergent strategy model of, 354–55
    fee-for-service reimbursement eliminated by
        Medicaid, Medicare, and, *xxii*
    revolution in healthcare services created by,
        xxi–xxii
    shift in focus of, 355
Management styles, inclusive and exclusive, 282
Manager
    authority of, 240
    collaboration in developing performance
        goals of employee and, 207
Martin, Edward, hospital standardization pro-
    moted by, *xvii–xviii*
Massachusetts General Hospital
    clinical privileges reiterated to medical staff
        by, *xiii*
    first disease/procedure index by, *xii, xiii*
Material safety data sheet
    care study for, 189
    content of, 174
    example of, 175–79
Mean, definition of, 36
Median
    definition of, 36
    middle value for, 36–37
Medicaid
    administration of, 244
    establishment in 1965 of, *xii, xx*
    evolution of healthcare delivery under, 303
    fee-for-service reimbursement eliminated by
        managed care, Medicare, and, *xxii*
    organizations receiving funds from, 244,
        250–51
Medical director among leaders for PI activities,
    232
Medical education reforms, *xv*
Medical equipment management plan, 172, 180
    focus of, 186–87
    quality control activities combined with, 187
Medical errors, Institute of Medicine report on,
    128, 353
Medical executive committee
    organizational relationships of, 235
    review of medical staff application and docu-
        mentation by, 210

Medical necessity for treatments and care, drive
    for, 88
Medical records
    conversion from paper-based to electronic,
        case study in, 300–301
    early U.S., *xviii*
    in managing risk exposure, 131
    paper-based, sources of information in, 268
    performance expectations of, 161
    review of PI initiatives using, 345
Medical staff
    organization of, 232
    as self-policing, 232
    standing committees of, as PI program
        resource, 235, 238
Medical staff bylaws
    due process for assessment of denied staff
        privileges, 210
    initial and reappointment of privileges defined
        in, 208
    provisional period for new appointments
        specified in, 210
Medical staff committee, documentation of
    reviews by, 140
Medical staff leaders, among leaders for PI
    activities, 232
Medical staff officers, elected, 232, 233
Medicare
    administration of, 244
    establishment in 1965 of, *xii, xx*
    evolution of healthcare delivery under, 303
    fee-for-service reimbursement eliminated by
        managed care, Medicaid, and, *xxii*
    organizations receiving funds from, 244,
        250–51
Medication errors
    in medical malpractice, 128, 155–56
    reviewed in drug use monitoring, 157, 158
Medication management, five rights of, 156
Medication systems and processes
    evaluations of, 155–58, 159
    patient and family education about, 155
Meetings of PI teams
    advance preparation for, 282
    case study of, 286–87
    ground rules for, 281–82
Microsoft Project software, project application
    for scheduling using, 301
Midwives, training for, *xvi*
Minimum Data Set for Long-Term Care, 163,
    251
Minimum Standard, The (ACS), *xviii–xix*
Mintzberg, Henry, strategic planning model of,
    354–55

Minutes of PI team progress and activities, 53–57
   conclusion section of, 57
   CRAF method for, 54
   follow-up section of, 57
   sample, 55–56
Mission statement of PI team, 24, 291
   sample, 24
Monitoring of QI results, ongoing, 12
Monthly delinquent medical record rate, 6
MSDS. *See* Material safety data sheet
Mutual assent to contract, 340

National Committee for Quality Assurance, 162–63
   accreditation of managed care organizations by, 245
National Fire Protection Association, 173, 188
National Library of Medicine journal index, 269
National Patient Safety Goals, 172, 249
National Practitioner Data Bank, 209, 210
NCQA. *See* National Committee for Quality Assurance
Near misses, intense investigation of, 139
Negligence, definition of, 344
Newtonian model of traditional education, 360
NFPA. *See* National Fire Protection Association
Nominal (categorical) data, uses of, 35
Nominal group technique to select important ideas, 17
Normal distribution
   bell-shaped curve appearance for, 38
   calculated mean at center of, 38–39
Nosocomial infection acquired in healthcare facility, 113
NPDB. *See* National Practitioner Data Bank
Nurse training programs, *xv–xvi*
Nurses' licenses to practice, validation of, 140
Nursing, behaviors exhibited by caring, 325
Nursing practice
   growth of body of knowledge and skills in, 360
   quality improvements in, *xv–xvi, xvii*

Observations of continuous measure
   bell-shaped curve of normal distribution for higher numbers of, 38
   initial, 37
Occupational Safety and Health Act of 1970, 195
   standards for hazardous materials and waste management by, 173
   standards for utility management by, 188
Occupational Safety and Health Administration, 115

Occurrence reports, 130, 131–37
   employee training for, 206
   example of, 131–37
   reviewed in drug use monitoring, 157
   tracked to identify incidents involving hazardous material and waste management, 173
Office of Inspector General, 1999 report of, 353
Ongoing measurements described in PI program evaluation report, 332
Open-ended questions, for surveys, 81
Operating costs of PI project, 293
Optimal spending for optimal health, 91–92
Ordinal (ranked) data, use of, 35
Organization chart, sample hospital, 233
Organizational improvements described in PI program evaluation report, 332
Orientation of new employees, 199, 206
ORYX external benchmark (JCAHO)
   to compare results of care across hospitals, 32, 332
   core measure sets of, 160, 249
   indicators for patient care in, 162
   sample report of, 33
OSHA. *See* Occupational Safety and Health Administration
Outcomes of care
   as class of quality assessment, 363
   reporting, 70

P and T committee. *See* Pharmacy and therapeutics committee
Pareto chart
   to determine priorities in problem solving, 42
   differences among uses of histogram, pie chart, and, 40
   example of, 44
Park, Roswell, antisepsis principles disseminated by, *xvii*
Pathology laboratory, real-life example of internal and external customers of, 76, 79
Patient advocacy function, managing, 130, 140–41
Patient care, optimizing, 147–70
   accountability for, 239–40
   case study in, 168
   critical factor in, 147
   ongoing developments in, 161–64, 165–67
   project application in evaluating customer satisfaction for, 168
   QI toolbox technique of criteria set for, 164, 168
   real-life example of, 164
   steps in, 154–61
Patient care model for nursing, 325

Patient care outcomes review, 159, 160–61
Patient care process cycle, 154, 155
Patient care quality data, review by board of directors of, 237
Patient evacuation procedures, 186
Patient health information, physical safeguards of, 343
Patient rights standards by JCAHO for disclosure of outcomes, 131
Patient safety goals established by JCAHO, 129
Patient Safety Improvement Act of 2003, 128
Patient safety program defined by JCAHO, 129
Patient satisfaction survey, 69
    data collection using, 325
    example of, 77–78
    vendors for, table of, 71
Patient-identifiable information, access to, 345
Patient-specific data, 260
    JCAHO standards for, 273
Patients
    as external customers, 68
    health risks in healthcare facilities created by, 127
Payment systems, 358–59.
PCEs. *See* Potentially compensable events
Peer review, reporting findings of, 238
Peer review committee
    admissibility as evidence of documentation from, 348–49
    definition of, 347
    establishing, 347–48
    institutional, 348
    referrals of practitioners from, 348
Peer review organizations
    case study of, 351
    formation of, *xii, xx*
Pennsylvania Hospital as first U.S. hospital, *xiii*
Perceived quality of healthcare services, 73
Performance appraisal process, managing, 199, 207
Performance goals, development of, 207
Performance improvement
    accountabilities for, 234–35
    as cyclical process, 4–5, 147, 234
    future of, predicting, 353–67
    goal of, 147, 344
    human side of, managing, 303–11
    identifying opportunities for, 9, 11–19, 74, 290–91, 326
    as information-intensive activity, 259
    as intradepartmental, 330
    leadership commitment to, 231
    legal implications of, 337–51
    organizing for, 231–42
    reasons for emphasis on, *xxiii–xxiv*
    synonyms for, *xxiii,* 3

Performance improvement council
    organizational relationships of, 235
    representatives to, 234
    risk manager as member of, 139
    structure, process, and knowledge issues examined by, 130, 139
Performance improvement data, 31–52
Performance improvement information management tools, 259–76
Performance improvement model, 1–63
    based on continuous monitoring and assessment of performance measures, 259
    cycles in, 4–5, 282, 327
    defining, 3–10
    implementing, 8
    as improvement opportunity, example of, 7
Performance improvement plan
    areas addressed in, 326
    case study for, 327
    design of, 314–26
    developing, 313–28
    implementing, 326–27
    sample, 315–22
    setting priorities for, 314, 323–24, 326
Performance improvement processes
    board of directors' oversight of, 235–37
    organizationwide, 4, 239
    seven steps of, 291
    team-based, 5, 9–10
Performance improvement program
    case study for review of, 334
    components of review for, 330–2
    empowerment of employees to contribute to, 231
    evaluating, 329–35
    implementing, 314
    leading activities in, 232–35
    outcomes as major focus of, 363
    recommendations for, 334
    redesign of, 330
    resources for, 237–38
    Six Sigma philosophy applied to, 359
    steps in implementing, 8–9
    systematizing, 329
Performance improvement projects. *See* Health-care performance improvement projects
Performance improvement report, example of, 261–67
Performance improvement resource department, developing, 239–40
Performance improvement teams
    charters of, 278–80
    composition of, 22–23, 289, 290–91
    convening, 9
    copyright violations by, avoiding, 340
    cross-functional, 21

Performance improvement teams (continued)
    defining performance expectations for, 9
    developing effective, 277–87
    disbanding, 10, 292
    example of effective, 25–27
    expectations of, 277
    flow of information from, 53–54
    formed when end products of process do not
        meet customer requirements, 15
    group dynamics of, 292
    information sharing among, 269
    life cycle of, 289, 292
    meetings of, 281–82
    mission of, 289
    mission statement for, 24, 291
    objectives of, 283, 285
    people issues in, 283–86
    program evaluation report highlighting activi-
        ties of, 332–33
    project application for goals and purposes of,
        29
    recorder or secretary of, 54
    as reinforcing beginning of new process, 309
    roles within, 23–24
    size recommended for, 22
    sources of information for, 268
    vision statement for, 24–25
Performance measures
    activities assessed using, 268
    identification and monitoring of, 5, 7, 12, 73
    measurements from, 259
PERT. *See* Program Evaluation and Review
    Technique
Pharmacy and therapeutics committee, 156
    functions of, 156–57
    organizational relationships of, 235
    therapeutic interchanges by, 158
Pharmacy department
    advisory role to medical staff of, 156
    P and T committee as advising, 157
Photocopying copyrighted material, 339
Physician index summary, 212, 216–24
Physician performance profiles
    blank, 225–26
    case study for, 212–26
    content of, 212
    sample, 213–15
Physicians and surgeons
    credentialing of, 140, 156
    as internal and external customers, 68–69
Physicians' role in healthcare system, forces
    shaping, 360–61
PI. *See* Performance improvement
Pie chart
    creating, 44
    differences among uses of histogram, Pareto
        chart, and, 40

    examples of, 45, 50
    to show relationship of parts to whole, 42
Pisacano, Nicholas J., family practice specialty
    promoted by, *xvii*
Plagiarism in violation of copyright law, 338
Point-of-service plans, *xxii*
Policies on standard practices, review of, 159,
    161
Policy and procedure committee, 161
POS plans. *See* Point-of-service plans
Position descriptions, content of, 199
Position requisitions to initiate recruitment, 197
    sample, 198
Postdischarge planning, 94, 96
Postprogram assessments
    to evaluate employee knowledge after train-
        ing, 189
    example of, 190–92
Potentially compensable events, 128
    communicating relevant data and information
        for, 130, 141
    documenting, 131
PPOs. *See* Preferred provider organizations
PPS, Medicaid and Medicare. *See* Prospective
    payment system for Medicare and
    Medicaid
Preadmission care planning, 93, 94
Precertification of medical procedures, 362
Preferred provider organizations, *xxii*
Primary source verification of staff qualifica-
    tions, 208–9
    real-life example of, 211–12
Privacy of health information, policies on, 345
Privileges, clinical, 199
    delineation of, 209
    history of conferring, *xii*
    initial appointment for, 208, 209–10
    reappointment for, 208, 210, 238
    sources of information about applicants for,
        209–10
Proactive error reduction activities, 129
Process of identifying PI opportunities, steps in,
    12–15
Process redesign, steps in, 9–10
Program Evaluation and Review Technique, 294,
    301
Program evaluation report
    description of performance review structure
        in, 331–32
    executive summary of, 331
    improvement opportunities described in, 332
    overview of, 331
    PI team activities highlighted in, 332–33
    review topics in, 333–34
    staff survey questions for, 331
Progress, means of accomplishing, *xi*

Progress note, sample, 138
Progress of care, reviewing, 94
Project application
    for applying teamwork in PI, 29
    for assessing outcomes of training sessions,
        189
    for creating best graphic presentation for QI
        project data, 48
    for fishbone diagram and root-cause analysis
        in storyboard projects, 145
    for flowchart use in student projects, 125
    for human resources issues and staffing
        recommendations, 212
    for identifying customer expectations, 83
    for identifying improvement opportunities, 18
    for optimizing patient care using standardized
        criteria, 168
    for performance measures of continuum of
        care, 107
Project portfolio, establishing comprehensive,
    297, 299
Projects, performance improvement. *See* Health-
    care performance improvement projects
Prospective payment system for Medicare and
    Medicaid, 92
    effects of, 358
    establishment during 1980s of, *xii, xxi*
    evolution of healthcare delivery under, 303
Proximate cause, of sentinel event, 143–44
Psychiatric facilities, CMS and CARF surveys
    of, 247, 252
Public Law 89-97, Medicare and Medicaid cre-
    ated by, *xx*
PubMed search engine (National Library of
    Medicine), 269

QI toolbox techniques, 10, 16, 27–28, 295
    experts in use of, 239
    in flowcharting care paths in infectious
        disease control, 122–23
    in identifying improvement opportunities, 18
    in measuring customer satisfaction, 80–83
    in optimizing continuum of care, 98–106
    in optimizing patient care using JCAHO Core
        Measure data set, 164, 168
    in postprogram assessments of employee
        knowledge of care environment, 189–92
    in root-cause analysis using cause-and-effect
        or fishbone diagram, 142–45
QIOs. *See* Quality Improvement Organizations
Qualifications for staff, credentials and licenses
    as, 199
Quality assurance
    retrospective efforts for, *xx*
    synonyms for, 3

Quality improvement, *xxiii,* 3
Quality Improvement Organizations, *xxiii,* 246
Quality improvement philosophies, healthcare
    areas of focus for, 11–12
Quality improvements, early healthcare, *xi–xix*
Quality management
    ISO 9000 standards for, 363
    synonyms for, 3
Quality measures for each product and service,
    identifying, 73, 74
Questions for employment interviews
    appropriate and discriminatory, 200, 205
    structured and unstructured, 200
Questions in customer satisfaction interviews, 83
Questions in PI team meetings
    power of, 283
    types of, descriptions of, 284

RBRVS system. *See* Resource-based relative
    value scale system
Real-life examples
    of graphic presentation for data, 47–50
    of improving care environment, 188–89
    of managing continuum of care, 96–98
    of measuring customer satisfaction, 74–79
    of optimizing patient care, 164
    of PI project for implementing revised
        JCAHO standard, 295–97
    of preventing and controlling infectious
        disease, 116–21
    of recruitment of medical staff, 211–12
    of sentinel event, 141–42
    of standards review to meet multiple stan-
        dards, 253–55
    of storyboards of PI project, 60–61
Recall logs for medical equipment, 187
Recruitment practices of healthcare
    organizations, 195
Recruitment process
    initiating, 197–98
    interviews during, 200–205
    managing, 197–98, 199–205
    real-life example of, 211–12
Refrigerators, temperature logs of, 114
Rehabilitative care facilities, CARF surveys of,
    247, 252
Reimbursement process
    DRGs as driving, 361
    with electronic billing systems, 358–59
Relative frequency, definition of, 36
Reliability in healthcare performance, 70
Reports of PI team, 57
    example of quarterly, 58
    final, 291
Research phase of PI process, 9

Resource-based relative value scale system, *xxi*
Responsiveness to unanticipated service needs,
    72
Retention of employees
    managing process of, 199, 208
    variations in healthcare delivery affecting,
        195
Retrospective payment system, *xx–xxi*
Risk
    definition of, 128
    unnecessary, detecting, 144
Risk exposure, 127–46
    case study of, 145
    program evaluation report review of, 333
    project application of, 145
    QI toolbox technique for root-cause analysis
        of, 142–45
    real-life example of, 141–42
    sources of, 127–28
    steps in decreasing, 128–41
Risk management and procedure policies, devel-
        opment of, 129–30
Risk management educational programs, 130–31
Risk management program, measuring impact of,
        140
Risk managers
    as leader in developing insurance strategy,
        130, 141
    as members of PI council, 139
    as patient advocates, 140
    proactive error reduction activities of, 128–29
    review of occurrence reports by, 131
    review of operational policies and procedures
        by, 130
    trends in risk occurrences identified by, 139
Root-cause analysis, 14
    definition of, 349
    fishbone diagrams in, 142–45
    lessons learned about safety as findings from,
        129
    of sentinel events, 139, 350–51
    of transfusion reactions, 160

Safety plan, 172
    monitoring and improving, 180, 188
Satisfaction scales for each product and service,
        identifying, 73, 74
Screening indicators in measuring staff effective-
        ness (JCAHO), 197
Scribe, functions of PI team's, 278
Seclusion, restraints, and protective devices
    evaluating use of, 158–59
    orders for, 161
Secretary of PI team as record keeper, 24
Security management plan, 172–73, 180

Security rule of HIPAA, 342–43
Senge, Peter, systems approach of, 359–61
Sentinel events, 349–51. *See also* Potentially
        compensable events
    Accreditation Watch following, 250
    causes of, 206
    definition of, 349
    as dramatic information about healthcare
        services' end products, 13
    investigation of, 130, 139
    JCAHO definition of, 14, 139, 349
    lessons learned from, goals and recommenda-
        tions based on, 129
    real-life example of, 141–42
    reportable to JCAHO, 349–50
    reported to state licensing agencies, 139
Shared vision in PI systems, 360
Side effects of medications, patient and family
        education about, 155
Site visit for accreditation survey, CARF, 252
Site visit for accreditation survey, JCAHO
    exit conference to summarize, 249
    interviews with leaders, staff, and patients
        during, 249
    opening conference of, 248–49
    resources surveyors bring to, 249
    trigger issues for, 249
Six Sigma philosophy applied to PI programs, 359
Skills testing during employment interviews, 200
Software standardization for information sharing,
        269
Special cause variation in control chart, 46
Special treatment procedures, 158–59
Sponsorship of PI project, 290
Spreadsheet applications to perform statistical
        analysis, 260
Staff development, 195–227
    identifying needs for, 207
Staff education, for new processes, 10
Staff education plan, to train employees in PI, 8
Staff self-development and lifelong learning,
        196, 208
Staff survey questions for PI program review,
        331
Standard deviation
    calculating, 37
    defined by percentage of frequencies
        contained in portions of normal
        distribution, 39
    in Six Sigma processes, 359
Standardization of software
    data confidentiality and, 356
    for information management tools, 269–70
Standards of care
    changes in, 345
    evaluation of, 159, 161
    evidence of, 344

State departments of health, certification and licensure of long-term care facilities by, 250–51
State licensing agencies, 243
    authority granted to healthcare facilities by, 246
    compliance surveys for *Conditions of Participation* conducted by, 253
    focus of, 247
    history of, *xii, xiv*
    sentinel events reported to, 139
State nursing associations, *xvi*
States, early healthcare performance improvements in, *xii*
Statistical analysis, 36–40
    advanced, 47
    spreadsheet applications to perform, 260
Storyboards (storytelling) to report PI activities, 57–60
    benefits of, 58–59
    case study of, 60–62
    introduction of technique of, 57
    keys to successful, 59
    real-life example of, 60–61
    sample, 60, 61
STPs. *See* Special treatment procedures
Strategic plan
    definition of, 313
    example of, 323–24
    modifications to, 234
    process of developing, 325
Stress interviews, 200
Structured interviews, 200
Summary profiles of physician performance
    content of, 212
    sample, 213
Supervisor, collaboration in developing performance goals of employee and, 207
Surgical site infections associated with hospital care, 113
Survey process for accreditation and licensure reviews, 246–53
    CARF, 252, 256
    JCAHO, 247–50, 253–56
    keys to successful, 246–47
Survey team, JCAHO, 248
Surveys
    case study of customer satisfaction, 83–84
    check sheets to tally responses for, 35
    customer satisfaction, 70–71, 74–79, 80–84
    in data collection for customer satisfaction, 73
    project application for customer satisfaction, 83
    question types in, 81
    reading level of, 81
SWOT analysis, 313

Systems approach
    of continuous improvement in future accreditation process, 249
    for evidence-based management, 362
Systems thinking in quality improvement, 359–61
    diagnosis-related groups as example of, 361
    disciplines mastered for, 359–60

Team charters, 278
    example of, 279–80
Team facilitator, responsibilities of, 278
Team leader of PI team, 293
    responsibilities of, 23, 292
Team member of PI team, functions of, 23
Team problem-solving, 277
Team recorder (scribe), functions of, 278
Teamwork, in developing care plans, 148
Teamwork, in performance improvement, 21–30. *See also* Performance improvement teams
Theory of constraints, 363–65
    case study for, 365
Therapeutic interchange program in pharmacy and therapeutics program, 157, 158
Timekeeper for PI team, 281
TOC. *See* Theory of constraints
Top-down improvement initiative, 234
Tort, definition of, 343
Tort law, standard of care and, 343–45
Tort of negligence, 344
Total quality management (TQM)
    adoption by U.S. healthcare during 1990s, *xii, xxii–xxiii*
    synonyms for, 3
Tracer method used for onsite surveys, 249
Training programs
    to address improvement opportunity, 21, 22
    for change management, 306
    in creative problem solving and team building, 309
    in data analysis and statistical package use, 270
    in emergency response, 186
    in fire safety, 186
    HIPAA requirements for, 343
    for JCAHO surveyors, 248
    by leaders in performance improvement, 239
    for new employees, 199, 206
    for nurses, *xv–xvi*
    postprogram assessment tools to evaluate, 189–92
    variations in healthcare delivery affecting, 195
Transfusion reaction, investigation of, 160

Transitional period, in change process, 304, 305
  avoiding conflicting messages during, 309
  managing, 308–9
Treatment plan
  assessment process for, 148
  cornerstone of patient care as establishing,
    147
Trends
  in healthcare PI, emerging, 354–55
  in risk occurrences identified by risk manager,
    130, 139
Trigger issues considered during site visits for
  accreditation, 249, 251
Tuberculosis prevention plan, state requirements
  for, 115

Unannounced surveys by accrediting and licens-
  ing agencies, 247
Units, case study of combining, 310–11
Universal precautions, 112–13
UR programs. *See* Utilization review programs
Utilization and documentation standards commit-
  tee, organizational relationships of, 235

Utilization review programs, *xx*
Utility management plan, 172, 180
  control areas labeled in, 188
  focus of, 187
  occurrence reporting system for, 188

Vaccines, development of, 111
Viable project process, developing, 297
Viral infections, antibiotics as ineffective in
  treating, 114
Vision statement for PI team, 24
  sample, 25
Voluntary reviews for healthcare facility accredi-
  tation, 244

Watson, Jean, theory of human caring of, 325
Web browser user interfaces, access to informa-
  tion warehouses using, 270
Wellness centers, 116
Workplace environment, federal legislation
  affecting, 195